Genome Analysis
in Domestic Animals

© VCH Verlagsgesellschaft mbH, D-6940 Weinheim (Bundesrepublik Deutschland), 1990.

Vertrieb:
VCH, Postfach 101161, D-6940 Weinheim (Bundesrepublik Deutschland)
Schweiz: VCH, Postfach, CH-4020 Basel (Schweiz)
Großbritannien und Irland: VCH (UK) Ltd., 8 Wellington Court, Wellington Street.
     Cambridge CB 1 1HZ (England)
USA und Canada: VCH, Suite 909, 220 East 23rd Street, New York, NY 10010–1606 (USA)

ISBN 3-527-28097-9   ISBN 0-89573-968-2

Genome Analysis in Domestic Animals

Edited by
H. Geldermann and F. Ellendorff

 VCH   Weinheim · New York · Basel · Cambridge

Editors:

Prof. Dr. Hermann Geldermann
Institute for Animal Husbandry
University Hohenheim
Garbenstraße 17
D-7000 Stuttgart 70
Federal Republic of Germany

Prof. Dr. Dr. Franz Ellendorff
Institute for Small Animal Research
Federal Research Center of Agriculture
Dörnbergstraße 25
D-3100 Celle
Federal Republic of Germany

Published jointly by
VCH Verlagsgesellschaft mbH, Weinheim (Federal Republic of Germany)
VCH Publishers Inc., New York, NY (USA)

Editorial Director: Dr. Hans-Joachim Kraus
Production Manager: Dipl.-Wirt.-Ing. (FH) H.-J. Schmitt

Library of Congress Card No.: 90-12874

British Library Cataloguing-in-Publication Data:
Genome analysis in domestic animals.
   1. Organisms. Genes. Mapping
   I. Geldermann, H. II. Ellendorff, F
   575.12
   ISBN 3-527-28097-9
   ISBN 0-89573-968-2

Deutsche Bibliothek Cataloguing-in-Publication Data:
Genome analysis in domestic animals / ed. by H. Geldermann
and F. Ellendorff. – Weinheim ; New York ; Basel ; Cambridge : VCH, 1990
   ISBN 3-527-28097-9 (Weinheim ...)
   ISBN 0-89573-968-2 (New York)
NE: Geldermann, Hermann [Hrsg.]

Printing: betz-druck, D-6100 Darmstadt.
Bookbinding: Großbuchbinderei J. Schäffer, D-6718 Grünstadt.
Printed in the Federal Republic of Germany.

Preface

From the beginning of domestication, man has improved selection in animals paying attention and care to those characteristics in animals that met his needs for nutrients, labor, protection etc.. Advances in molecular biology are providing knowledge and concepts which are helpful not only to enhance our basic understanding of the genome of domestic animals but also to assess the impact of modern biotechnology in animal production by evaluating its chances and risks. This book is intended to show the different aspects of a growing area in animal breeding largely as reviews, i.e. analysis of genes and their functions, molecular mechanisms of disease, patentability of genetic inventions. Ranging from basis to application, the book should stimulate progress in the field and interaction between scholars of various disciplines.

The editiors are deeply indebted to the publisher. The development of the book has mainly been supported by Dr. H.J. Kraus, VCH Publishers, Weinheim (FRG), and we are thankful for his valuable help in all stages of preparation. We also thank the authors for submitting their chapters. Moreover, we express our gratitude for all the assistance we obtained from our co-workers in the process of editing. Amongst them, the significant support of Bioengineer Stefan Neander is singled out who performed the computer-assisted text preparation and printing. For computer-aided graphics we acknowledge the help of Gudrun Geldermann. The support in data procession by Dr. J. Buitkamp, Ingrid Wittwer and Ingo Schwan is gratefully mentioned.

The book is a proceeding of a symposium, held in Hannover and financially supported by a number of institutions. All of them are gratefully acknowlegded, amongst them the authors are specially indebted to the Volkswagen-Stiftung and the H.Wilhelm Schaumann Stiftung. Finally, the editors recognize that the Biotechnology Committee of the German Society for Animal Breeding responded favourably to their request to make the symposium part of its activities.

Hannover, March 1990
H. Geldermann
F. Ellendorff

Table of Contents

5.   Techniques, Strategies and Stages of DNA Sequencing   49

P. Heinrich and H. Domdey

9.   How to Demonstrate Genetic Individuality in Any Human and Animal Subject   135

H. Zischler, I. Nanda, C. Steeg, M. Schmid
and J.T. Epplen

10.   Information on RFLPs and VNTRs in Farm Animals   143

C.-M. Prokop

List of Contributors

L. Andersson, Department of Animal Breeding and Genetics, Swedish University of Agricultural Sciences, Husargatan 3, Box 595, S-675124 Uppsala, Sweden

D. Baum, Abteilung für Biochemie, Institut für Anatomie, Physiologie und Hygiene der Haustiere, Universität Bonn, Katzenburgweg 7-9, D-5300 Bonn 1, FRG

M.P. Baur, Institut für Medizinische Statistik, Dokumentation und Datenverarbeitung der Universität Bonn, Sigmund-Freud-Str. 25, D-5300 Bonn 1, FRG

F. Clerget-Darpoux, INSERM U155, Chateau de Longchamp, Bois de Boulogne, 75016 Paris, France

H. Domdey, Laboratorium für Molekulare Biologie, Genzentrum der Ludwig-Maximilians-Universität München, D-8033 Martinsried, FRG

J.T. Epplen, Max-Planck-Institut für Psychiatrie, Am Klopferspitz 18A, D-8033 Martinsried, FRG

H. Feldmann, Institut für Physiologische Chemie, Physikalische Biochemie und Zellbiologie der Universität München, Schillerstr. 44, D-8000 München 2, FRG

H. Geldermann, Institut für Tierhaltung und Tierzucht der Universität Hohenheim, Garbenstr. 17, D-7000 Stuttgart 70, FRG

M. Georges, Department of Genetics, Faculty of Veterinary Medicine, Université de Liège, Rue de Vétérinaires, 45, B-1070 Bruxelles, Belgium

R.C. Gorewit, Department of Animal Science, Cornell University, Ithaca, NY 14853, USA

G. Graser, Abteilung für Biochemie, Institut für Anatomie, Physiologie und Hygiene der Haustiere, Universität Bonn, Katzenburgweg 7-9, D-5300 Bonn 1, FRG

W. Hecht, Institute for Animal Breeding and Genetics, Department for Veterinary Genetics and Cytogenetics, Justus-Liebig- University, Hofmannstr. 10, D-6300 Giessen, FRG

List of Contributors

M. Heib, Abteilung für Biochemie, Institut für Anatomie, Physiologie und Hygiene der Haustiere, Universität Bonn, Katzenburgweg 7-9, D-5300 Bonn 1, FRG

P. Heinrich, Consortium für Elektrochemische Industrie, D-8000 München 70, FRG

R. Ivell, Institute of Hormone and Fertility Research, Grandweg 64, D-2000 Hamburg 54, FRG

R.S. Jack, Institut für Genetik der Universität zu Köln, Weyertal 121, D-5000 Köln 41, FRG

H. Kräußlich, Chairman of DGfZ-Arbeitsausschuß "Methoden der Biotechnologie in der Tierproduktion", Institut für Tierzucht und Tierhygiene der Ludwig-Maximilians-Universität München, Lehrstuhl für Tierzucht, Veterinärstr. 13, D-8000 München 22, FRG

G. Krampitz, Abteilung für Biochemie, Institut für Anatomie, Physiologie und Hygiene der Haustiere, Universität Bonn, Katzenburgweg 7-9, D-5300 Bonn 1, FRG

J.-C. Mercier, Laboratoire de Génétique Biochimique, Institut National de la Recherche Agronomique, INRA-CRJ, Domaine de Vilvert, F-78350 Jouy-en-Josas, France

R. Moufang, Max-Planck-Institute for Foreign and International Patent, Siebertstr. 3, D-8000 München 80, FRG

I. Nanda, Institut für Humangenetik, Koellikerstr. 2A, D-8700 Würzburg, FRG

C.-M. Prokop, Institut für Tierzucht und Vererbungsforschung der Tierärztlichen Hochschule Hannover, Bünteweg 17p, D-3000 Hannover 71, FRG

C. Provot, Laboratoire de Génétique Biochimique, Institut National de la Recherche Agronomique, INRA-CRJ, Domaine de Vilvert, F-78350 Jouy-en-Josas, France

K.H. Scheit, Max-Planck-Institut für Biophysikalische Chemie, Abteilung Molekulare Biologie, Postfach 2841, D-3400 Göttingen, FRG

M. Schmid, Institut für Humangenetik, Koellikerstr. 2A, D-8700 Würzburg, FRG

S. Schüler, Abteilung für Biochemie, Institut für Anatomie, Physiologie und Hygiene der Haustiere, Universität Bonn, Katzenburgweg 7-9, D-5300 Bonn 1, FRG

C. Steeg, Max-Planck-Institut für Psychiatrie, Am Klopferspitz 18A, D-8033 Martinsried, FRG

G. Stranzinger, Institut für Nutztierwissenschaften, Gruppe Tierzucht, ETH-Zentrum, Tannenstr. 1, CH-8092 Zürich, Schweiz

R. Studer, Max-Planck-Institut für Psychiatrie, Am Klopferspitz 18A, D-8033 Martinsried, FRG

S. Suhai, Molecular Biophysics Group, German Cancer Research Center, Im Neuenheimer Feld 280, D-6900 Heidelberg 1, FRG

J.-L. Vilotte, Laboratoire de Génétique Biochimique, Institut National de la Recherche Agronomique, INRA-CRJ, Domaine de Vilvert, F-78350 Jouy-en-Josas, France

H. Zischler, Max-Planck-Institut für Psychiatrie, Am Klopferspitz 18A, D-8033 Martinsried, FRG

1. Introduction

H. Kräußlich

Chairman of DGFZ-Committee "Methoden der Biotechnologie in der Tierproduktion"

The introduction of efficient methods to characterise and map genes has dramatically increased the number of homologous genes that have been mapped in more than one species. Comparative data are presently available for more than 25 species of mammals and the number of genes mapped in more than one species ranges from less than 20 for several pairs of species (including farm animal species) to more than 250 for mouse and man [1]. Thus, comparative mapping is of considerable importance to establishing saturated maps in many species, but especially in domestic animals where chromosomal locations have been determined for a relatively small number of defined genes.

Fundamental to DNA diagnostics in animal breeding is the determination of nucleic acid sequences that are informative in important characters and the development of techniques that permit to monitor these sequences in individual animals at low cost. It is a challenge to the academic animal breeder to reach the standard, that is already available for the human and mouse geneticists, as soon as possible. It is the aim of the documentation of the papers of the symposium on "Genome Analysis in Domestic Animals" to provide valuable information to this end.

Knowledge of structure, function and organization of major genes directly allows application of molecular biology methods to animal breeding using genomic selection or gene transfer. But, most traits important to animal breeding undergo continuous variation due to multiple gene function that is modified by the evironment. Since decades quantitative geneticists where engaged to determine the location and number of genes that condition quantitative traits and to estimate the magnitude of individual gene effects. Paterson et. al. [2] demonstrated for the first time experimentally the dissection of genetic factors that condition a quantitative trait (fruit characters of tomato). They used a detailed linkage map and appropriate analytical methods. This work may ultimately lead to precise biochemical and molecular analysis of individual genes underlying quantitative characteristics of plants and animals.

Valid mapping experiments and marker based selection experiments are expensive and may in many cases not yet be justified by the technology currently available. Thus, national and international cooperation between scientists, breeding organisations and practical animal breeders are necessary to develop new methods, new strategies and new programmes in animal breeding. The symposium on "Genome Analysis in Domestic Animals" will provide valuable information for all participants and stimulate young scientists to work in this field.

References

[1]   Nadeau, J.H., *TIG*. *1989*, *5*, 82-86.
[2]   Paterson, A.H., Lander, E.S., Hewitt, J.D., Peterson, S.P., Lincoln, S.E. and Tanksley, S.D., *Nature* *1989*, *335*, 721-726.

2. Chromatin Organization in Interphase

R. S. Jack

2.1 Introduction

Eukaryotic nuclear genomes are packaged as chromatin in chromosomes. The resulting compaction of the DNA allows it to be fitted into the nucleus in a form which the cell can handle. However, packaged DNA is less accessible to sequence specific DNA binding proteins and hence the chromatin must be properly organized so as to permit regulated gene expression. With this in mind I will briefly review what is known about the structure of chromatin in interphase.

Both meiotic and mitotic chromosomes are sufficiently condensed to permit their structure to be examined by microscopy. Because of their much lower degree of condensation interphase chromosomes have proved more difficult to study. Nevertheless, over the past few years, considerable progress has been made. Two types of approach have been used to study chromatin structure. The first involves starting at the level of the simplest repeating unit and working towards greater complexity - working from the bottom up. The second takes the complementary approach of attempting to extract information from the intact chromosome - working from the top down.

2.2 Working from the Bottom Up

A list of statements about chromatin structure for which sustainable arguments could be presented would consist almost entirely of those derived from the strategy of working from the bottom up. This approach has involved the study in detail of the nucleosome - the smallest structural element of chromatin. A major milestone was reached with the solution of the nucleosome crystal structure at 7 A resolution [1]. An open array of such nucleosomes forms a fibre with a diameter of 10 nm [2]. The next higher structural level involves the condensation of the 10 nm fibre into one of 30 nm. This is achieved by winding of the 10 nm fibre into a contact helix or solenoid [3]. The formation of this solenoid is a self-assembly process which takes place spontaneously at moderate salt concentration [2, 3, 4, 5]. The solenoid has been shown to be present in high molecular weight preparations of native chromatin [6]. The great success of the strategy of working from the bottom up has quite properly not discouraged attempts to work from the top down. It is this latter approach which we will consider in some detail.

2.3   Working from the Top Down

The first problem in working with intact interphase chromosomes is that they are large delicate structures which are easily induced to precipitate into a tangled insoluble mess. This problem is generally avoided by keeping the chromosomes safely packaged inside the nuclei. In general in this type of work the nuclear membrane is removed with a mild non-ionic detergent treatment to increase the accessibility of the interior. Such detergent treated nuclei are bound by the lamin meshwork [7] which maintains the structural integrity of the organelle.

When considering experimental manipulations carried out on isolated nuclei it is important to bear in mind the stupendous macromolecular solute concentrations which are involved. DNA is present inside a eukaryotic nucleus at a concentration of around 100 mg/ml, there is about the same amount of RNA, and the concentration of protein is even higher. Exposure of nuclei to extremely non physiological conditions may therefore result in the generation of artefactual associations which do not tell us anything about the *in vivo* organization of interphase chromosomes. Two examples should suffice to act as an awful warning. The c-myc protein is localized in nuclei from which it can be quantitatively extracted with 200 mM salt. If however the isolated nuclei are exposed to temperatures above 35°C then a rapid and irreversible change takes place such that c-myc becomes tightly bound within the nucleus [8]. This curious effect is by no means restricted to the c-myc protein. In rat liver nuclei virtually all of the extractable proteins become trapped in this way after a brief exposure to 37°C [9]. The molecular basis for this effect is unknown.

The second example concerns the isolation of a protein from rat liver which bound with high specificity to single stranded DNA. Such proteins are often referred to as helix destabilizing proteins (HDP). Antisera to this protein cross-react strongly with preparations of HDPs from mouse, human and *Drosophila* cells. Immunofluorescence on *Drosophila* salivary gland polytene chromosome preparations showed that the protein was localized in puffed regions [10]. Subsequent work identified this protein. It is nothing more than lactate dehydrogenase, a protein of the intermediary metabolism which binds nicotinamide adenine dinucleotide as a cofactor [11]. Not only does the enzyme come through the screen used to isolate HDPs, it artefactually binds to polytene chromosomes during the preparation of the squashes. These examples illustrate a problem which bedevils the whole field. Interphase chromosomes are complex structures. In order to obtain interpretable information most investigative strategies start off by reducing the degree of complexity. Unfortunately the steps taken to achieve this are frequently not gentle. All are intrusive - some bizarrely so. The results obtained must then be viewed in the light of the experimental steps taken to obtain them. Do they really reflect chromosome structure or are they merely demonstrating for us the ease with which artefacts can be induced? For these reasons we will first of all critically consider the principle methodologies involved and the results obtained using them. We will then try to see to what extent these results form a picture of the way in which interphase chromatin is organized in the chromosome.

2.4   Methodologies

2.4.1 Ethidium Bromide Titration

Ethidium bromide is an intercalating dye which introduces positive supercoiling into closed circular DNA molecules. Titration of a negatively supercoiled plasmid with increasing concentrations of ethidium bromide therefore progressively removes the naturally occurring negative supercoils. A point is reached at which the amount of positive supercoiling introduced by ethidium binding exactly counters the negative supercoiling introduced *in vivo*. At this point the plasmid is a relaxed circle. Increasing the ethidium concentration beyond this point results in the molecule being wound up as a positive supercoil. The change from supercoil to open circle can be experimentally visualized as a change in sedimentation rate. Experiments of this sort have been carried out on carefully prepared high molecular weight eukaryotic chromatin [12, 13]. The results suggest that the chromosome is divided up into a series of independent topological domains. Values in the range of 80 kb were obtained both for *Drosophila* domains [13] and for those from rat liver [14]. The interphase chromosome is therefore organized, at least in part, into closed topological domains of about 80 kb. This is one of the very few statements of structural consequence for the chromosome which might command general agreement.

2.4.2 2M Salt Extraction

This protocol, developed to try to obtain information about the complex, probably labile and certainly dynamic structure of the interphase chromosome, involved extracting nuclei with 2M salt and trashing the genome with DNAseI [15]. Not surprisingly most of the DNA is released from the nuclei but some always remains inside. That which remains is deemed to represent the fraction of the genome which interacts with the nuclear matrix [16]. The nuclear matrix is operationally defined as the fibrous array visible in EM thin sections of nuclei which have been extracted with high salt and treated with DNAseI. This matrix is composed of RNA and protein [17]. It has been suggested that the matrix forms a structural network within the nucleus with which the genome interacts. Transcribed sequences are found to be specifically associated with it [18]. However exposure of nuclei to high salt precipitates RNPs [19] as well as some proteins [20] so that a strong suspicion exists that the nuclear matrix is nothing more than a precipitation artefact. This is not to say that a nuclear matrix does not exist but merely that it is a concept for which no compelling experimental justification is currently available. The proposition that anything useful can be learned about the structure of interphase chromosomes after 2M salt extraction must be viewed with extreme caution.

2.4.3 Detergent Extraction

The next protocol was developed specifically with a view to avoiding the artefacts associated with high salt extractions. This procedure is carried out at low salt but requires instead the use of the powerful detergent lithium diiodo salicylate (LDIS). Nuclei are prepared and then exposed to 37°C to "stabilize" the chromosomes. This heat treatment is an essential prerequisite of the protocol. The nuclei are then extracted with LDIS. The effect of this extraction is to destroy transcription complexes and to quantitatively strip the histones off the genome [20, 21]. The extracted nuclei are then treated with restriction enzymes which release the bulk of the genome. Some DNA does however remain associated with the nuclei and this is considered to represent those sequences which interact with structural components within the nucleus.

The results obtained using this protocol differ from those obtained with the 2M salt procedure in two important respects. First, those sequences retained inside the nuclei after high salt extraction are released in the LDIS procedure. Thus, transcribed sequences score as matrix attached after high salt extraction but are - with few exceptions - released in the LDIS protocol [21, 22, 23]. Second, in the LDIS procedure specific restriction fragments are found to be attached irrespective of the developmental status of the cells used to prepare the nuclei. This result has been interpreted as demonstrating that there exists a fixed structural framework or "scaffold" and that specific sequences direct attachment to it. Analysis of a collection of such scaffold bound fragments has revealed the presence of three conserved sequence elements. The first of these is an A rich sequence termed the A box. The second is a T rich sequence known as the T box. The third is a group of sequences which have a passing resemblance to the rather vaguely defined Topoisomerase II consensus sequence and which are collectively termed the Top 2 box [22].

A detailed analysis of one such scaffold binding site has been undertaken. This is the site which lies upstream from the adult promoter of the *Drosophila* alcohol dehydrogenase gene. Judicious choice of restriction sites was used to demonstrate that fragments lacking A or T boxes are scaffold bound, while a fragment containing five copies of the Top 2 box was released. This indicates that neither the A nor the T boxes are necessary for binding and that multiple copies of the Top 2 box are not sufficient [22]. Apart from the fact that all attached fragments contain highly AT rich regions within them, the sequence requirements for attachment remain, for the present, obscure.

Sequences which are bound in the LDIS protocol share the property that when added as end labelled DNA fragments to extracted and restriction digested nuclei, they bind with great specificity [22, 23]. In contrast, when the same experiment is carried out using fragments which do not contain an attachment site, no specific binding takes place. Thus, scaffold binding is an autonomous property of these small restriction fragments. It follows that the genomic fragments which score as scaffold bound may in fact have associated with the scaffold only after LDIS extraction.

This may well provide the explanation for the fact that scaffold attachment sites have turned out to be remarkably frequent. In addition, binding induced during the course of the experiment would neatly explain the inherently unlikely proposition that a functioning attachment site is present within the mouse immunoglobulin kappa sequence in myeloma cells within which this gene is being heavily transcribed [23]. Were attachment to contribute to gene regulation, one might well expect that the genome would contain many more sites than are attached in any given cell type. Unfortunately the LDIS protocol cannot tell us whether any particular site was cryptic or attached at the start of the experiment.

Although the LDIS protocol is clearly a great advance on the 2M salt extraction procedure, the absolute requirement for a heat treatment of the nuclei prior to extraction is certainly disquieting [21]. Furthermore the possibility that exposure of the nuclei to a powerful detergent may result in the generation of artefacts remains a problem which must be kept in mind when assessing the results.

2.4.4 Encapsulation

A sophisticated means of avoiding some of the potential artefacts has recently been introduced. In this procedure cells are first encapsulated within agarose beads [24]. The cells may then be grown for a period to allow them to recover from this manipulation after which the preparation of nuclei may be carried out inside the beads. The great advantage of this procedure is that the preparation of the nuclei can be done under isotonic conditions in which free nuclei are difficult to handle. Digestion of the genome under mild conditions followed by electroelution of the liberated *chromatin* fragments has permitted the visualization of a matrix-like structure within the nucleus [25]. This nucleoskeleton is spread throughout the nucleus, is composed of elements similar in appearance to cytoplasmic intermediate filaments, and is associated with residual clumps of chromatin.

2.4.5 High Resolution *in situ* Analysis

These experiments have made use of the polytene chromosomes present in the salivary glands and other tissues of *Drosophila* third instar larvae. In these interphase chromosomes extra rounds of replication of the paired homologous result in the formation of bundles of precisely aligned chromosomes whose structure is amenable to analysis under the light microscope. Fixed and squashed salivary gland polytene chromosomes have been used to carry out a high resolution *in situ* analysis of the *Notch* locus [26]. Probes from various positions along the molecular map were hybridized to polytene chromosome preparations. Where this region of the genome to be organized as a closed loop lying perpendicular to the chromosome axis then the signals derived from the various probes could not be resolved at the cytological level. The results, however, demonstrate that the molecular and cytological maps are collinear even within the band. This therefore tells us that the *Notch* locus is not organized as a chromatin loop. The transcription unit itself lies within the band and a comparison of its cytological versus its molecular dimensions suggests that the chromatin in this band has a packing ratio compatible with it being in the form of a 30 nm fibre. The control region of the gene lies in the adjacent interband which has a more open structure and may exist as a 10 nm fibre.

2.4.6 Optical Sectioning

Optical sectioning is without doubt the least intrusive non-genetic approach used to date. A fluorescence study in which Nomarski optics was used to generate optical sections of polytene nuclei provided the first information on the distribution of the chromosome within the nucleus. The results demonstrate that the chromosomes interact with the nuclear envelope at points along their length. The chromosome arms are not intertwined. The nucleolus is situated in - but does not fill - the central region. The fused chromocentres are situated close to the nuclear envelope and lie at the opposite side of the nucleus from the telomeres [27, 28]. A comparison of polytene nuclei from different tissues indicated that the spatial organization of the chromosome arms is not entirely constant [29, 30]. However, in none of the tissues studied does the relative positioning of the chromosome arms provide any evidence to support the notion that the spatial organization of the nucleus is rigidly controlled. Furthermore, the sites of interaction of the chromosome arms with the nuclear envelope do not obviously relate to the pattern of transcriptional activity.

2.4.7 Genetic Analysis

Genetic analysis provides a means of probing the higher order structure of interphase chromosomes in a way which is non-intrusive and non-destructive. The information which it gives is not directly interpretable in molecular terms but it does yield functional information that no other currently available methodology can.

The correct temporal and spatial expression of genes contained within many chromosomal rearrangements tells us that the precise overall order of genetic loci along the chromosome is unimportant for regulated gene expression. This is confirmed by the successful production of transgenic animals in which genes are expressed in the correct tissue specific manner even when inserted at "inappropriate" positions. However, position effects do occur. The precise details of expression of any particular construct vary from line to line. A good example of this is the collection of experiments involving the *Drosophila white* gene. Mutations at this gene result in alterations in the amount or distribution of pigment in the adult eyes, ocelli, and testis sheath, and in the larval malpighian tubules. Deletion of this non-essential locus abolishes pigmentation of these tissues. A large number of transgenic lines have been produced in which insertion of the gene into ectopic chromosomal locations allows normal pigmentation in animals carrying a deletion of the wild type *white* locus [31, 32, 33].

Powerful position effects may arise when a chromosomal rearrangement places a gene close to centromeric heterochromatin. When this happens the gene may be inactivated. The inactivation need not take place in every cell, in which case the phenomenon of position effect variegation results (for review see Spofford [34]). Among the *white* transgenic lines a number were recovered in which the pigmentation of the eye was variegated. In two cases the fact that the variegation was indeed due to a position effect was proved by mobilizing the transgene and recovering new inserts which gave homogeneously colored eyes [35].

A second type of position effect is seen in transvecting systems [36, 37]. These situations involve a curious form of complementation between alleles. Briefly, two alleles one of which is mutant within the transcribed region and the other mutant in an upstream controlling sequence , may complement each other. What seems to be happening is that the intact promoter element upstream of the disrupted transcription unit can be recruited to run the intact transcription unit of the allele whose promoter is inactive [38]. This form of complementation works only if the two alleles are so situated that they can pair. Rearrangements which locally disrupt pairing also disrupt the transvection effect. It has long been recognized that the major conclusion to be drawn from the existence of transvection effects is that *Drosophila* interphase chromosomes are paired. This fact alone is of great importance since it must have a structural basis and any model of the *Drosophila* interphase chromosome must be able to account for it. Whether interphase pairing is a general phenomenon in higher eukaryotes or whether it is a peculiarity of the fly remains a critical unanswered question.

Chromosome pairing could only modulate gene expression if it is at least locally reversible. The only well studied example of transcriptional regulation of wild type alleles which is pairing dependent involves the interaction between the genes *zeste* and *white*. The wild type product of *zeste* is a DNA binding protein which has been shown to bind *in vitro* to an enhancer like element lying upstream of the *white* promoter [39]. A mutation at *zeste* (z^1) depresses transcription at the *white* locus [40, 41]. This transcriptional down-regulation takes place only if there are two copies of the *white* gene present which may pair. Thus the result of the interaction of the *zeste*[1] product with paired and unpaired copies of the wild type allele of white are quite different.

The *zeste* - *white* interaction is the clearest demonstration that chromosome pairing may indeed affect gene expression. However it must be kept in mind that while this principle is established we still do not know whether pairing dependence really makes a significant contribution to gene regulation. The problem is that *zeste*[1] is a mutation. We simply do not yet know whether *zeste*[1] tells us something fundamental about gene expression or whether it is merely an exotic genetic curiosity.

While in general transgenes introduced into *Drosophila* function remarkably well this may not necessarily be the case in other systems. Genes introduced into the mouse genome by transformation are frequently found to be expressed at rather low levels. Furthermore the level of expression achieved is quite strongly influenced by the integration position. Experiments in which part of the human ß globin locus were introduced into mice permitted the identification of a so-called **D**ominant **C**ontrol **R**egion (DCR). The DCR has the ability to work like a very strong tissue specific enhancer elements [42]. A globin gene introduced on a vector carrying the DCR is expressed at a very high level. Furthermore this high level of expression seems now to be independent of the integration position. Sequences with similar properties have been found associated with the gene for the human T cell marker CD2 [43] and the chicken lysozyme gene [44]. Though it is not known how these DCR sequences function it will be clear that they have considerable potential value in facilitating high levels of expression from transgenes.

2.5   What Can One Make of All This?

The mere fact that a rapidly moving field may present a confusing picture is no cause for disappointment still less for despair. Although the structure of the interphase chromosome is very far from being solved the relevant questions can now be posed in a more focussed form. A few of the obvious questions are considered below.

2.5.1   Basic Organization of Chromatin in Interphase

The winding of the 10 nm nucleosomal filament into the 30 nm solenoid provides the basic structural element of the interphase chromosome. This 30 nm fibre has been demonstrated to be present in preparations of high molecular weight interphase chromatin [6]. Less directly, the *Notch* transcription unit has been shown to be contained in chromatin compacted to the level predicted for a 30 nm fibre [26]. If the picture which has emerged from the analysis of *Notch* turns out to be general then the basic structural motive of the interphase chromosome will be of large domains folded into 30 nm fibres separated by shorter stretches of more open chromatin.

The *Notch* gene is not transcribed in third instar salivary glands. For this reason the 30 nm fibre may serve to package those sequences which are not being transcribed. Certainly the conversion of polytene chromosome bands into puffs at transcriptionally active sites indicates that active chromatin is much less condensed than the chromatin in an inactive band. The physical basis for deciding which sequences are compacted to which level is entirely unclear at the moment. Equally there is no information as to the molecular mechanisms by which a change of compaction level is effected.

2.5.2 Interphase Domains

The interphase chromosome is divided up into a series of topologically closed domains [12, 13, 14]. The formation of a domain within a linear chromosome requires that the ends of the domains be held fixed with respect to each other. How is this achieved? It might be done by having specific proteins cross link the ends of the domain so as to form a loop. The notion that interphase chromosomes are formed in a series of loops is attractive since this is a structural motive seen in metaphase. If this is what happens then, merely to form the domains, there is no requirement for a matrix or scaffold. Alternatively we can dispense with loops by fixing the ends of a linear domain to a rigid matrix. The results from the high resolution *in situ* analysis on the *Notch* locus were interpreted as showing that this part of the chromosome is not organized into a loop [26]. If this turns out to be generally true then we require no loops but do require a fixed structural element to organize the chromatin.

The closed domains within the interphase chromosome are estimated to be about 80 kb in length. This is within the same range as the estimated length of sequence in a polytene chromosome band. Are closed domains and bands the same thing? If yes, then the sequences which lie at the band interband borders at *Notch* would be expected not only to define these borders but they should have the capacity to direct the formation of the limit of a domain of topology. Though experiments to establish these points will not be easy, they are nevertheless crucially important if the concepts arising from the diverse collection of current methodologies are to be usefully synthesized.

2.5.3 Supercoiling

The DNA is wrapped in a left handed superhelix around the surface of the nucleosome. If this is carried out within a closed domain then the negative supercoiling which this process introduces will be equilibrated by the introduction of positive supercoiling in the inter nucleosomal linker regions. The negative supercoiling caused by the association of the DNA with the nucleosome is said to be constrained. This is because it is trapped by the tight interaction between the DNA and the histones. A break in the DNA helix introduced by a topoisomerase will not release it. In contrast the positive supercoiling in the linker regions is not constrained and will be readily released by topoisomerase. The domain is thereafter negatively supercoiled but all of the supercoiling is constrained. The closed chromatin domains defined in the ethidium titration experiments [13] have been aptly described as being "topologically stressed but energetically relaxed" [45]. This is in contrast to the situation in prokaryotes where unconstrained supercoiling can and does contribute to gene regulation [46].

2.5.4 The Interphase Nuclear Matrix

The idea that chromosomes possess a central structural element is derived from work on metaphase chromosomes. The scaffold - like the nuclear matrix - is an operationally defined structure. Like the nuclear matrix doubts have been expressed as to its reality [47]. Nevertheless, the biochemical and electron microscopy data in favor of a metaphase chromosome scaffold is persuasive [48, 49, 50, 51], though not unchallenged [52]. The question that concerns us is whether there is any solid evidence to support the idea that a structure exists within the interphase nucleus which organizes the interphase chromosome. Evidence in favor of an interphase scaffold or matrix comes from the work referred to above in which scaffold attachment sites are mapped using LDIS extracted nuclei. Scaffold preparations obtained from interphase nuclei contain the enzyme Topoisomerase II [53]. This enzyme has been localized on *Drosophila* polytene chromosomes [54]. The enzyme is also a major component of metaphase chromosome scaffolds [51, 55]. Fragments which score as scaffold attached in the LDIS assay contain sequences resembling the Topoisomerase II consensus sequence [56]. The conclusion drawn from this is that Topoisomerase II is a major component of the chromosomal scaffold both during interphase and metaphase and that it is involved in binding the genome. There is no serious argument with the idea that Topoisomerase II is associated with the chromosomes. The crucial point is whether it is organized into an insoluble scaffold *in vivo* or whether this is something which is merely an artefact of the preparation procedures used. The evidence available at the moment is insufficient to resolve this point. Evidence in favor of a nuclear skeleton comes from the work using encapsulated cells [25]. Biochemical analysis of the components of this skeleton may go a long way to resolving many of the problems which have so far dogged attempts to define structural elements within the nucleus.

Whatever the status of the matrix/scaffold/skeleton, the evidence from optical sectioning does strongly suggest that interactions take place between the chromosomes and the nuclear envelope. There is, however, no evidence in this data that these interactions are directly related to the transcriptional status of particular genes. There is however suggestive biochemical data which indicates that expressed genes may indeed tend to be in close proximity to the nuclear envelope [57].

2.5.5 Chromosome Structure and Gene Expression

Gene expression in eukaryotes, like that in prokaryotes, is largely carried out by the interaction of specific transcription factors with the genome. It is unlikely that this would take place efficiently on segments of the genome which are wrapped up as a 30 nm fibre. It therefore seems almost inevitable that the structural realities of the chromosome are bound to affect the ways in which gene regulation is carried out. However, it would be unwise to try to explain too much on a purely structural basis. After all, *in vitro* transcription systems which use naked DNA as template, have been enormously successful in mimicking the regulation of specific genes. It may well be that the contribution of dynamic chromosome structure to gene regulation lies in modulating the availability of the genomic targets for transacting transcription factors. The *Notch* data suggest that the controlling region of this transcription unit is somehow prevented from forming a 30 nm fibre. An elucidation of the way in which this is achieved will tell us much about how chromosome structure and gene regulation are integrated. Furthermore, the controlling regions of genes are usually associated with short hypersensitive sites. Some of these sites are present all the time and are said to be constitutive, but many of them are inducible. These sites are generally thought of as being nucleosome free regions within a stretch of chromatin. Their generation and maintenance are poorly understood though they are fundamental to the control of eukaryotic genes. Beyond this, nuclear architecture may contribute to the control of gene expression by regulating access of factors and substrates.

2.6   Perspectives

It will be clear from the foregoing that progress in understanding the structure of the interphase chromosome will depend to a large degree on the development and exploitation of those methodologies which are least intrusive. Where this is not possible then the results must be taken with a pinch of salt until they, or predictions flowing from them, are testable by procedures which are artefact free. The molecular basis of domain formation, the organization of chromatin beyond the level of the 30 nm fibre, the precise nature of a nuclear scaffold, and the interplay between chromosome structure and gene expression are areas in which rapid progress can be expected.

2.7   References

[1] Richmond, T.J., Finch, J.T., Rushton, B., Rhodes, D., and Klug, A., *Nature* **1984**, *311*, 532-537.
[2] Thoma, F., Koller, T., and Klug, A., *J.Cell Biol.* **1979**, *83*, 403-427.
[3] Finch, J.T., and Klug, A., *Proc.Natl.Acad.Sci.USA* **1976**, *73*, 1897-1901.
[4] Butler, P.J.G., and Thomas, J.O., *J.Mol.Biol.* **1980**, *140*, 505-529.
[5] Widom, J., *J.Mol.Biol.* **1986**, *190*, 411-424.
[6] Widom, J., and Klug, A., *Cell* **1985**, *43*, 207-213.
[7] Aebi, U., Cohn, J., Buhle, L., and Gerace, L., *Nature* **1986**, *323*, 560-564.
[8] Evan, G.I.and Hancock, D.C., *Cell* **1985**, *43*, 253-261.
[9] Izaurralde, E., Mirkovitch, J., and Laemmli, U.K., *J.Mol.Biol.* **1988**, *200*, 111-125.
[10] Patel, G.L., and Thompson, P.E., *Proc.Natl.Acad.Sci.USA* **1980**, *77*, 6749-6753.
[11] Williams, K.R., Reddigari, S., and Patel, G.L., *Proc.Natl.Acad.Sci.USA* **1985**, *82*, 5260-5264.

[12] Cook, P.R.and Brazell, I.A. *J.Cell Sci.* *1975, 19,* 261-279.

[13] Benjayati, C., and Worcel, A., *Cell 1976, 9,* 393-407.

[14] Igo-Kemenes, T., and Zachau, H.G., *Cold Spring Harbor Symp.Quant.Biol. 1976, 42,* 109-118.

[15] Bekhor, I., and Mirell, C.J., *Biochemistry 1979, 18,* 609-616.

[16] Berezney, R., and Coffey, D.S., *Biochem.Biophys.Res.Commun. 1974, 60,* 1410-1417.

[17] Nelson, W.G., Pienta, K.J., Barrack, E.R., and Coffey, D.S., *Ann.Rev.Biophys.Chem. 1986, 15,* 457-475.

[18] Robinson, S.I., Nelkin, B.D.and Vogelstein, B., *Cell 1982, 28,* 99-106.

[19] Kirov, N., Djondjurov, L., and Tsanev, T., *J.Mol.Biol. 1984, 180,* 601-614.

[20] Roberge, M., Dahmus, M.E., and Bradbury, E.M., *J.Mol.Biol. 1988, 201,* 545-555.

[21] Mirkovitch, J., Mirault, M-E., and Laemmli, U.K., *Cell 1984, 39,* 223-232.

[22] Gasser, S.M., and Laemmli, U.K., *Cell 1986, 46,* 521-530.

[23] Cockerill, P.N., and Garrod, W.T., *Cell 1986, 44,* 273-282.

[24] Jackson, D.A., and Cook, P.R., *EMBO J. 1985, 4,* 913-918.

[25] Jackson, D.A., and Cook, P.R., *EMBO J. 1988, 7,* 3667-3677.

[26] Rykowski, M.C., Parmelee, S.J., Agard, D.A., and Sedat, J.W., *Cell 1988, 54,* 461-472.

[27] Agard, D.A., and Sedat, J.W., *Nature 1983, 302,* 676-681.

[28] Hochstrasser, M., Mathog, D., Gruenbaum, Y., Saumweber, H., and Sedat, J.W., *J.Cell Biol. 1986, 102,* 112-123.

[29] Hochstrasser, M., and Sedat, J.W., *J.Cell Biol. 1987, 104,* 1455-1470.

[30] Hochstrasser, M., and Sedat, J.W., *J.Cell Biol. 1987, 104,* 1471-1483.

[31] Gehring, W.J., Klemenz, R., Weber, U., and Kloter, U., *EMBO J. 1984, 3,* 2077-2085.

[32] Hazelrigg, T., Levis, R., and Rubin, G.M., *Cell 1984, 36,* 469-481.

[33] Pirotta, V., Steller, H., and Bozzetti, M.P., *EMBO J. 1985, 4,* 3501-3508.

[34] Spofford, J.B., in; *The Genetics and Biology of Drosophila.* Ashburner, M., and Noviski, E. Eds. Academic Press, *1976.*

[35] Levis, R., Hazelrigg, T., and Rubin, G.M., *Science 1985, 229,* 558-561.

[36] Lewis, E.B., *Am. Nat. 1954, 88,* 225-239.

[37] Lewis, E.B., *Genetics 1956, 41,* 651.

[38] Biggin, M.D., Bickel, S., Benson, M., Pirotta, V., and Tjian, R., *Cell 1988, 53,* 713-722.

[39] Benson, M., and Pirotta, V., *EMBO J. 1987, 6,* 1387-1392. [40] Jack, J.W., and Judd, B.H., *Proc.Natl.Acad.Sci.USA 1979, 76,* 1368-1372.

[41] Bingham, P.M., and Zachar, Z., *Cell 1985, 40,* 819-825.

[42] van Assendelft, G.B., Hanscombe, O., Grosfeld, F., and Greaves, D.R., *Cell 1989, 56,* 969-977.

[43] Greaves, D.R., Wilson, F.D., Lang, G., and Kioussis, D., *Cell 1989, 56,* 979-986.

[44] Steif, A., Winter, D.M., Strätling, W.H., and Sippel, A.E., *Nature 1989, 341,* 343-345.

[45] Futcher, B., *TIG 1988, 4,* 271-272.

[46] Higgins, C.F., Dorman, C.J., Stirling, D.A., Waddel, L., Booth, I.R., May, G., and Bremer, E., *Cell 1988, 52,* 569-584.

[47] Hadlaczky, G., Sumner, A.T., and Ross, A., *Chromosma* **1981**, *81*, 557-567.

[48] Adolph, K.W., Cheng, S.M., Paulson, J.R., and Laemmli, U.K., *Proc.Natl.Acad.Sci. USA* **1977**, *74*, 4937-4941.

[49] Paulson, J.R., and Laemmli, U.K., *Cell* **1977**, *12*, 817-828.

[50] Marsden, M.P.F., and Laemmli, U.K., *Cell* **1979**, *17*, 849-858.

[51] Gasser, S.M., Laroche, T., Falquet, J., Boy de la Tour, E., and Laemmli, U.K., *J.Mol.Biol.* **1986**, *188*, 613-629.

[52] Labhart, P., and Koller, T., *Cell* **1982**, *30*, 115-121.

[53] Mirkovitch, J., Gasser, S.M., and Laemmli, U.K., *J.Mol.Biol.* **1988**, *200*, 101-109.

[54] Berrios, M., Osheroff, N., and Fisher, P.A., *Proc.Natl.Acad.Sci.USA* **1985**, *82*, 4142-4146.

[55] Earnshaw, W.C., Halligan, B., Cooke, C.A., Heck, M.M.S., and Liu, L.F., *J.Cell Biol.* **1985**, *100*, 1706-1715.

[56] Gasser, S.M., and Laemmli, U.K., *EMBO J.* **1986**, *5*, 511-518.

[57] Hutchinson, N., and Weintraub, H., *Cell* **1985**, *43*, 471-482.

3. On the Organization of Animal Genomes: Ubiquitous Interspersion of Repetitive DNA Sequences

R. Studer and J.T. Epplen

3.1 Introduction

Most eukaryotic genomes are orders of magnitude larger than those of prokaryotes. However, much of the eukaryotic DNA appears to be in excess of the minimum required in a given group of species [1]. Repetitive sequences [2] constitute a major part of this "secondary DNA". Secondary DNA has been supposed to have structural meaning, i.e. in the architecture of the interphase nucleus, meiotic chromatin and/or the metaphase chromosomes. In interphase nuclei the chromatin fibers are organized into domains with DNA loops of some 100 kilobases (kb) in length. To form mitotic chromosomes the individual DNA molecule of the uninem chromatid has to be compacted over 10,000-fold [3]. Though repetitive DNA sequences (and protein components of the chromatin fiber) appear as candidates for attachment sites to accomplish the compactation in a radial loop/"scaffold" model [4], their involvement in this process has not been proven positively. Similarly during meiotic prophase, the chromatin undergoes remarkable organizational changes. The meiotic arrangement must sequentially position the genes in a linear order measured by genetic linkage. Meiotic recombination takes place between homologous chromosomes which are partly parallel-aligned in the configuration of the synaptonemal complex. Only recently the first sequences of synaptonemal complexes have been isolated and preliminarily characterized: Clearly these elements are of repetitive nature but they share no homology with known sequences [5].

Enormous amounts of information have accumulated on repetitive elements and their organization in animal genomes. Consequently there is a considerable range of opinions on their meaning (see Fig. 3-1). Apart from that discussion it has been useful to classify repetitive sequences according to their structure, distribution and reiteration frequency [6, 7]. Here we discuss different aspects of selected repetitive elements that are *interspersed* in eukaryotic genomes. First we concentrate on interspersed simple, tandemly repeated DNA and subsequently we deal with long and short interspersed nucleotide elements - LINE's and SINE's [8].

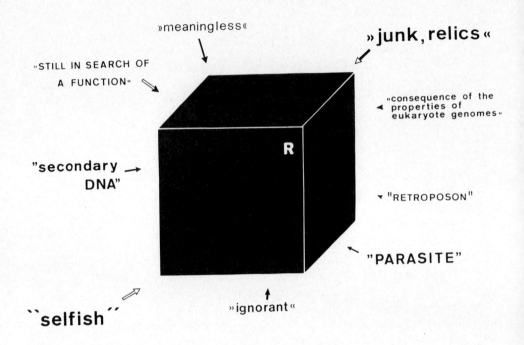

Fig. 3-1 Repetitive DNA (R) is simplistically symbolized as a black box. Since many different types of repetitive elements are artificially combined in this black box, numerous opinions have been expressed concerning the nature and meaning of this purely kinetically defined category of DNA. The collection of vocabulary given is by no means exhaustive; it represents to some extent only the extremes of a broad range of judgements [49, 53, 114, 115, 116, 117, 118].

3.2 Interspersed Simple Tandemly Repeated Sequences

"Typical satellite DNA is composed of short, tandemly repeated sequences each usually less than 1000 nucleotide base pairs in length and present in more than 10^5 copies per haploid genome" [9]. Large blocks of these sequences are often concentrated in specific parts of the chromosome complement, the heterochromatin [10]. This subject was extensively reviewed by John and Miklos [11] leaving little to be added to the field of typical satellite DNA's. Simple repetitive sequences are of even lower complexity with short sequence motives of less than 10 base pairs (bp) repeated over and over in a tandem fashion. Different mechanisms have probably contributed simultaneously to generate the tandem sequence arrays. These mechanisms could include: several consecutive unequal crossing-over events between initially repetitive or at least duplicated sequences [12], slipped-strand mispairing [13, 14, 15, 16], a "rolling circle" mechanism [17] and "aberrant *in situ* replication" [18]. Any of the aforementioned random mechanisms can produce two kinds of the simplest possible repeats when the unit is one base (e.g. A and G; T and C are on complementary strand). Obviously poly (dA)$^+$ stretches are adequately explained by reinsertion into the genome of poly (A)$^+$ mRNAs after their reverse transcription. Poly (dG)$^+$ could initially be polymerized by terminal nucleotidyl transferase [19] as exemplified in the antigen recognition portions of immunoglobulin and T lymphocyte antigen receptor genes. Redistribution and amplification requires additional mechanisms such as the ones mentioned above (for retroposition see also below). Apart from the homopolymers, how many different simple repeats are principally possible? If the basic simple repeat unit is two bases long, six different permutations are unique (AA, AC, AG, AT, CC, CG). All the other possibilities with two bases are covered by shifting the phase by one nucleotide or are represented on the complementary strand, respectively. Using three bases there are 18 different base patterns possible out of the 64 different base triplets. It seems obvious that the formula $4^{n-1} + 2$ (for $n \geq 2$) applies [20]. For tetrameric repeats 66 ($4^{4-1}+2$) different simple tandem units exist. Recently a specific hexamer repeat has gained widespread attention: The $(TTAGGG)_n$ elements are present at the telomeres of a variety of animal chromosomes [21]. Their similarity to functional telomeres isolated from lower eukaryotes suggests that these sequences are functional animal telomeres. In lower eukaryotes like *Tetrahymena* hypothetical "telomerases" synthesize the ends of chromosomes in a non-templated way one base at a time [22].

Higher eukaryote genomes have so far not been screened systematically for the presence of each of the theoretically possible sequence permutations. Those simple sequences that are present in an animal species often give rise to polymorphism. By carefully hybridizing with a panel of simple repeat oligonucleotide probes it seems possible to select one or a few to individualize animals in all species [23]. Yet it is safe to state that many of these simple repeat combinations are not present in the investigated eukaryote genomes in very substantial amounts [20]. Furthermore sequence specific amplification mechanisms for selected nucleotide combinations are not known with the exception of the telomeric simple repeats described above. The spreading of certain simple sequences may be the result of chance in any given species, e.g. by being included in a retroposable unit.

3.2.1 $(CA/TG)_n$ Simple Repeats

One of the most frequently occurring simple DNA sequences is the alternating purine pyrimidine co-polymer $(CA/TG)_n$ first identified in spacer DNA of the murine rRNA [24]. Members of the $(CA/TG)_n$ family are present in all eukaryotic DNAs [23, 25, 26]. Rogers [27] has suggested that $(CA/TG)_n$ sequences may have arisen at staggered breaks in the target DNA by 3'-polymerization of $(CA)_n$ onto one strand or by insertion and ligation of preexisting $(CA)_n$ stretches. Once invented and present the genome $(CA/TG)_n$ blocks may spread and insert by a retroposon-like mechanism (see below). A "retrogene" generation for the $(CA/TG)_n$ family is not generally accepted [28], although the evidence in favor of such a mechanism is manifold [29; for a sequence compilation see 27, 30]. The transcriptional activity of the $(CA/TG)_n$ simple repeats is limited and due to RNA polymerase II. As for other simple repeated sequences it has been suggested that elements are adventitiously included in longer transcription units and thus may survive RNA processing [28]. $(CA/TG)_n$ blocks are indeed often seen in the vicinity of genes including those belonging to the immunoglobulin gene superfamily [31]. We also stumbled upon several additional $(CA/TG)_n$ blocks in the human and mouse T-lymphocyte receptor alpha- and ß-chain loci [32, and own unpublished observations]: $(TG)_{21}$ was found in an intron of a human variable $(V_ß)$ element, $(CA)_{24}$ is present in the vicinity of a novel mouse joining (J_α) segment. The meaning of these sequence associations remains elusive so far, especially with respect to the regulation of transcription. It has been demonstrated, however, that $(CA/TG)_n$ repeats have in certain expression vectors a weak enhancer-like activity [33].

Isolated CACA tetramers could also serve as precursor elements for the GACA simple quadruplet repeats. A potential additional transition mutation in GACA generates the GATA element. Again amplification and redistribution modes are required to multiply and spread the initial simple repeats: The inclusion of $(CA/TG)_n$ sequences into otherwise unrelated retrogenes would perhaps be the most straight forward explanation for dispersion in the genome.

3.2.2 GATA/GACA Simple Repeats

Initially GATA/GACA simple quadruplet repeats were isolated from female Colubrid snakes in the form of sex-specific satellites (*Bkm* DNA from the *Banded krait* minor satellite; [34]). The identification of the GATA/GACA components in the satellite was first possible in cloned *Elaphe radiata* DNA [35]. Tandem arrays of GATA/GACA repeats have been demonstrated in every eukaryote investigated so far [36]. The smallest and fundamental unit of amplification is the tetrameric repeat, but the final stretches of tandemly repeated sequences may be extremely large. The possible mechanisms for this amplification have been discussed above. Independent mechanisms may contribute to the phenomenon of random distribution (see below). As a result of the spreading mechanisms, the GATA/GACA elements are randomly interspersed throughout the autosomal chromosomes in several mammalian and reptile species [20]. Often, the organization of GATA/GACA repeats varies between the heteromorphic sex chromosomes within a certain species. Unequal crossing-over between homologous loci on the heteromorphic sex chromosomes cannot account for the difference in GATA/GACA repeat length, since these sequences are located outside the meiotic pairing regions. The gonosomal GATA/GACA distributions in mice, snakes, fish [37] and several additional species are suggestive that these sequences are involved in primary sex determination [34]. Rather than determining sex themselves, GATA/GACA sequences may be closely linked to sex-determining genes [38].

Accumulation of GATA/GACA sequences on the heteromorphic sex chromosomes is not the only interesting phenomenon observed in the investigation of these elements. In the mull-vole *Ellobius lutescens* for example the sex-chromosomes had not been identified. This species is endowed with 17 chromosomes whereby the singular chromosome number 9 does not pair during meiosis. Initially we had entertained the suspicion that chromosome 9 is the X chromosome because GATA/GACA repeats were significantly reduced on the presumptive X chromosome as demonstrated by oligonucleotide hybridization *in situ*. Recently this view could be confirmed by somatic cell genetic methods [39]. In another rodent, *Microtus agrestis,* the extremely large X chromosome contains an enormous block of constitutive heterochromatin. Here, in contrast to *E. lutescens,* we found a very high concentration of GATA/GACA sequences in the X-chromosomal heterochromatin by hybridization *in situ* [40]. In rats and hamsters no substantial GATA/GACA accumulations were found on the sex chromosomes (unpublished data). Thus in rodent evolution there are mechanisms at work that result in extreme variations of gonosomal GATA/GACA sequences as compared to the more or less even distribution on autosomal chromosomes.

GATA/GACA are ubiquitously interspersed on the autosomes in many reptile, bird and mammalian species though there are again exceptions to the rule. Recently Traut [41] showed the presence of significant GATA stretches in only very few blocks or restriction fragments in an insect, the moth *Ephestia kuehniella*. This result was confirmed with base-specific GATA/GACA oligonucleotide probes (unpublished results). Ubiquitous autosomal interspersion of GATA/GACA sequences in man and mouse on the other hand allows the production of individual-specific DNA fingerprints in these species [42, 43, 44]. Although several mechanisms for tandem sequence amplification have already been discussed, we favor an additional possibility to generate homogeneous GATA/GACA sequence distribution over the whole genome. It has been shown that GATA/GACA sequences are transcribed in many mouse tissues [45]. Sequence analysis of several independent cDNA clones revealed flanking inverted repeats, in addition to the demonstration of GATA/GACA elements. Furthermore, these cDNA clones show several sequence characteristics of transposable elements such as inverted repeats and consensus ends [45]. Since several GATA/GACA repeats are transcribed and polyadenylated they are potential candidates for reverse transcription and may reenter the vertebrate genome to generate their random interspersion pattern.

With respect to sequence-specific functions of the simple GATA/GACA elements, studies on the major histocompatibility complex genes of the mouse should be critically evaluated: $(GACA)_6$ sequences are spatially connected to hotspots of recombination [46, 47]. In an unrelated investigation on the sex-reversed mutation (sxr) of the mouse, we found an exceedingly high rate of unequal recombination of the murine Y chromosome in fertile XY[sxr] carriers [43]. The sxr region is tightly linked to long GATA/GACA-containing DNA stretches. Nevertheless we cannot imagine the significance of GATA/GACA repeats themselves as being preferred sites of recombination simply because their preponderance in eukaryote genomes (more than 10,000 copies per haploid mouse genome; [48]) would predict excessive recombinatorial events also perhaps between non-homologous chromosomes. Presumably only one of the GATA/GACA stretches is involved in the unequal recombination in XY[sxr] carrier mice [43].

3.3   Interspersed Nucleotide Elements

The interspersed nucleotide elements comprise a family of non-tandemly organized repetitive DNA sequences found in many if not all eukaryotic genomes. Simply on the basis of their physical lengths two classes of repetitive elements can be categorized: the short and the long interspersed nucleotide elements (SINE's < 500 bp; LINE's, several kb) [8]. Both, SINE's and LINE's, occur very abundantly in animal genomes and have several structures in common: Among these are the A-rich tracts at their 3'-ends, the RNA polymerase promoters and the flanking short direct repeats. These characteristics point to RNA intermediates which can be reverse-transcribed and reinserted into the genome by staggered breaks [49, 50, 51, 52, 53, 54]. In order to differentiate such elements from other transposons, the term of retroposons has been coined [55].

3.3.1   SINE's

Most of the primate SINE's contain a restriction endonuclease cleavage site for Alu I at a defined position. These SINE's are members of the so-called Alu family which serves here as a prototype for the whole class of these elements. Alu sequences are very abundantly interspersed throughout the chromosomes as has been shown by renaturation studies revealing a copy number of about 500,000 per haploid human genome. Thus the average distance between two such elements is less than 10 kb in man. The human Alu sequences are approximately 300 bp long and consist of two tandem monomer units. The 3'-monomer of approximately 150 bp is longer than its 5'-equivalent because of a 31 bp addition. A short (A)-rich sequence of varying length defines the 3'-end of each monomer. No monomeric Alu elements have ever been found in man. The rodent equivalents of the so-called B1 family share homology only to the 5'-part of the dimeric human Alu element; they are generally monomeric [7, 49, 53, 56]. Both, the dimeric Alu and the monomeric B1 elements are most likely derived from 7 SL RNA [57, 58]. The latter is a constituent of the signal recognition particle (SRP) which comprises six different polypeptides and one 7 SL RNA molecule. The SRP mediates translocation of secretory and membrane proteins across the endoplasmic reticulum [59]. The assumption that Alu and B1 sequences are derived from the 7 SL sequence, rather than vice versa, rests primarily on the remarkable evolutionary conservation of 7 SL RNA. Comparison of the human Alu consensus sequence with the human 7 SL RNA revealed an internal deletion of 155 nucleotides to produce the 3'-Alu monomer. In addition some 30 bp are lacking adjacent to the deletion in the 5'-Alu monomers. Therefore, one could argue that the putative ancestral 3'-Alu monomer could have been generated by differential RNA splicing (excision of an internal stretch) and acquisition of a poly (A)+ tract at the 3'-end of the native RNA sequence. In principle the Alu sequence could also have been produced at the DNA level through a non-homologous recombination event within or between 7 SL RNA genes [57, 58]. On this basis the 5'-Alu monomer could have been derived from the putative ancestral 3'-Alu monomer by tandem duplication and deletion. But independent splicing or recombination processes in the 7 SL RNA genes could also have generated the 5'-Alu monomer.

Most of the Alu sequences [60, 61, 62] and some of the B1 elements [63] can be transcribed by RNA polymerase III *in vitro* initiated by a promoter sequence at their 5'-ends. This RNA polymerase promoter exhibits an A and B consensus block typical for the split internal promoter of tRNAs. The A blocks lie close to the 5'-end of the element and appear to increase both the strength and accuracy of initiation [64, 65, 66]. However, only the B blocks seem to be essential for transcription [64, 65]. Several Alu members are transcribed by RNA polymerase II colinearly with protein-coding sequences [67] and they are later removed during mRNA maturation [68]. Alu sequences are found only exceptionally in mature mRNA, e. g. in the 3'-untranslated region of a human low density lipoprotein (LDL) receptor transcripts [69]. Another characteristic of Alu sequences are their flanking short terminal repeats. Short terminal repeats are usually found at the ends of transposable elements suggesting a mechanism of staggered breaks at the site of insertion with the result of target site duplications surrounding the mobile element [68]. Consistent with these structural and functional features Jagadeeswaran et al. [70] and van Arsdell et al. [71] have proposed a mechanism of transposition by self-primed reverse transcription to explain the high abundance of Alu sequences. This model requires an internal RNA polymerase III promoter which is located in such a way that transcription of the element begins at the 5'-end and runs through the entire element including the 3'-(A) rich tract. Transcription terminates at variable sites in adjacent non-repetitive DNA within a U-rich sequence which is characteristic for RNA polymerase III termination sites. If the 3'-oligo (U) sequence of the RNA transcript pairs with the most 3' internal oligo (A) stretch to prime cDNA synthesis by reverse transcription, the flanking non-repetitive DNA would be looped out in a hairpin structure. The result would be a reverse transcript without flanking non-repetitive DNA that conceals its origin. Since the promotor would be transposed with the repeat, each new copy has the potential to be transcribed, reverse-transcribed and retroposed [53, 72]. A separate model (Fig. 3-2) to explain the self priming process for reverse transcription and subsequent integration mechanism in the genome makes use of the frequent $(GT_x)_y$ tails predominantly within the 3'-(A) rich region of SINE's [49]. This model includes 3'-tailing of a nick in chromosomal DNA with a sequence, e.g. $(CA_x)_y$ which is complementary to the $(GU_x)_y$ part present in the RNA intermediate of a repetitive element. Subsequent pairing of RNA and chromosomal DNA would suffice to prime cDNA synthesis (5' to 3') on the 3'-(A) rich part of the RNA. A staggered second nick on the complementary DNA strand is necessary to link the mRNA sequence on chromosomal DNA. After ligation of newly synthesized cDNA into chromosomal DNA, repair enzymes could replace the RNA template with DNA. Recently Lin et al. [73] presented direct evidence for transposition of an Alu element. They transfected lung carcinoma cells with a bacterial target gene to search for transposable element insertions. Indeed one subline revealed an Alu element within the target gene. The element shows all the characteristics of an Alu sequence (270 bp; 97% homology to the Alu element Blur 8 [74]; RNA polymerase III promoter, poly (A)$^+$ stretch at the 3'-end of each monomer, flanking direct repeats). Although it has been suggested that as much as 10% of the genome of many species may have arisen by this process [75], the question of reverse transcriptase remains

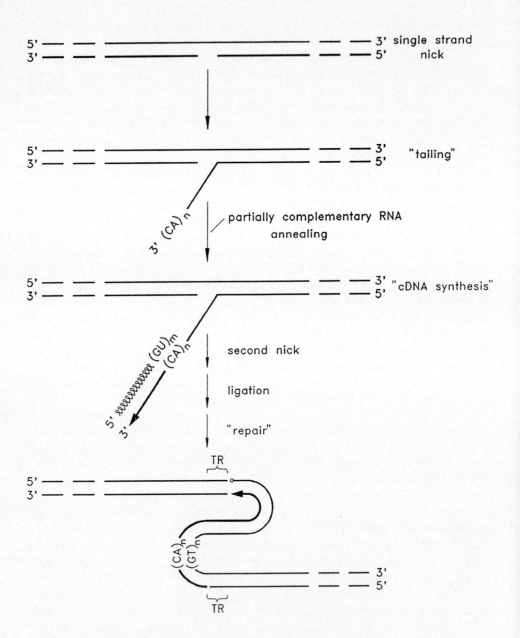

Fig. 3-2 Model for insertion of retroposons (simplified according to Rogers [49]). The 3'-end of nicked DNA is e. g. $(CA)_n$-tailed; RNA ($\ell\ell\ell$) containing $(GU)_m$ in the 3' untranslated portion anneals and cDNA is synthesized. Introducing a second nick, ligating the ends and repairing the product generates the new copy of the retroposon flanked by terminal repeats (TR).

still open: A plausible source in mice are the so-called intracisternal A-particles (IAP's) which have reverse transcriptase activity. IAP's are endogenous retrovirus like but non-infectious structures. These proviral elements are actively transcribed in many mouse tumor cells and are also found in oocytes and preimplantation mouse embryos [76, 77]. The first discovery of 7 SL RNA as a component of avian and murine retroviral particles points into the same direction [57]. But yet unknown enzymes with similar properties could also be responsible for reverse transcription in animals. Taken together these results indicate that Alu sequences are legitimate transposons that continue to shape the human genome. Such alterations could have deleterious effects as shown by two examples of Alu-Alu recombination: Rouyer et al. [78] described a XX male in whom an abnormal crossing over event between a Y and a X-chromosomal locus had occurred during paternal meiosis. A second case of Alu-Alu recombination was found in studies of familial hypercholesterolemia [79]. This disease is caused by a defective LDL receptor that binds cholesterol-carrying lipoproteins. The reason for the defective LDL receptor gene is a duplication of exon 2 to 8 caused by an unequal crossing over event between homologous Alu sequences from intron 1 and intron 8. This pathogenic event could also reflect a model for the evolution in which homologous recombination between repetitive elements in introns leads to exon duplication. Most of the breakpoints caused by homologous or non-homologous recombinations at Alu sequences within globin and LDL receptor genes are located between repetitive elements in introns leads to exon duplication. Most of the breakpoints caused by homologous or non-homologous recombinations at Alu sequences within globin and LDL receptor genes are located between the A and B blocks of the putative RNA polymerase III promotor sequence. These findings suggest a role for transcriptionally active Alu sequences in genetic recombination [78, 79].

From evolutionary studies the beginning of the Alu sequence amplification process has been dated back some 60 million years ago. However, there are some Alu family members which have probably been transposed more recently (200 000 to a few million years ago) because they are exceptionally closely related. Their close relation suggests also that only one or a few Alu members are able to be amplified and the vast majority has little if any retroposon activity (ineffective transcription units, truncated sequences or other still unknown reasons [80]). Similar to the Alu family other mammalian SINE's may have evolved from tRNA genes. They share significant sequence homology (about 70%). These tRNA-like SINE's include the C family of artiodactyls (cows and goats), the Monomer family of the bushbaby Galago, the rodent B2 or Type II family and the rat identifier (ID) or R.dre.1 sequences. Theoretically, homology may still allow folding into a secondary structure similar to that of typical tRNA molecules. These tRNA-like secondary structures are potentially recognized by reverse transcriptase, as tRNAs serve as primers for reverse transcription in retroviruses [50, 72, 81].

3.3.2 LINE's

In the past many repetitive DNA sequences, e.g. the primate Kpn I fragments or the mouse BamH I fragments (MIF-I = mouse interspersed fragment I) had been named independently until sequence data revealed that they are parts of longer coherent units, the LINE's [8, 52]. For further simplification this family is called L1 (LINE family 1) and a two letter suffix identifies the genus and species, e.g. L1Hs for Homo sapiens [82]. Additional letters and numbers can be used to specify individual family members.

The largest known L1 family members are 6-7 kb and appear as the maximum unit length of these elements. However, most of the L1 elements are randomly truncated at their 5'-ends. Therefore sequences of the 5'-end are much less frequently dispersed throughout the genome (about 10^4 copies per haploid genome) than the sequences of the 3'-end (about 10^5 copies) [49, 52]. Some L1 elements, including both full-length and truncated ones, have structural characteristics in common with retroposons. These are 3'-(A) rich tails usually preceded by a polyadenylation signal AATAAA and the surrounding target-site duplications of varying lengths. Moreover, a variety of polymorphisms in several mammalian genes are caused by L1 insertions or deletions [53, 83, 84]. These characteristics suggest a multiplication mechanism similar to that proposed for Alu-like sequences: transcription of the L1 element, reverse transcription into cDNA and insertion of the cDNA into staggered breaks of a target site.

Most of the RNA transcripts complementary to L1 elements that have been detected in a number of murine and primate cell lines and tissues [85, 86, 87, 88, 89, 90] are confined to the cell nucleus. In general these transcripts are synthesized by polymerase II [87], are very heterogenous in size [91], receive no poly $(A)^+$ tail and are likely to be primed outside the L1 element as a "read-through" part of a coding transcript [85, 87, 89, 92]. Nevertheless, cytoplasmic poly $(A)^+$ RNAs that contain full length L1 sequences have been detected in the human teratocarcinoma cell line N Tera 2D1 as a 6.5 kb RNA [91] as well as in different murine lymphoid tumor cells which expressed 8 and 18 kb transcripts [93]. Both, human and murine transcripts represent only one L1 strand, the strand containing the open reading frames (ORF's, defined by the substantial distance between two termination codons in frame). Moreover, the human N Tera 2D1 cell line expressed the highest amount of L1 transcripts when it exhibited stem cell character with embryonal carcinoma morphology. Yet it expressed no L1 RNA when differentiation was induced by retinoic acid. Skowronski and Singer [91] hypothesized that these N Tera 2D1 cells behave like cells in the cleavage-stage human embryos. In analogy one could imagine that putative functional genes of the L1 family are active in cleavage-stage mammalian embryos. Interestingly, the B2 murine short retroposons are also abundantly transcribed in early embryos [94]. In this context it is noteworthy that those mammalian embryos, which begin to transcribe the zygotic genome at the two-cell stage, contain numerous retroposon copies. Those animals, which do not begin transcription until later in development, have only few retroposons. Therefore, the embryonal cleavage stage could be the phase in which new retroposons appear [49]. The extraordinary abundance of L1 sequences in mammalian genomes also suggests that an active dispersal process may exist in germ line cells or in their progenitors in early embryos [91].

Hybridization studies showed that **L1 sequences** are present throughout the marsupial and placental mammalian orders [95]. A detailed comparison of full-length L1 sequences from four mammalian orders (carnivore, lagomorph, rodent and primate) revealed a conserved internal region of about 4-5 kb whereas their 5'- and 3'-ends are different [84, 96]. The sequences of many randomly cloned genomic L1 elements indicate the presence of ORF's covering more or less this conserved internal region [97, 98]. However, only a few L1 segments harbor a complete ORF of about 5 kb that is split into two parts. Such an element has been described and termed L1 Md-A2 (Loeb et al. [99]). This mouse L1 segment is split into two frames of 1137 bases at their 5'-end (ORF1) and 3900 bases at their 3'-end (ORF2). Both parts of the ORF's overlap by 14 bases. The consensus sequence of several cDNA clones gained from cytoplasmic poly (A)$^+$ L1 RNA of the human teratocarcinoma cell line N Tera 2D1 contain ORF's of similar length (ORF1: 1122 nucleotides, ORF2: 3852 nucleotides). Both ORF's are in the same reading frame separated by 33 bp containing two conserved stop codons. Moreover, in the majority of the cDNA's ORF 2 is interrupted by at least one additional randomly positioned stop codon [100].

Comparison of the L1 Md-A2 sequence (6851 bp) with a consensus sequence compiled from 20 overlapping segments revealed a large sequence homology [52]. At the protein level the ratio of replacement to silent mutations was calculated: silent sites (mutations resulting in no alteration of the amino acids) are more frequent than replacements (amino acid exchanges). On the basis of this analysis, Loeb et al. [99] conclude that essentially the entire 3900 bp long ORF region is evolving as under selection for protein function. Using probes of this ORF region for hybridization, Burton et al. [95] showed homology in a wide cross-section of mammalian orders. Taken together, most of the full-length L1 sequence data comprise a coding potential of one or more proteins. Indeed, recent comparisons between the conserved ORF's of L1 elements with sequences of several retroviruses and retrotransposons (e.g. the yeast Ty 1 elements, *Drosophila copia* like elements, *Rous sarcoma* virus genes) on the DNA as well as on the protein level revealed that L1 family members are closely related to reverse transcriptase genes [99, 101]. The putative gene located within the ORF 2 includes several regions successively termed A, B, C, D, E, F and G which are homologous to reverse transcriptase protein sequences [96]. Two other conserved regions at the 3'-end of ORF 2 (termed H and I) are markedly similar to the transferrin (and lactotransferin) proteins [101] and nucleic acid binding proteins [96], respectively. Segment I shares a striking homology with the conserved domains common to many retroviral nucleic acid binding proteins. These proteins, encoded within the gag genes of retroviruses, contain less than 100 amino acids and are able to bind to single-stranded DNA and to RNA. The amino acid sequence of this putative nucleic acid binding protein is similar to the repeated zinc-binding domains of the transcription factor III A which are considered to serve as DNA-binding "fingers" [102, 103 104]. On the basis of these data, it has been suggested that putative gene products of L1 elements have the ability to bind nucleic acids and may function as a reverse transcriptase [96, 99, 101].

Singer et al. [105] have demonstrated that one of the cDNA's (cD 11) obtained from N Tera 2D1 transcripts could be transcribed and translated *in vitro*. A polypeptide of about 42 kD encoded by ORF 1 has been detected but no translation product of ORF 2. Only when ORF 1 is deleted from the constructions used for translation *in vitro,* a polypeptide appeared corresponding in size to that predicted for ORF 2. This suggests that the ORF 2 of cD 11 is also in principle translatable. However, a read-through translation product involves some problems because the ORF's are either in different frames and overlap in their sequences as demonstrated for the L1Md-A2 elements or the two L1 ORF's are in the same frame but are separated by a short region including two conserved in-frame stop codons as demonstrated for a number of N Tera 2D1 cDNA clones. Nevertheless, there are examples that overcome these problems. A translational frame-shift has been demonstrated in *Rous sarcoma* virus, where the gag and pol ORF's are in different overlapping frames. This is also necessary for mouse L1 elements to produce a fusion precursor polypeptide. In the case of human N Tera 2D1 cDNA's the production of an ORF 1/2 encoding fusion protein would require suppression of at least two stop codons. Suppression of one stop codon between the gag and pol gene have been proved for Moloney murine leukemia virus. Another possible translation process involves termination of ORF 1 and reinitiation in ORF 2 without prior dissociation of ribosomes [100].

But how do these putative translation products interact with each other? Fanning and Singer [96] postulated that L1 RNA is packaged into an intracellular ribonucleoprotein particle in which L1 RNA is reverse transcribed and subsequently reintegrated into the cellular genome. In this context it is interesting to note that mouse IAP retrotransposons produce particles with reverse transcriptase activity that is particularly pronounced in early embryos and lymphoid tumor cells [76, 77]. If L1 elements are indeed transcribed through their own translation products, reverse transcripts could enter the genome and establish new elements within germ line cells. In the absence of any specific excision and destruction mechanism this could explain why L1 elements are so abundant. The typically truncated 5'-ends of L1 elements may be due to early termination of reverse transcription rather than aberrant initiation by polymerase II. Premature termination is a well known feature associated with reverse transcriptase activity *in vitro* [83, 98]. There is good evidence that L1 elements occur throughout the eukaryotic genomes. Their strong conservation and their characteristic features argue that several invertebrate movable elements belong to the L1 retroposons: the *Drosophila I* [106], F [107], G [108] and jockey element [109, 110], the *Trypanosoma ingi* elements [111, 112] and the R 2 element of *Bombyx mori* [113]. Nevertheless, it is not clear which functional role the L1 family members play in the genome. The coding potential for putative nucleic acid binding proteins and reverse transcriptases together with their retroposon features led to the suggestion that they merely maintain their self-propagating potential. If self-propagation is the primary function of L1 elements also secondary effects in consequence of this process could be suggested, e.g. influence of the expression of neighboring genes and support of rearrangements within neighboring genomic regions [52].

3.4   Summary

Animal genomes harbor different classes of DNA sequences, single copy and repeats. The *de-novo* generation, amplification, contraction and/or persistence of repetitive sequences in eukaryotes is only partly understood. Different categories of repetitive sequences depend on the degree of repetition, length of the element and other structural criteria. The interspersed repeats have been studied in relation to structural genes, recombination, transcription and retroposition. Interspersed $(CA/TG)_n$, GATA/GACA and related simple sequences are evolutionarily conserved from "primitive to most advanced" animals (or at least the mechanism of their generation). Among the complex repetitive sequences, the so-called SINE's and LINE's are abundantly found in higher vertebrates. Both are nearly ubiquitously interspersed; they differ in length and other sequence characteristics. It is not yet clear, however, which part of the repetitive elements has present-day functions. Or do they simply exist on the basis of their evolutionary potential as efficient self-propagating elements?

Acknowledgements

We thank Ursula Qreini for typing and Viktor Steimle as well as Hans Zischler for reading the manuscript. Work in the authors' laboratory is supported by the Stiftung Volkswagenwerk, the BMFT and the Deutsche Forschungsgemeinschaft.

3.5   References

[1] Hinegardner, R., in: *Molecular Evolution:* Ayala, F.J. (ed) Sunderland, Ma.: Sinauer Associates, *1976*, pp. 179-199.

[2] Britten, R.J., and Kohne, D.E., *Science 1968, 161*, 529- 540.

[3] Earnshaw, W.C., *Bio Essays 1988, 9*, 147-150.

[4] Paulson, J.R., in: *Chromosomes and Chromatin:* Adolp, K.W. (ed.) Boca Raton, Florida, CRC Press, *1988*, Vol. 3, pp. 3-36.

[5] Moens, P.B., and Pearlman, R.E. *Bio Essays 1988, 9*, 151-153.

[6] Schmidtke, J., and Epplen, J.T., *Hum Genet. 1980, 55*, 1-18.

[7] Jelinek, W.R., and Schmid, C.W., *Ann. Rev. Biochem. 1982, 51*, 813-844.

[8] Singer, M.F., *Cell 1982, 28*, 433-434.

[9] Macgregor, H.C., and Sessions, S.K., *Phil. Trans. R. Soc. 1986, 312*, 243-259.

[10] Miklos, G.L.G., and Gill, A.C., in: *Evolution of Genetic Systems:* Smith, H.H. (ed.) New York, Gordon and Breach *1982*, pp. 366.

[11] John, B., and Miklos, G.L.G., *Int. Rev. Cytol. 1979, 58*, 1-114.

[12] Smith, G.P., *Science 1976, 191*, 528-535.

[13] Streisinger, G., Okada, Y., Emrich, J., Newton, J., Tsugita, A., Terzaghi, E., and Inouye, M., *Cold Spring Harbor Symp. Quant. Biol. 1966, 31,* 77-84.

[14] Kornberg, A., Bertsch, L.L., Jackson, J.F., and Khorana, H.G., *Proc. Natl. Acad. Sci. 1964, 51,* 315-323.

[15] Tautz, D., and Renz, M., *Nucl. Acids. Res. 1984, 12,* 4127-4138.

[16] Levinson, G., Marsh, J.L., Epplen, J.T., and Gutman, G.A., *Mol. Biol. Evol. 1985, 2,* 494-504.

[17] Hourcade, D., Dressler, D., and Wolfson, J., *Proc. Natl. Acad. Sci. 1973, 70,* 2926-2930.

[18] Schimke, R., in: *Gene Amplification:* Schimke, R. (ed.) New York: Cold Spring Harbor Press *1982*; pp. 317-333.

[19] Alt, F.W., and Baltimore, D., *Proc. Natl. Acad. Sci. USA 1982, 79,* 4118-4122.

[20] Epplen, J.T., *J. Hered. 1988, 79,* 409-417.

[21] Moyzis, R.K., Buckingham, J.M., Cram,L.S., Dani, M., Deaven, L.L., Jones, M.D., Meyne, J., Ratliff, R.L., and Wu, J.-R. *Proc. Natl. Acad. Sci. USA 1988, 85,* 6622-6626.

[22] Roberts, L., *Science 1988, 240,* 982-983.

[23] Zischler, H., Nanda, I., Steeg, C., Schmid, M., and Epplen, J.T., *This Volume, 1990.*

[24] Arnheim, N., Seperack, P., Banerji, I., Land, R.B., Miesfeld, R., and Marcu, K.B., *Cell 1980, 22,* 179-185.

[25] Miesfeld, R., Krystal, M., and Arnheim, N., *Nucl. Acids Res. 1981, 9,* 5931-5947.

[26] Hamada, H., Petrino, M.G., and Kakunaga, T., *Proc. Natl. Acad. Sci. USA 1982, 79,* 6465-6469.

[27] Rogers, J. *Nature 1983, 305,* 101-102.

[28] Sun, L., Paulson, K.E., Schmid, C.W., Kadyk, L., and Leinwand, L., *Nucl. Acids Res. 1984, 12,* 2669-2690.

[29] Gebhard, W., and Zachau, H.G., *J. Mol. Biol. 1983, 170,* 255-270.

[30] Greaves, D.R., and Patient, R.K., *EMBO J. 1985, 4,* 2617-2626.

[31] Hunkapiller, T., and Hood, L., *Nature 1986, 323,* 15-16.

[32] Hinkkanen, A., Kempkes, B., Steimle, V., Stockinger, H., Weltzien, H.U., and Epplen, J.T., in: *The T Cell Receptor:* Davis, M.M., Kappler, J. (eds.) New York: CRC Press, *1988*; pp. 253-262.

[33] Hamada, H., Seidman, M., Howard, B.H., and Gorman, C.M., *Mol. Cell Biol. 1984, 4,* 2622-2630.

[34] Singh, L., Purdom, I.F., and Jones, K.W., Cold Spring Harbor *Symp. Quant. Biol. 1980, 45,* 805-814.

[35] Epplen, J.T., McCarrey, J.R., Sutou, S., and Ohno, S., *Proc. Natl. Acad. Sci. USA 1982, 79,* 3798-3802.

[36] Epplen, J.T., and Ohno, S., in: *Selected Topics in Molecular Endocrinology:* Lau, Y.C.F. (ed.) New York: Oxford University Press, *1988*; pp. 169-180.

[37] Shapiro, D.Y., *Environmental Biol., 1989,* in press.

[38] McLaren, A., Simpson, E., Epplen, J.T., Studer, R., Koopmann, P., Evans, E.P., and Bourgoyne, P.S., *Proc. Natl. Acad. Sci. USA 1988, 85,* 6442-6445.

[39] Vogel, W., Steinbach, P., Djalali, Mehnert, K., Ali, S., and Epplen, J.T., *Chromosoma* **1988**, *96*, 112-118.

[40] Nanda, I., Neitzel, H., Sperling, K., Studer, R., and Epplen, J.T., *Chromosoma* **1988**, *96*, 213-219.

[41] Traut, W., *Genetics* **1987**, *115*, 493-498.

[42] Ali, S., Müller, C.R., and Epplen, J.T., *Hum. Genet.* **1986**, *74*, 239-243.

[43] Epplen, J.T., Studer, R., and McLaren, A., *Genet. Res.* **1988**, *51*, 239-246.

[44] Epplen, J.T., Cellini, A., Romero, S., and Ohno, S., *J. Exp. Zool.* **1983**, *228*, 305-312.

[45] Schäfer, R., Böltz, E., Becker, A., Bartels, F., and Epplen, J.T., *Chromosoma* **1986**, *93*, 496-501.

[46] Kobori, J.A., Strauss, E., Minard, K., and Hood, L., *Science* **1986**, *234*, 173-179.

[47] Steinmetz, M., Stephan, D., and Fischer Lindahl, K., *Cell* **1986**, *44*, 895-904.

[48] Schäfer, R., Ali, S., and Epplen, J.T., *Chromosoma* **1986**, *93*, 502-510.

[49] Rogers, J., *Int. Rev. Cytol.* **1985**, *93*, 187-279.

[50] Rogers, J., *Nature* **1985**, *317*, 765-766.

[51] Rogers, J., *Nature* **1986**, *319*, 725.

[52] Singer, M.F., and Skowronski, J., *Trends Biochem. Sci.* **1985**, *10*, 119-122.

[53] Weiner, A.M., Deininger, P.L., and Efstratiadis, A., *Ann. Rev. Biochem.* **1986**, *55*, 631-661.

[54] Hardman, N., *Biochem. J.* **1986**, *234*, 1-11.

[55] Rogers, J., *Nature* **1983**, *301*, 460.

[56] Schmid, C.W., and Jelinek, W.R., *Science* **1982**, *216*, 1065-1070.

[57] Ullu, E., and Tschudi, C., *Nature* **1984**, *312*, 171-172.

[58] Brown, A.L., *Nature* **1984**, *312*, 106.

[59] Walter, P., and Blobel, G., *Nature* **1982**, *299*, 691-698.

[60] Duncan, C., Biro, P.A., Choudary, P.V., Elder, J.T., Wang, R.R.C., Forget, B.G., De Riel, J.K., and Weissman, S.M., *Proc. Natl. Acad. Sci.USA* **1979**, *76*, 5095-5099.

[61] Duncan, C.H., Jagadeeswaran, P., Wang,R.R.C., and Weissman, S.M., *Gene* **1981**, *13*, 185-196.

[62] Elder, J.T., Pan, J., Duncan, C.H. and Weissman, S.M., *Nucl. Acids Res.* **1981**, *9*, 1171-1189.

[63] Young, P.R., Scott, R.W., Hamer, D.H., and Tilghman, S.M., *Nucl. Acids Res.* **1982**, *10*, 3099-3116.

[64] Fuhrman, S., Deininger, P., La Porte, P., Friedmann, T., and Geiduschek, E.P., *Nucl. Acids Res.* **1981**, *9*, 6439-6456.

[65] Perez-Stable, C., Ayres, T., and Shen, C.-K.J., *Proc. Natl. Acad. Sci. USA* **1984**, *81*, 5291-5295.

[66] Paolella, G., Lucero, M.A., Murphy, M.H., and Baralle, F.E., *EMBO J.* **1983**, *2*, 691-696.

[67] Allan, M., Lanyon, W.G., and Paul, J., *Cell* **1983**, *35*, 187-197.

[68] Sharp, P.A., *Nature* **1983**, *301*, 471-472.

[69] Yamamoto, T., Davis, C.G., Brown, M.S., Schneider, W.J., Casey, M.L., Goldstein, J.L., and Russell, D.W., *Cell* **1984**, *39*, 27-38.

[70] Jagadeeswaran, P., Forget, B.G., and Weissman, S.M., *Cell* **1981**, *26*, 141-142.

[71] Van Arsdell, S.W., Denison, R.A., Bernstein, L.B., Weiner, A.M., Manser, T., and Gesteland, R.F., *Cell 1981, 26*, 11-17.

[72] Deininger, P.L., and Daniels, G.R., *Trends Genet. 1986, 2*, 76-80.

[73] Lin, C.S., Goldthwait, D.A., and Samols, D., *Cell 1988, 54*, 153-159.

[74] Deininger, P.L., Jolly, D.J., Rubin, C.M., Friedmann, T., and Schmid, C.W., *J. Mol. Biol. 1981, 151*, 17-33.

[75] Baltimore, D., *Cell 1985, 40*, 481-482.

[76] Kuff, E.L., in: *Eukaryotic Transposable Elements as Mutagenic Agents:* Lambert, M.E., McDonald, J.F., Weinstein, F.B. (eds.) New York: Cold Spring Harbor Laboratory, **1988**; Vol. 30, pp. 79-89.

[77] Callahan, R., in: *Eukaryotic Transposable Elements as Mutagenic Agents:* Lambert, M.E., McDonald, J.F., Weinstein, I.B. (eds.) New York: Cold Spring Harbor Laboratory, **1988**; Vol. 30, pp. 91-100.

[78] Rouyer, F., Simmler, M.Ch., Page, D.C., and Weissenbach, J., *Cell 1987, 51*, 417-425.

[79] Lehrman, M.A., Goldstein, J.L., Russell, D.W., and Brown, M.S., *Cell 1987, 48*, 827-835.

[80] Deininger, P.L., and Slagel, V.K., *Mol. Cell. Biol. 1988, 8*, 4566-4569.

[81] Daniels, G.R., and Deininger, P.L., *Nature 1985, 317*, 819-822.

[82] Voliva, Ch.F., Jahn, C.L., Comer, M.B., Hutchison, C.A., and Edgell, M.H., *Nucl. Acids Res. 1983, 11*, 8847-8859.

[83] Voliva, Ch.F., Martin, S.L., Hutchison, C.A., and Edgell, M.H., *J. Mol. Biol. 1984, 178*, 795-813.

[84] Skowronski, J., and Singer, M.F., *Cold Spring Harbor Symp. Quant. Biol 1986, 51*, 457-464.

[85] Kole, L.B., Haynes, S.R., and Jelinek, W.R., *J. Mol. Biol. 1983, 165*, 257-286.

[86] Lerman, M.I., Thayer, R.E., and Singer, M.F., *Proc. Natl. Acad. Sci. USA 1983, 80*, 3966-3970.

[87] Shafit-Zagardo, B., Brown, F.L., Zavodny, P.J., and Maio, J.J., *Nature 1983, 304*, 277-280.

[88] Schmeckpeper, B.J., Scott, A.F., and Smith, K.D., *J. Biol. Chem. 1984, 259*, 1218-1225.

[89] Sun, L., Paulson, K.E., Schmid, C.W., Kadyk, L., and Leinwand, L., *Nucl. Acids Res. 1984, 12*, 2669-2690.

[90] Jackson, M., Heller, D., and Leinwand, L., *Nucl. Acids Res. 1985, 13*, 3389-3403.

[91] Skowronski, J., and Singer, M.F., *Proc. Natl. Acad. Sci. USA 1985, 82*, 6050-6054.

[92] Di Giovanni, L., Haynes, S.R., Misra, R., and Jelinek, W.R., *Proc. Natl. Acad. Sci. USA, 1983, 80*, 6533-6537.

[93] Dudley, J.P., *Nucl. Acids Res. 1987, 15*, 2581-2592.

[94] Vasseur, M., Condamine, M., and Duprey, P., *EMBO J. 1985, 4*, 1749-1753.

[95] Burton, F.H., Loeb, D.D., Voliva, Ch.F., Martin, S.L., Edgell, M.H., and Hutchison, C.A., *J. Mol. Biol. 1986, 187*, 291-304.

[96] Fanning T., and Singer, M., *Nucl. Acids Res. 1987, 15*, 2251-2260.

[97] Potter, S.S., *Proc. Natl. Acad. Sci. USA, 1984, 81*, 1012-1016.

[98] Martin, S.L., Voliva, Ch.F., Burton, F.H., Egdell, M.H., and Hutchison, C.A., *Proc. Natl. Acad. Sci. USA.*, *1984*, *81*, 2308-2312.

[99] Loeb, D.D., Padgett, R.W., Hardies, S.C., Shehee, R.W., Comer, M.B., Edgell, M.H., and Hutchison, C.A., *Mol. Cell Biol.* *1986*, *6*, 168-182.

[100] Skowronski, J., Fanning, T.G., and Singer, M.F., *Mol. Cell. Biol.* *1988*, *8*, 1385-1397.

[101] Hattori, M., Kuhara, S., Takenaka, O., and Sakaki, I., *Nature 1986*, *321*, 625-628.

[102] Miller, J., McLachlan, A.D., and Klug, A., *EMBO J.* *1985* *4*, 1609-1614.

[103] Berg, J.M., *Science 1986*, *232*, 485-487.

[104] Evans, R.M., and Hollenberg, S.M., *Cell 1988*, *52*, 1-3.

[105] Singer, M.F., Skowronski, J., Fanning, T.G., and Mongkolsuk, S., in: *Eukaryotic Transposable Elements as Mutagenic Agents:* Lambert, M.E., McDonald, J.F., Weinstein, F.B. (eds.) New York: Cold Spring Harbor Laboratory, *1988*; Vol. 30, pp. 71-78.

[106] Fawcett, D.H., Lister, C.K., Kellett, E., and Finnegan, D.J., *Cell 1986*, *47*, 1007-1015.

[107] DiNocera, P.P., and Casari, G., *Proc. Natl. Acad. Sci. USA 1987*, *84*, 5843-5847.

[108] DiNocera, P.P., Graziani, F., and Lavorgna, G., *Nucl. Acids Res.* *1986*, *14*, 675-691.

[109] Mizrokhi, L.J., Georgieva, S.G., and Ilyin, Y.V., *Cell 1988*, *54*, 685-691.

[110] Priimägi, A.F., Mizrokhi, L.J., and Ilyin, Y.V., *Gene 1988*, *70*, 253-262.

[111] Kimmel, B.E., Ole-Moiyoi, O.K., and Young, J.R., *Mol. Cell. Biol. 1987*, *7*, 1465-1475.

[112] Murphy, N.B., Pays, A., Tebabi,P., Coquelet, H., Guyaux, M., Steinert, M., and Pays, E., *J. Mol. Biol. 1987*, *195*, 855-871.

[113] Burke, W.D., Calalang, C.C., and Eickbush, T.H., *Mol. Cell. Biol.* *1987*, *7*, 2221-2230.

[114] Ohno, S., *Brookhaven Symp. Biol.* *1972*, *23*, 366-370.

[115] Doolittle, W.F., and Sapienza, C., *Nature 1980*, *284*, 601-603.

[116] Orgel, L.E., and Crick, F.H.C., *Nature 1980*, *284*, 604-607.

[117] Lewin, R., *Science 1982*, *217*, 621-623.

[118] Doolittle, W.F., *Trends Genet. 1987*, *3*, 82-83.

4. Screening and Cloning of Genes for Application in Animal Genetics and Animal Breeding

H. Feldmann

4.1 Introduction

The ultimate objective in animal breeding is the improvement of livestock for mankind by genetic change. So far, research in animal breeding has concentrated on the recognition of genetic merit to best manipulate the genetic make-up by selection both among and within populations. Achievements of gene technology and recent advances in micro-manipulation have now made it realistic to improve livestock by direct gene transfer. Virtually the same methodology may be applied to animal molecular genetics as to human genome research. In both cases large and complex genomes have to be handled, but the objectives differ considerably.

Major research goals on the human genome relate to prediction of risk, prenatal diagnosis, prognosis, and therapy of genetic abnormalities or defects, which by definition include multifactorial diseases like coronary heart disease or cancer. A more detailed genetic map of the human genome, ordered clone libraries of the whole genome or single chromosomes in combination with cDNA sequences, and an ordered library of DNA probes are needed for further insight. Broader family studies, centralized storage of material and data, and access to these data for all researchers are of importance. Further necessities consist in the improvement of current gene technology and development of new techniques: for example, methods for cloning and handling of large DNA molecules, DNA transfer techniques (transfer of chromosomes, site-directed integration), new *in vitro* gene vectors (containing defined regulatory sequences), and methods for gene amplification.

From the list of problems to be solved in human genome research, it becomes evident that animal molecular genetics will be confronted with similar tasks. A major area of current interest involves gene mapping in domestic animals, the concept of genetic approaches to the control of major diseases and productivity traits, and other methods such as vaccines and vector control. Undoubtedly, genome analysis in domestic animals will profit from progress that has been made by genome research in man or in laboratory animals. On the other hand, animal research provides models for genetic aspects of human disease as well. Technical improvement that comes from any of the fields will also be beneficial to the others.

4.2   Methods for Screening and Cloning
- New Developments

4.2.1   Standard Procedures

A large set of standard procedures allows, in principle, screening of any gene of a given organism and its subsequent cloning [1]. A particular protocol to be followed depends on the knowledge, either of the gene itself, the gene product(s), the function(s) of the gene product(s), or at least on phenotypical consequences that arise from malfunction. The most common approach for cloning a mammalian gene involve the use of one of the gene products as a tool in the selection procedure, that is either the messenger RNA or the corresponding protein, in combination with appropriate DNA libraries. Such libraries can be constructed either from genomic DNA or from complementary DNA (cDNA) that is prepared from total cellular messenger RNA, using lambda or cosmid vectors as cloning vehicles [1]. In most cases, screening is first performed in cDNA libraries. If a specific messenger RNA is abundant, it can be isolated and may serve as a probe in screening DNA libraries by hybridization techniques.

Special libraries can be constructed by the use of vectors that will allow expression of the cloned cDNA. Then, antibodies are generated against the protein in question and used to detect corresponding clones by immuno-blotting techniques. Whenever the expression of a cloned gene leads to a functional protein, an appropriate test for function can be set up and used in the screening procedure. For example, human interferon genes have been cloned by such methods.

4.2.2   From Proteins to Genes

For the isolation of genes of rare products (e.g., peptide hormones, growth factors, receptors, or regulatory proteins) synthetic oligonucleotides became powerful instruments in hybridization assays. A specific oligonucleotide (or a set of degenerate oligonucleotides) is easily prepared by retro-translating a short characteristic amino acid sequence of the desired protein: micro-sequencing techniques allow performance of such an analysis with minimal amounts of material [2]. This concept may even be extended towards a more or less systematic analysis of all coding sequences of a given species. High-resolution two-dimensional electrophoresis has reached a technological level that allows most of the numerous protein species of a mammalian organism to be resolved if appropriate strategies are used. In favorable cases, more than 5000 protein species have been resolved. Based on this approach, complex cellular protein databases for a number of organisms (including *E.coli, Bacillus subtilis, Salmonella typhimurium, Saccharomyces cerevisiae, Zea mays*) and several human and rodent cell lines or tissues are already available or in preparation [3]. Proteins from such gels can be transferred to membranes and directly subjected to automated micro-sequencing. Amino acid sequences from the N-terminus then render sufficient information to synthesize corresponding oligonucleotides. A newly developed and versatile approach to prepare such screening probes is the "mixed oligonucleotide primed amplification of cDNA" (MOPAC) [4], taking advantage of the "polymerase chain reaction" (PCR) that is explained below.

4.2.3 RFLPs and VNTRs

Once a cDNA clone has been obtained, it can be used to screen a genomic library in order to obtain clones of the chromosomal gene including all regulatory regions, etc. Naturally, in many cases, it is also feasible to directly screen genomic libraries with oligonucleotide probes. This has successfully been carried out for a variety of human and mouse genes.

However, one has to consider that most mammalian genes extend over very large genomic regions and cannot be obtained as a singular cloned piece of DNA by the use of conventional vectors. Cosmid vectors, for example, will accommodate some 35 to 45 kb of DNA, but there are many genes that far exceed this size. The largest region mapped to date is the human Xp21 locus [5] that carries the dystrophin gene (ca. 2000 kb in length), defects of which cause Duchenne's or Becker's muscular dystrophy. The Duchenne gene and several other loci involved in inherited diseases (e.g., cystic fibrosis, polycystic kidney disease, Huntington's chorea) are examples [6] to illustrate a further possibility for screening, namely the use of probes that are connected to "restriction fragment length polymorphism" (RFLP). Originally, RFLP was observed in an attempt to identify mutant human ß-globin genes by the use of diagnostic cDNA probes [7]. Two important principles were established in a single observation: (i) natural variation in DNA structure (polymorphisms) could readily be detected by RFLPs; and (ii) the presence of RFLP alleles could be correlated with disease alleles which would then facilitate disease diagnosis. Furthermore it became apparent that in many cases RFLPs do not necessarily imply phenotypical modifications and often are located outside of functional regions. Nevertheless, they can serve as reliable genetic markers, as they are inherited strictly to the Mendelian rules. It was found that a particular disorder in a family is associated with the presence of an RFLP, and the successful approaches to map even diseases whose gene defects were unknown has set in motion this new approach - DNA linkage analysis - for several applications in genome research.

Two types of DNA variation represent the molecular basis of RFLPs (Fig. 4-1): single base alterations and repeated sequences; in some cases, the RFLP has components of both. Single base RFLPs are recognized by restriction endonuclease cleavage and detected by hybridization analysis. Jeffreys et al. [8] were the first to isolate probes that had a highly polymorphic character because of a variable number of tandem nucleic acid repeat sequences (VNTRs), and Nakamura et al. [9] utilized selected oligonucleotides from tandem repeat regions of the myoglobin gene, the zeta-globin pseudogene, the insulin gene, and the X-gene of hepatitis B virus for the determination of new human polymorphisms and the isolation of cosmid clones. VNTRs have proven clinically useful in the diagnosis of several

H. Feldmann

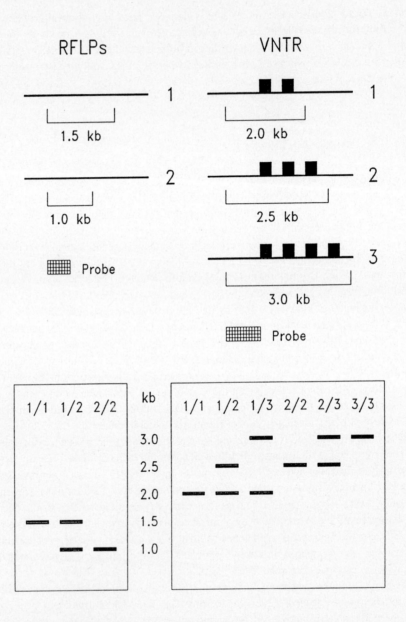

Fig. 4-1 RFLP mechanisms. RFLPs associated with base changes are distinguished by the difference in restriction endonuclease cleavage sites yielding fragments of different length. RFLPs associated with tandem repeat sequences lead to restriction fragments of different sizes; these are due to the presence of variable numbers of repeats flanking the region of interest. The size variations that would appear after electrophoresis and Southern hybridization are schematized below each diagram [6].

genetic diseases and they are of general importance as markers in genome analyses. Furthermore, the use of RFLPs predominantly based on VNTRs has become applicable in forensic analysis [10] as well as in animal population genetics [11]. This method can be considered as a general means for genetically identifying individuals by DNA variation. The great advantages of this approach are at hand: (i) The patterns obtained by hybridization with the probes are highly informative. The probability that the patterns of two individuals are identical has been calculated to be in the range of 10^{-13}. (ii) The tests require minimal amounts of DNA. The "classical" hybridization technique can now be supplemented by the polymerase chain reaction (PCR, see below) that allows amplification of the DNA material. In this way, DNA from the cells adhering to a single hair [12] or even from a single body or sperm cell [13] is sufficient for an analysis. (iii) The same oligonucleotides can be used to probe DNAs from different species, because similar patterns of repetitive sequences do occur.

4.2.4   Manipulation of Large DNA Molecules - PFG and YACs

The examples mentioned above have clearly demonstrated that a major problem arising in the analysis of mammalian genomes is to overcome large distances (in terms of base-pairs). One approach still in use is "chromosomal walking": once a particular clone has been established, probes can be prepared from its ends and used to screen the library for extending clones (i.e., those containing overlapping DNA segments). Although this is a tedious method it seems straightforward. In reality, however, one can encounter considerable difficulties, mainly because of the obstacle that mammalian genomes are loaded with repetitious sequences.

One of the most powerful tools for manipulation of large DNA molecules, developed during recent years, is pulsed field gel electrophoresis (PFG) [14, 15]. This new technique uses exactly the same kind of agarose gels as in conventional gel electrophoresis, but alternating fields are applied to separate DNA molecules up to several million base-pairs. For lower eukaryotes such as *Trypanosomes, Saccharomyces cerevisiae, Schizo-saccharomyces pombe,* etc.) it is possible to fractionate whole, intact chromosomes.

Care has to be taken, of course, not to break the DNA in preparing the gels. For this purpose, cells are embedded and lysed in small blocks of agarose which are then transferred to the separating gels. This procedure is also convenient to restrict the DNA by the use of 'rare cutter enzymes' [16] prior to electrophoresis and in this way to produce large defined DNA fragments. After electrophoresis, the DNA can be transferred to filters and subjected to hybridization with labelled probes as done with conventional gels. In combination with appropriate probes (RFLPs), PFG has been shown to be an indispensable tool for mapping and directed cloning of several large genes as well as whole genomes [17]. This technique will always be extremely important in any genome analysis.

A recent development that might facilitate the cloning of large DNA segments is the construction of yeast artificial chromosomes (YACs) [18]. These constructs contain all the components necessary for the maintenance of eukaryotic chromosomes: a replication origin, a centromere for the distribution of sister chromatids during mitosis, and two telomeres, elements responsible for the stability and complete duplication of the chromosomes. These vectors can accommodate up to several hundred kilo-bases of foreign DNA and still be transformed into yeast cells in a stable manner. PFG is then applied to distinguish and to isolate these extra chromosomes from the rest of the yeast chromosomes.

4.2.5 DNA Amplification

A method that has become very popular because of its versatility, is the polymerase chain reaction (PCR) designed to amplify DNA. The principle is based on repeated primed synthesis [19]: two oligonucleotides, each one complementary to one end of the matrix, are synthesized and annealed to the separated strands. After a first round of DNA synthesis with polymerase, the strands are separated again by heat denaturation, the same oligonucleotides (which are present in excess) re-annealed, a second round of synthesis is performed, etc. The reaction can be conducted with labeled nucleoside triphosphates and carried out for multiple rounds. Since the DNA is doubled each time, one can expect a million times of DNA after 20 rounds. In the beginning, Klenow polymerase was used; a recent improvement is to employ *Taq*-polymerase which is stable up to $65°C$, so that addition of new enzyme after each round is no longer required [20]. Some applications of PCR have already been mentioned above, such as in forensic medicine, in obtaining diagnostic DNA probes, or in synthesizing oligonucleotides for gene screening. One word of caution is necessary regarding the quality of the probes obtained by PCR. Since this is a highly sensitive method, any impurity in the starting material will be co-amplified and eventually lead to wrong interpretation of the results obtained with PCR probes.

4.3   DNA Sequences of Interest for Animal Genetics and Animal Breeding

DNA sequences of interest for application in domestic animals can be grouped into different categories: (i) DNA sequences - mainly cDNAs of mammalian genes - that are biotechnically used to obtain compounds valuable for animal breeding; (ii) DNA sequences that serve analytical or diagnostic purposes; and (iii) genes that can be introduced into the germ line to obtain transgenic animals.

4.3.1 DNA Sequences for Biotechnical Application

4.3.1.1 Hormones

One application that immediately comes to mind is the biotechnical production of peptide hormones. In practice, any species specific peptide hormone or releasing hormone should be amenable to biosynthesis, as soon as the corresponding gene has been cloned. Often it turns out that these genes are simple and lead to unmodified gene products, features that comply with the use of prokaryotic systems for their biosynthesis. For animal breeding applications, there is the example of hormones that stimulate growth or affect other body functions in a favorable way. The first experiments with biosynthetic somatotropin (growth hormone) were promising. However, in general, it turns out to be extremely difficult to predict all the effects that may be induced by applied hormones, since very complex regulatory circuits are involved, not forgetting the problems that may arise from regulations or prejudice of consumers.

In any case, the isolation and characterization of hormone genes from domestic animals offer the possibility of further investigating their regulation and interplay. The following list briefly summarizes some studies on different hormones for which at least cDNAs are available:

(1) Somatotropin (growth hormone, GH): genomic DNAs isolated from rat and man; biotechnically produced from rat (rGH), man (hGH), calf (bGH), pig (pGH) cDNAs.
(2) Somatoliberin (somatocrinin, somatotropin releasing factor)
(3) Prolactin, structurally similar to somatotropin; broad spectrum of effects on lactogenesis, growth, etc.
(4) Prolactoliberin: prolactin stimulating factor.
(5) Somatomedines and insulin-like growth factors (IGFs) [21, 22]: several components, synthesized mainly in the liver, mediators of different effects on growth and metabolism; synthesis stimulated by GH, probably by prolactin. Synthesis of transport proteins for IGFs also controlled by GH. Somatomedin C is equivalent to IGF-I; IGF-II is probably identical to multiplication-stimulating factor.
(6) Gonadotropins: LH and FSH; active only as glycoproteins. FSH is composed of two chains, FSHα and FSHß.
(7) Gonadoliberin (gonadotropin releasing-hormone, GnRH): GnRH is a decapeptide, processed out of a precursor. Human cDNA is used for biosynthesis [23]. The last 56 amino acids of the precursor are GAP (gonadotropin-releasing hormone associated peptide), probably an inhibitor of prolactin excretion (i.e., identical to PIF, prolactin-release inhibiting factor).
(8) Inhibin: Inhibitors of FSH, produced via porcine and bovine cDNAs, respectively. They control differential secretion of gonadotropins. Heterodimers from α- and ß-subunits which have larger precursors [24]; their genes are related, and show considerable homology with human ß-TGF (transforming growth factor).
(9) Activin: stimulating FSH secretion, counterpart of inhibin [25].

4.3.1.2 Vaccines

The biotechnical preparation of appropriate vaccines to animal pathogenic viruses or parasites affords a detailed structural analysis of their genomes. In order to achieve an active immunization, it is extremely important to elucidate which proteins or protein domains of a pathogen will act as antigenic determinants (e.g. surface antigens). In this regard, the analysis of several important animal viruses (bovine leukemia virus; foot-and-mouth disease virus; animal hepatitis B viruses; rabies virus; bluetongue virus; vaccinia virus) has made considerable progress. One possibility of preparing active vaccines by recombinant DNA techniques, is to cut defined segments from the viral genome and let appropriate micro-organisms synthesize virus-specific proteins ('subunit vaccines'). Such an approach has been successful for hepatitis B virus but less so for foot-and-mouth disease. Also in other cases, initial experiments along these lines were not too encouraging, mainly because the immunogenicity of synthetic vaccines turned out to be lower than of inactivated whole virus particles. A promising variant for immunization, vaccination with 'live vaccines', is currently explored [overview: 26]. In this approach, the gene of an immunoactive structural protein from a pathogenic virus is incorporated into the genome of another 'attenuated' (non-pathogenic) virus. Studies are concentrating on bovine vaccinia virus as a recombinant carrier, especially trying to explore vaccinia virus gene organization and to characterize vaccinia virus specific promoters which are necessary to express foreign genes [e.g., 27].

4.3.2 Genetic Markers and DNA Probes

One major area of current interest involves gene mapping in domestic animals [28]. The genetic approach to control production traits and major diseases is increasing in popularity. In this respect, the greatest contribution of gene technology to animal improvement may lie in the identification of appropriate genes and the mapping of polymorphic loci in the genomes of livestock. As mentioned above, the approach with RFLPs has revolutionized human genome research by providing marker-assisted diagnosis and counseling for a number of inherited diseases. Up to now, only a small number of RFLPs in farm animals have been reported, and considerable interest has arisen to intensify these studies. On the one hand, decisions have to be taken on which farm animals to concentrate. On the other hand, the analysis might be facilitated by the fact that the modern techniques now applied in animal breeding (such as artificial insemination and embryo transfer) will restrict the contribution of a few types of individuals to the gene pool, which in turn could have the effect of reducing the frequency of DNA polymorphism in livestock species. Undoubtedly, an important role for RFLP analysis will be in monitoring the loss or the conservation of variety within the gene potential. Marker-assisted selection for economically important animals will ultimately depend on the progress of deriving their gene maps with RFLPs. Furthermore, these approaches will permit the isolation of interesting new genes or even complex loci and make them amenable to functional analysis outside the animal body.

Specific DNA probes can be developed for any kind of diagnostics. In animal agriculture, three areas may be of outstanding importance: (i) DNA probes that identify infectious diseases in animals. (ii) DNA probes that are designed for the study of the breeding structure of populations or for the identification of parentage in domestic species. RFLPs are highly informative, and VNTRs meet all the criteria for DNA fingerprinting. Both types of probes have been successfully applied to laboratory as well as domestic animals and populations of wild animals. (iii) A specific area of application and obvious commercial value of DNA probes will lie in the ability to control the sex ratio of offspring in breeding populations of livestock. Successful manipulation of sex ratios will probably come through the sexing of embryos prior to transfer: embryo splitting will permit karyotype analysis. A small number of cells (or even one cell) should be sufficient to reveal Y-chromosome specific DNA sequences, and such probes should become available for the most important animals in agriculture.

4.3.3 Gene Transfer and Transgenic Farm Animals

Gene transfer into farm animals [29, 30] can serve different purposes: (i) improvement of productivity by introducing new metabolic pathways, or improvement of productivity by interfering with regulation of metabolism; (ii) introduction of tolerance or disease resistance; and (iii) gene farming. Although in the majority of cases, transferred genes were found to be integrated and expressed, there are still a number of problems to be solved to make gene transfer more precise and gene expression more specific.

In animal systems, DNA introduced by any of the currently available transfer techniques [31] (micro-injection of recombinant DNA directly into fertilized eggs [e.g., 32, 33]; exposure of early embryos to retroviral vectors; transfection of DNA into totipotent teratocarcinoma cells, followed by injection of selected cells into the blastocyte; electroporation) is integrated into the genome at random sites and mostly in multiple copies (except with retroviral vectors). One complication arising from this fact is that insertional mutagenesis of endogenous genes may occur [34]. A further complication is that the level of expression from one transgenic animal to another is unpredictable and usually does not correlate with gene copy number. These phenomena are related to the fact that proper function of a gene is highly dependent on its genomic environment. It would be advantageous, therefore, to develop techniques that are capable of targeting integration to predetermined chromosomal sites. There are, however, considerable difficulties to overcome. For example, the exchange of a chromosomal copy of a given gene by an (*in vitro*) modified copy -an approach that would be necessary in gene therapy - is an intricate procedure for mammalian cells. In yeast, where homologous recombination events are prevalent, gene replacement is a standard technique, whereas higher eukaryotes have a preference for illegitimate recombination and do not permit a similar approach. Hence, the success of targeted gene transfer experiments in mammalian systems largely depends on efficient selection procedures. Mansour et al. [35] have described a generally applicable approach that is based on the transformation of embryonic stem cells with linearized DNA and a selection procedure that takes advantage of two markers (*tk,* thymidine kinase, and *NEO,* neomycin resistance, respectively) which are contained in the vectors. In the case of the human HPGRT gene it was possible to make this selection highly specific. However, how such selections could be done in fertilized eggs, is extremely difficult to conceive.

A problem closely related to site specific integration is the regulated expression of a transferred gene. Although a lot has been learned about the principles that govern tissue-specific gene expression, hormone-controlled gene regulation, and gene expression during development, practical experience clearly shows that we are far from optimal solutions [36].

4.3.3.1 Interference with Metabolic Pathways

The introduction of new metabolic pathways [29] seems to offer the possibility of improving livestock in two cases: (i) There is a shortage or a lack of metabolic compounds; this may be particularly true for certain amino acids. Increase of cysteine, for example, in sheep is thought to augment the quantity and to improve the quality of wool. (ii) Unusual, and hitherto unused, nutritional compounds could be metabolized by animals if appropriate metabolic circuits were introduced. For example, acetate or propionate, major products of cellulose or fat degradation, do not normally contribute to the net synthesis of glucose in mammals; these compounds could be channeled into gluconeogenesis by transferring pathways from other organisms such as micro-organisms or plants (e.g., glyoxalate pathway enzymes) into farm animals. However, the enormous difficulties encountered in expressing these exogenous genes in mammals are obvious.

Activation of growth and productivity through the application of hormones have already been discussed. In the long term, it may be desirable to maintain hormonal effects directly by implanted hormone genes rather than by regular administration of hormones. In pilot experiments that have been carried out, for example, with fusion products of mouse metallothioneine promoter and human growth hormone gene, different animal species revealed different rates of integration and expression, and more severe, multiple aberrations in progeny of transgenic animals were observed. Furthermore, problems may arise from the fact that heterologous hormones (differing in amino acid sequence from genuine products) will not be functional or even cause undesired effects. Answers to some of the technical problems will certainly come from experiments with transgenic laboratory animals. However, species specific requirements can only be approached by research with livestock animals themselves.

More straightforward results could be expected in those cases in which (single) genes directly responsible for the quality of animal products can be manipulated. This may, for example, apply to genes of milk proteins or genes involved in muscling.

4.3.3.2 Tolerance and Resistance Genes

Up to now there have been only a few cases in which tolerance or resistance is known to be encoded by a single gene (*Mx* gene of the mouse, resistance to influenza virus; porcine stress resistance gene). It is, therefore, not surprising that these genes are under investigation in transgenic animals. As soon as such properties are determined by complex genotypes, it will be necessary to proceed via appropriate genetic markers; such approaches are under way, for example, for the trypano tolerance of West African N'Dama beef.

4.3.3.3 Gene Farming

From several points of view, gene farming appears to be one of the most promising applications of transgenic animals. The rationale is to develop expression vectors that allow targeting specific gene expression to a particular tissue, preferably, for example, to body fluids, from which a product could be isolated more easily than from solid tissue. Moreover, many proteins of biomedical importance are secreted into body fluids. A favorite model for this approach is the (lactating) mammary gland: milk protein genes can be expressed very specifically and to high levels; if it is possible to fuse these promotors to other genes, the corresponding proteins should be produced in comparable quantity. Problems may arise from the fact that in the mammary gland proteins are processed or modified differently compared to other tissues, that stability of proteins other than milk proteins is not high enough, or that purification becomes tedious. Pilot projects along these lines have been started, and some perspectives of this interesting approach have recently been reviewed [29].

4.4 Outlook

In this study, we have described several novel developments in recombinant DNA technology and their possible application to animal breeding. Biotechnology allows us to produce pharmaceuticals which are not easily available by conventional procedures, such as hormones or vaccines. Particularly, the production of recombinant vaccines against various animal diseases that are caused by viral infections or infections from parasites is a field of active research. The technology that enables stable and transmissible integration of cloned DNA segments into the mammalian genome is now being established for agricultural animals. Transgenic livestock may be generated to improve food conversion efficiency, growth, fertility, or overall economic merit; the introduction of specific resistance genes seems possible. Another opportunity is to use transgenic animals for the production of valuable biomedical compounds. The progress in this exciting and important area will depend on the fruitful cooperation of scientists from those many disciplines that contribute to molecular biology and animal agriculture.

4.5 References

[1] Maniatis, T., Fritsch, E.F., and Sambrook, J., *Cold Spring Harbor Laboratory,* Cold Spring Harb., N.Y., *1989.*

[2] Hunkapiller, M., Kent, S., Caruthers, M., Dreyer, W., Firca, J., Griffin, C., Horvath, S., Hunkapiller, T., Tempst, P., and Hood, L., *Nature 1984, 310,* 105-111.

[3] Celis, J.E., Protein Databases in Two-dimensional Electrophoresis, *Electrophoresis, 1989,* vol. 10.

[4] Lee, C.C., Wu, X., Gibbs, R.A., Cook, R.G., Muzny, D.M., and Caskey, C.T., *Science 1988, 239,* 1288-1291.

[5] van Ommen, G.J.B., Verkerk, J.M.H., Hoker, M.H., Monaco, A.P., Kunkel, L.M., Ray, P., Worton, R., Wieringa, B., Bakker, E., and Pearson, P.L., *Cell 1986, 47,* 499-504.

[6] Caskey, C.T., *Science 1987, 236,* 1223-1229.

[7] Kan, Y.W. and Dozy, A.M., *Proc. Natl. Acad. Sci. USA, 1978, 75,* 5631-5636.

[8] Jeffreys, A.J., Wilson, V., and Thein, S.L., *Nature 1985, 314,* 67.

[9] Nakamura, Y., Leppert, M., O'Connell, P., Wolff, R., Holm, T., Culver, M., Martin, C., Fujimoto, E., Hoff, M., Kumlin, E., and White, R., *Science 1987, 235,* 1616-1622.

[10] Jeffreys, A.J., *Biochem. Soc. Transactions 1987, 15,* 309- 317.

[11] Hill, W.G., *Nature 1987, 327,* 98-99.

[12] Higuchi, R., Van Beroldingen, C.H., Sensabaugh, G.F., and Erlich, H.A., *Nature 1988, 332,* 543.

[13] Li, H., Gyllensten, U.B., Cui, X., Saiki, R.K., Erlich, H.A., and Arnheim, N., *Nature 1988, 335,* 414-417.

[14] Smith, C.L., and Cantor, C.R., *TIBS 1987, 12,* 284-287.

[15] Anand, R., *Trends in Genet. 1986, 2,* 278-283.

[16] Barlow, D.P., and Lehrach, H., *Trends in Genet. 1987, 3,* 167-171.

[17] Smith, C.L., Econome, J.G., Schutt, A., Klco, S., and Cantor, C.R., *Science 1987, 236,* 1448-1453.

[18] Burke, D.T., Carle, G.F., and Olson, M.V., *Science 1987, 236,* 806-811.

[19] Mullis, K.B., and Faloona, F.A., *Meth. Enzym. 1987, 155,* 355.

[20] Saiki, R.K., Gelfand, D.H., Stoffel, S., Scharf, S.J., Higuchi, R., Horn, G.T., Mullis, K.B., and Erlich, H.A., *Science 1988, 239,* 487-491.

[21] Froesch, E.R., Schmid, C., Schwande, J., and Zapf, J., *Ann. Rev. Physiol. 1985, 47,* 443-467.

[22] Blundell, T.L., Bedarkar, S., and Humbel, R.E., *Fed. Proc. 1983, 42,* 2592-2597.

[23] Nikolics, K., Mason, A.J., Szönyi, E., Ramachandran, J., and Seeburg, P.H., *Nature 1985, 316,* 511-517.

[24] Mason, A.J., Hayflick, J.S., Ling, N., Esch, F., Ueno, N., Ying, Y.-S., Guillemin, R., Niall, H., and Seeburg, P.H., *Nature 1985, 318,* 659-663.

[25] Ling, N., Ying, S.-Y., Ueno, N., Shimasaki, S., Esch, F., Hotta, M., and Guillemin, R., *Nature 1986, 321,* 779-782.

[26] Brown, F., Schild, G.C., and Ada, G.L., *Nature 1986, 319,* 549-550.

[27] Bertholet, C., van Meir, E., ten Heggeler-Bordier, and Wittek, R., *Cell 1987, 50,* 153-162.

[28] Womack, J.E., *Trends in Genet. 1987, 3,* 65-68.

[29] Land, R.L., and Wilmut, I., *Theriogenology 1987, 27,* 169- 179.

[30] Clark, A.J., Simons, P., Wilmut, I., and Lathe, R., *TIB Tech. 1987, 5,* 20-24.

[31] Palmiter, R.D. and Brinster, R.L., *Cell 1985, 41,* 343-345.

[32] Hammer, R.E., Pursel, V.G., Rexroad, C.E.jr., Wall, R.J., Bolt, D.J., Ebert, K.M., Palmiter, R.D., and Brinster, R.L., *Nature 1985, 315,* 680.

[33] Brem, G. et al., *Theriogenology 1986, 25,* 143.

[34] Gridley, T., Soriano, P., and Jaenisch, R., *Trends in Genet. 1987, 3,* 162-166.

[35] Mansour, S.L., Thomas, K.R., and Capecchi, M.R., *Nature 1988, 336,* 348-352.

[36] Maniatis, T., Goodburn, S., and Fischer, J.A., *Science 1987, 236,* 1237-1245.

5. Techniques, Strategies and Stages of DNA Sequencing

P. Heinrich and H. Domdey

5.1 Introduction

The information contained in the genome of an organism is the fundamental description of this living system. Questions concerning gene expression, genome replication, etc. eventually depend on the knowledge of genomic sequences. Therefore genomic sequences are of primary interest for biologists.

One of the most bizarre characteristics of vertebrate genomes is the fact that they contain repeated DNA sequences, sequences of unknown function, and sequences which might have no function at all. These DNA sequences comprise nearly 90% of the total DNA content. Without the complete DNA sequences of several genomes it will hardly be possible to determine whether these sequences have a specific meaning (they could be, e. g., the play-ground for evolutionary processes) or if they are not more than ancestral junk sequences.

Genome sequences are also important for addressing questions concerning evolutionary biology. The reconstruction of the history of life on earth, the definition of gene families, and the search for a universal ancestor all require an understanding of the organization of genomes. A comparison of homologous sequences of different organisms will be important to reveal regions of high conservation and to indicate those regions with stringent evolutionary pressures.

Another important aspect is that genomic sequence data are essential tools for the identification of polymorphisms and of specific disease genes. These data will be of special interest for biologist working with classical breeding techniques as well as for those who intend to generate transgenic animals.

Of course the idea of sequencing entire genomes can be strongly questioned. Even the recently started project of the complete analysis of the human genome is discussed most controversially.

A survey of the published DNA-sequences shows that only a minute amount of DNA-sequence information is available for domestic animals. Fig. 5-1 shows a diagram of DNA-sequences stored and distributed in the GenBank release 58 (January 1989). A closer look into the databank reveals that, e. g., from chicken less than 100 partially or completely sequenced genes or cDNAs were contained in the data bank and - as another example - only 15 genes from sheep. Already these two examples demonstrate that sequences from domestic animals are extremely underrepresented in the sequence data bases. On the other hand, as mapping and sequencing technologies have been and are still being improved rapidly, sequencing of at least the fundamental genes and those which might be of specific interest for the investigated species can be expected in the near future also for domestic animals. For this purpose, the following chapters will bring a short overview about the current state in sequencing strategies and technologies starting with the methods how to obtain the genes of interest.

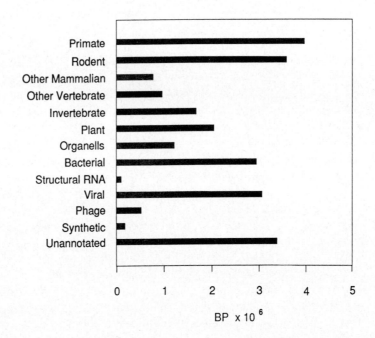

Fig. 5-1 Base pairs in GenBank release 58.
Release 58 of GenBank contains a total of 21,248 loci and 24,690,876 bases; this represents an increase of 8,000,000 bases and almost 6,000 loci over the past year. This release also incorporates data from EMBL release 15 and from DDBJ release 3.0.
The graph shows the number of bases in each taxonomic category. The largest group in terms of bases is the primate sequences, comprising more than 16% of GenBank. The largest group in terms of entries is the rodent sequences, with more than 17% of all the entries.

5.2 Generation of Recombinant DNA Libraries

A DNA library is a collection of independently isolated vector-linked DNA-fragments derived from a single organism. There are two basically different ways of preparing a library. In the first, the genomic DNA is enzymatically fragmented and ligated to the appropriate vector to produce a genomic library. In the second, the DNA fragments are obtained by reverse transcription from the cellular mRNA; this type of library is called a cDNA library.

Construction of recombinant DNA molecules by *in vitro* ligation of vector DNA and the DNA fragments of choice is a well established process. There may arise special problems, however, when the fragment of interest is only a very small part of the target DNA. In general these problems show up either when isolating single copy genes from a complex eucaryotic genome or when isolating rare cDNA clones out of a complex mRNA population. A successful isolation of genes highly depends on the quality of the library used. It has to be made sure that the library contains at least one copy of every DNA sequence of interest. An ideal library represents all of the DNA fragments within the smallest number of clones. This goal can be achieved only approximately. A mammalian genome contains about 3×10^9 base pairs; thus theoretically 2.5×10^6 clones containing a particular 5 kb DNA fragment are required to fully represent an entire mammalian genome with a 99 % probability, whereas around 2,500 clones already represent a bacterial genome of 3×10^6 bp. Similarly, mRNAs present in low amounts may comprise only one part in 106, highly abundant messages, however, can represent 10 % or more of the total mRNA.

5.2.1 Preparation of a Genomic DNA Library

The general main problem in making a genomic DNA library is to maximize the probability that all genes are represented at least once within the smallest number of clones. To approach this ideal it is advisable to minimize the number of clones by incorporating large DNA-fragments and to increase the cloning efficiency by using vectors based on bacteriophage λ.

5.2.1.1 Representation of the DNA Library

To construct a library with an as complete coverage as possible with as few clones as possible, the cloned DNA fragments should be randomly distributed on the DNA. The likelihood (p) that a sequence of interest is present in N clones of a random library is given by the Poisson distribution N = ln (1-P) / ln (1-f) in which f is the size of the fragment expressed as fraction of the genome [1]. This assumption is only true, if the distribution of the target DNA is completely random. If some DNA segments are preferentially represented in the library, the total number of plaques to be screened is larger [2]. Representation in genomic libraries is also influenced by two types of sequences, those which contain palindromes [3] and those which contain direct repeats. These types of sequences have been found either being deleted or lost in bacteriophage λ in an *E. coli* wild-type host [3, 4]. Palindromic sequences and direct repeats may be stabilized and efficiently propagated by plating genomic bacteriophage λ libraries under homologous recombination deficient conditions provided by rec A⁻, rec B⁻, rec C⁻, scb B⁻ hosts [5, 6].

For mammalian genomes, a reasonable library consists of 1-2 x 10^6 plaques. It is advantageous to amplify the library of choice and then to screen aliquots of the amplified library to have a libraries for future screenings. From the fact that some bacteriophage clones grow more efficiently during amplification than others it may be useful to screen a portion of the library before amplification.

5.2.1.3 Vectors for Constructing Genomic DNA Libraries

There are two types of vectors which combine high cloning efficiency and toleration of large insert sizes: (i) bacteriophage λ replacement vectors [6, 7, 8] which allow the substitution of internal, nonessential bacteriophage genes by a 15 - 20 kb eucaryotic DNA-fragment and (ii) hybrid plasmid vectors, called cosmids [9], which have been developed to allow the cloning of DNA-fragments up to 40 kb. The choice between phage and cosmid vectors is based on the size of the desired genomic DNA-fragment, but it seems easier to construct and screen λ replacement vector libraries.

5.2.2   Preparation of cDNA Libraries

cDNA clones differ from genomic DNA clones as they represent those parts of a gene that are expressed as RNA, in the simplest cases without introns and usually without regulatory sequences found in genomic DNA. In the last few years the techniques to generate cDNAs have been remarkably improved due to better enzymes and more straightforward strategies, so that the generation of cDNA copies from mRNAs up to 4 kb or even more is feasible [10, 11]. Nevertheless it is essential to start with high quality RNA, which has the main influence on the quality of the synthesized cDNA. If the RNA is obtained from a differentiated eukaryotic organism, the library will vary in its composition with the type of cells used. The composition also depends on the RNA source and the physiological state of the cell. In general, the cDNA clone has to be selected from a library of that cell type known to synthesize the desired gene product.

5.2.2.1 Representation of cDNA Libraries

A typical mammalian cell contains between 10^4 and 3×10^4 different types of mRNA and around 500,000 mRNA molecules per cell [12]. In order to ensure the cloning of any particular sequence, one can calculate the probability P that each mRNA will be represented at least once, with the equation $N = \ln (1-P) / \ln (1-f)$ in which N is the number of clones required, P the probability for isolating the clone of interest (usually set at 99 %), and n the fractioned proportion of the total mRNA population represented by a single mRNA species. Williams [12] has determined the number of clones needed to ensure successful cloning of a single rare mRNA from different cell types. The theoretical number of recombinants which is needed to ensure that also a very rare mRNA is contained within the library, can vary from 5×10^3 to 1×10^6 recombinants.

5.2.2.2 Vectors for Constructing cDNA Libraries

Before constructing a cDNA library one has to decide over the method with which the library will be screened. If the library will be screened with a nucleic acid probe, any insertion vector is appropriate. λgt 10 is particularly good for large libraries (> 10^5 clones) to be screened with this approach [13]. For small libraries (10^4 to 5×10^4 clones) the use of pUC vectors or plasmids with recognition sites for phage RNA polymerases are recommended. If the library is to be screened with antibody probes, appropriate *E. coli* expression vectors such as the pUC family of plasmids [14], the λgt 11 phage [15] or Lambda ZAP II phage [16] should be chosen. Such vectors are based on the expression of a fusion protein in which a segment of interest is fused to coding sequences of highly expressed and stable *E. coli* ß-galactosidase.

5.2.2.3 Subtracted cDNA Libraries

A subtracted cDNA library consists of cDNA clones corresponding to mRNAs present in a certain cell or tissue type but not present in another type. The advantage of creating a subtraction library is the enrichment of target cDNA clones. Briefly, mRNA from cell type A is reverse transcribed into cDNA. The radiolabelled cDNA is hybridized to an excess of RNA from cell type B and the non-hybridizing cDNA purified by chromatography. In this way it represents an enriched population of sequences expressed in cell type A but not in type B. This cDNA can be used to construct an enriched cDNA library. This subtraction approach allows the isolation of clones complementary to mRNAs which represent as little as 0.01 % of the total mRNA. This method was used, e. g., to isolate T-cell antigen receptor cDNAs [17] and the murine interleukin-4-receptor [18].

A good alternative to creating subtracted cDNA libraries is a screening method called "plus/minus". Here, for example, duplicate sets of filters from a tissue specific cDNA library are prepared first. One set is hybridized with RNA-derived probes from cell type A, and the other set is screened with probes derived from cell type B. The specific cDNA clones which are desired can be identified as plaques that hybridize with a type A cDNA probe but not with a type B cDNA probe. This approach was successfully used to identify the cDNA encoding the CD20 antigen of human B lymphocytes [19].

Furthermore, specific mRNAs can also be enriched by isolating RNA from nuclei [20], membrane-bound polysomes [21] and from polysomes with specific nascent proteins [22].

5.2.3 The Polymerase Chain Reaction in Gene Analysis and Cloning

The polymerase chain reaction (PCR) is a procedure for the enzymatic amplification of a specific target DNA sequence *in vitro*. Within a few hours, over 10^6 copies of the target DNA can be synthesized using specific oligonucleotide primers and DNA polymerase [23]. The specificity of PCR amplification is based on two oligonucleotide primers which flank the DNA segment to be amplified and which hybridize to opposite strands. The procedure involves repeated cycles of heat denaturation of the DNA, annealing of the primers to these complementary sequences, and extension of the annealed primers with DNA polymerase. Each primer becomes incorporated into the amplification products, so that the exponentially accumulating product is a discrete fragment whose termini are defined by the 5' ends of the PCR primers. PCR can also be used to introduce additional sequences such as restriction sites by incorporating these into the oligonucleotide primers. The original PCR protocols used the Klenow fragment of *E. coli* DNA polymerase I. This required the addition of fresh enzyme during each cycle. To overcome this problem, a thermostable DNA polymerase purified from the thermophilic bacterium *Thermus aquaticus* was introduced into the reaction [23]. Although the first published report appeared not more than 4 years ago [24], PCR amplification has already been applied for the diagnosis of genetic disorders [25, 26, 27], the detection of infectious disease pathogens such as viruses, parasites and bacteria [28], the study of activated oncogenes [29], the analysis of allelic sequence variations [24, 25, 30], direct genomic cloning of DNA up to 2.5 kb from genomic templates [27, 30], direct cDNA cloning [31], genetic fingerprinting of forensic samples [32], and direct genomic sequencing [33, 34]. One constraint of the method is that some sequence information at both ends of the target DNA must be available.

5.2.4 Screening DNA Libraries

The screening of DNA libraries involves the application of rapid and suitable assay procedures and also the decision what kind of library to screen, a genomic or a cDNA library. Recombinants within libraries can be screened for homology with a nucleic acid sequence, for expression of an antigen or for expression of a phenotype. A combination of selection schemes is recommended to make sure that a clone has been identified correctly and is not a false positive.

5.2.4.1 Screening Genomic Libraries

Genomic libraries can be constructed from DNA derived from any tissue of the target organism. The formula for predicting the number of clones that have to be screened for isolating the desired clone has been discussed above. The number is a function of the complexity of the genome and the average size of the inserts in the library. Usually about 10^6 bacteriophage clones need to be screened in order to identify a genomic clone. For amplified libraries the number of clones to be screened is up to 50 % larger than the number calculated with the formula.

The use of nucleic acid probes is generally applied in screening genomic libraries. Three different types of probes may be used:(i) heterogenous probes, which usually are heterogeneous mixtures of cDNAs complementary to mRNA populations reflecting cell- or gene-specific attributes supposed to be very similar to the gene desired; (ii) oligonucleotide probes based on protein sequence data; (iii) homogeneous cloned DNA fragments derived from a gene of interest from a different species or a related gene that has been isolated from the same species.

5.2.4.2 Screening cDNA Libraries

The optimal cDNA library is constructed from a particular tissue or cell type which expresses the mRNA searched for highly abundantly. The level of abundance of the specific mRNA is of critical importance for the selection of a cloning strategy. A typical eukaryotic cell contains approximately 1 pg of mRNA which is equivalent to about 10^6 molecules, transcribed from about 15,000 different genes [35, 36]. 3 different types of clones are generally obtained: clones derived (i) from low abundant genes, (ii) from medium abundant genes, and (iii) from high abundant genes. Thus the number of recombinant clones which have to be screened is determined by the abundance of the mRNA in the cell.

Several procedures for selecting genes of interest from cDNA libraries are available. Screening with nucleic acid probes for homology with a DNA sequence of an individual recombinant is an effective procedure that allows the identification of a single clone within a population of millions of other clones. Another common method for the identification of particular clones from expression libraries is the immunological screening method. Expression libraries are constructed using either plasmids [37] or bacteriophage vectors like λgt 11 [15] or λZAP II [16]. Inserts cloned in these vectors are expressed as part of a fusion protein and clones can be identified by using an antibody that is able to recognize the fused form of the desired protein. Polyclonal antibodies are generally used for this kind of screening since they recognize many epitopes. In contrast, monoclonal antibodies recognize only one specific epitope.

Another immunological screening procedure is the socalled "hybrid arrest of translation". Recombinant DNA clones can be used to select a specific mRNA from the total mRNA population for translation in a cell free protein synthesis system [38, 39]. Precipitation by homologous antibodies is used for recognition of the desired polypeptide. Alternatively, identification of a gene can be based on the inhibition of translation of a specific polypeptide by hybridization of the cloned DNA to its corresponding mRNA [40]. The immunodiagnostic steps can be replaced by monitoring the specific biological activity of the polypeptide synthesis in an *in vitro* translation system [41].

5.3 Strategies for DNA Sequencing

Two general methods are widely used to determine DNA sequences. One is the chemical method based on the cleavage with a base specific chemical reagent at one or two of the four nucleotide bases [42]. The other is the enzymatic method based on the incorporation of dideoxynucleotides for sequence-specific chain termination [43]. The latter has become used more frequently because of its simplicity and quickness. An important feature of this method is that it can be easily modified to apply several strategies designed to produce rapidly high amounts of data for large sequencing projects. Rapid methods for sequencing large fragments of DNA involve two basic general strategies, the random and the directed approach. The random approach is based on the presence of restriction sites or on the subcloning of randomly generated fragments. This procedure often leads to gaps, which have to be filled by employing additional approaches. The collected sequence data have to be put together by using one of different available computer program.

For large scale sequencing projects direct sequencing approaches will have to be employed for several reasons: Shotgun approaches would require considerably more individual sequence determination to cover long sequences completely; therefore the data storage and analysis problems would grow tremendously and with them the costs for DNA sequencing. The largest sequencing projects which have been published so far were performed by the random approach: The complete DNA-sequences of human mitochondrial DNA, 16 kb [44], of bacteriophage λ, 49 kb [45], of Epstein-Barr virus, 170 kb [46]. For sequencing entire genomes like *E. coli* (already in progress), *Saccharomyces cerevisiae* (also in progress), *Arabidopsis thaliana, Caenorhabditis elegans* or even the human genome more fundamentally new approaches have to be employed. One strategy for sequencing entire eukaryotic genomes or parts of it is to divide the purified individual chromosomes in an ordered manner into successively smaller pieces. These fragmented chromosomes are used to generate chromosome-specific cosmid libraries. An overlapping set of cosmid clones encompassing the entire chromosome is generated with the library, constituting a physical map of that specific chromosome. Each cosmid clone is individually fragmented into smaller pieces suitable for sequence analysis [47]. The efficiency of this approach depends upon the developmental status of the following techniques: 1) the automation of DNA sequence technology (discussed afterwards), 2) the automation of both clone preparation and DNA purification (commercially available machines already do exist), 3) automation of laser-activated chromosome sorting machines (Van Dilla et al., 1986), 4) availability of better procedures for the input of DNA sequences with the appropriate annotation, of better computers speeding up date analysis and of improved software for data analysis, 5) the physical map of the organism to be sequenced.

A highly promising technique (Multiplex sequencing) for efficient large scale sequencing analysis (described in chapter 5.4.8.) has been developed by Church and Kieffer-Higgins [49] and is presently applied for sequencing the entire *E. coli* genome.

It should be kept in mind that the sequencing procedures are not the rate-limiting factor in obtaining sequence data. The generation of subclones takes considerably more time. Therefore we will first describe briefly the strategies which are presently used for sequencing projects.

5.3.1   Random Approaches for DNA Sequencing

5.3.1.1   Sonification Method

At the beginning of a sequencing project, the DNA to be analyzed has to be isolated and then fragmented into a set of subsegments covering the entire region of interest. A random set of subfragments is generated by shearing the purified DNA to be sequenced by sonification. To increase the random nature of the process, the DNA should be self-ligated before sonification. It is an advantage of random cloning that the size of the subfragments desired can be controlled by altering the sonification parameters. The molecule ends resulting from sonification cannot be ligated directly into the cloning vectors. Their ends have to be repaired by generating enzymatically blunt ends which can be ligated directly to blunt-ended vectors. The cloning efficiency can be increased by adding restriction enzyme linkers to blunt ends and cloning into an appropriately cut vector. In practice, the lengths of the inserts to be sequenced should be at least as long as the maximum which can be resolved and read from a single gel run (300 - 600 bp). M13 mp vectors [50] or plasmid vectors providing several unique restriction sites for cloning can be chosen and the sequencing reaction can be performed by single stranded M13 or double stranded plasmid DNA sequencing [50, 51]. The sonification method results in a six to eight fold sequence determination of each base in a DNA fragment before a project is completed [45, 46, 52].

5.3.1.2 Enzymatic Generation of DNA Subfragments

If a detailed map of the DNA fragment to be sequenced is available, the subfragments can be generated by using restriction enzymes and the resulting fragments are subcloned into the appropriately cut vectors. This approach, however, exhibits some disadvantages. First, a wide range of subfragment sizes will be produced due to the uneven distribution of restriction sites within the target DNA. Second, small fragments show a much higher cloning efficiency so that some regions will be cloned many times. Third, several libraries and different enzymes have to be used in order to obtain overlapping fragments.

This approach was employed for sequencing the 8031 bp long cauliflower mosaic virus [53].

In addition, random DNA fragments can also be produced by the less sequence-specific DNA hydrolyzing enzyme DNase I [54]. Parts of the human mitochondrial genome were determined with another strategy [54]. Here a large DNA fragment is digested into smaller pieces by single and/or multiple digests with restriction enzymes and the pool of fragments deriving from one digest is cloned shotgun into M13 or appropriate plasmid vectors.

5.3.2 Directed Approaches for DNA Sequencing

5.3.2.1 DNase I Method

Employing DNase I for generating sequentially deleted subclones makes it possible to obtain sequence information progressively along the DNA fragment of interest [55, 56, 57]. The cloning vector (M13 or plasmid vector) must contain single restriction enzyme recognition sites B, C, A (in this order) adjacent to the primer binding site. The DNA to be sequenced is cloned into site B. The recombinant DNA is partially digested with DNase I in the presence of Mn2+ generating a mixture of linear double stranded molecules with various lengths of insert DNA on both ends. Treatment of the mixture with a corresponding restriction enzyme removes the DNA pieces at site A.

Circularization of these linear recombinants with T4 DNA ligase produces a clone mixture with insert DNA fixed at one end and sequentially shortened at the other end close to the sequencing primer binding site. DNase I generates breaks around the target molecule at random. Breaks on the side of the origin of replication (positioned at the left end of the insert) will lead mostly to nonviable bacteria (or phages). Breaks on the other side of the insert cannot be sequenced due to the lack of a primer binding site. Before transformation, the parental DNA molecules are "inactivated" by linearizing them with the restriction enzyme which cuts at site C. The final result of this procedure are clones, which are derived from breaks within the insert. Clones differing in length by about 250 nucleotides have to be sequenced from the primer binding site for covering the whole length of the insert DNA.

5.3.2.2 Bal 31 Exonuclease Method

This method is based on the ability of exonuclease Bal31 to progressively shorten a DNA fragment which, after generating blunt ends, can be ligated with T4 DNA ligase. Since Bal31 digests in a nearly linear way for up to 15 minutes, a nested set of deletions can be easily created. The number of nucleotides to be deleted is a function of the amount of Bal 31 used. This approach is suitable for M13 mp vectors and for a variety of plasmid vectors that contain appropriate restriction sites in a polylinker. The procedure involves the following steps [58, 59, 60]:

1) Digestion of cloned DNA with a restriction enzyme at a single site at one end of the insert.
2) Bal 31 digestion of the insert DNA for various lengths of time to create a nested set of deletions.
3) Digestion of the shortened inserts with a restriction enzyme at the other end of the insert and separation of the desired fragments from the vector by gel electrophoresis.
4) Elution of the shortened fragments from the gel and subcloning into an appropriate vector.

The nested set of deletions can be used for dideoxy sequencing using single or double stranded DNA templates. It has to be kept in mind that the Bal31 approach is designed for deletions of up to 2000 bp and that the rate of digestion of the DNA is not uniform as GC-rich regions are digested much more slowly than other regions.

5.3.2.3 Unidirectional Deletions with Exonuclease III

This strategy [61, 62] is based on the remarkably uniform rate at which Exonuclease III can digest DNA. This method can be used with M13 or similar bacteriophages, and plasmids containing the appropriate polylinker restriction sites. The procedure involves the following steps:

1) Cloning of the DNA fragment of interest into the polylinker region in such a way that at least two restriction sites remain between the insert and the sequencing primer binding site. The restriction enzyme that cuts at the site close to the insert has to leave a 5' protruding or blunt end; the enzyme that cuts farther away from the insert should leave a 4 nucleotides long 3'-protrusion.
2) Endonuclease III is added and digests synchronously in only the one desired direction, since it does not attack a 4-nucleotide long 3' protrusion.
3) Aliquots are removed at timed intervals and subsequently treated with S1 Nuclease or Mung bean nuclease, which remove the other strand of the DNA covered by the desired deletions.
4) The ends are filled up with Klenow polymerase and DNA ligase is added for recircularizing the molecules.

5) Transformation and plating of timed aliquots lead to clone families with clustered
   deletion breakpoints.

A potential limitation of the method is the requirement for appropriate unique restriction
sites between the insert and the primer binding site. The construction of vectors with 8
base recognition cites such as Not I or BstX I, however, should minimize this limitation,
because such sites are present so infrequently that almost any insert is likely to lack them.
    This approach appears to be more efficient, more rapid and easier to perform than
other nonrandom methods, because it yields breakpoints that are tightly clustered around
a particular point in the sequence, reducing the time-consuming steps required to fill gaps
within the sequence. Finally, there is no limitation concerning the fragment length to be
deleted, up to 10 kb fragments in length have been successfully deleted.

5.3.2.4 Transposon Method

Transposons show three commonly observed properties: (i) They carry selectable markers.
(ii) They insert at many different locations. (iii) They are mutagenic in regions adjacent
to the insertion site [63, 64]. These properties have been exploited for bacterial gene
mapping, for the isolation of new gene arrangements and also for the isolation of
transposon-promoted deletions that are very useful in sequencing long DNA fragments [65,
66]. The DNA fragment of interest is cloned in the vector pAA3.7x [66] and transferred
into the strain AA102. The DNA fragment is divided into a set of overlapping segments
by transposon T9-promoted deletions. These deletions can be isolated by positive selection
for galactose resistance (Gal)^R. From Gal^R colonies the plasmids are purified and
fractionated by gel electrophoresis to select the clones with the deletion intervals desired.
Overlapping clones are directly sequenced by the plasmid dideoxy sequencing method with
a sequencing primer derived from Tn9 [67]. This method has several advantages to other,
especially random methods: there is no accumulation of redundant sequences; DNA
fragments with several kb in length can be progressively shortened from one end; there is
no requirement for specific restriction sites; sequencing data are produced in an ordered
way facilitating the assembly of the DNA fragment sequenced.
    The method has been modified for cloning and sequencing long (40 kb) DNA
fragments [67]. The fragment is cloned in the sequencing cosmid pAA113x and subdivided
by a series of overlapping IS1-promoted deletions, which are isolated by positive selection
for galactose resistance. Plasmids from several thousand galactose-resistant colonies are
fractionated on an agarose gel, and DNA from each fraction is restricted with the
appropriate enzymes to shorten each deletion from the opposite end, resulting in a series
of short overlapping segments, spread across the entire length of the fragment which are
fused to IS1. The purified plasmids are used for supercoil sequencing with an IS1 primer.
The procedure may be useful in the analysis of large genomes.

5.3.2.5 The T4 Polymerase I / Terminal Transferase Tailing Approach

This method creates a nested set of deletions for sequencing using single stranded DNA and complementary DNA oligomers to generate specific cleavage and ligation substrates [68, 69]. The first step involves the creation of a fixed end of the deletion, by employing a double strand specific enzyme and a short oligonucleotide that hybridizes to single stranded vector DNA adjacent to the target DNA. Digestion with the appropriate enzyme linearizes the circular molecule. The second step uses the 3'- 5'- specific exonuclease activity of T4 DNA polymerase to digest the linearized DNA progressively at a constant rate. Removing aliquots at intervals, a series of overlapping deletions encompassing up to 3 kb may be generated within 1 hour. The third step uses Terminal Transferase to add a short homopolymeric tail complementary to the 3' end of the oligonucleotide. Fresh oligomer is annealed to the deletion products joining the two ends of the molecule. Treatment with T4 DNA ligase will seal the remaining nicks. After transformation of the deletion products in *E. coli*, clones with appropriate deletions are selected on agarose gels and used for single stranded DNA sequencing.

The method is restricted to single stranded DNA. It has the advantage that the whole procedure can be carried out in one reaction tube without phenol extractions and ethanol precipitations.

5.3.2.6 The T4 DNA Polymerase / Thiodeoxynucleotide Method

This method is useful for the preparation of deletions that extend in one direction from a fixed end and is based on the ability of deoxynucleoside [1-thio] triphosphates to be incorporated into DNA by DNA polymerase I (Klenow fragment) and the fact that α-thiophosphate-containing phophodiester bonds are resistant to hydrolysis by the 3'-5' exonucleolytic activity of phage T4 DNA polymerase [70]. Therefore, linear duplex DNA molecules blocked at one 3'-terminus with a thiophosphate are prepared and then degraded from the other end with exonunclease. After digestion for different lengths of time the DNA molecules are treated with Nuclease S1 and T4 DNA Ligase resulting in a nested set of deletion mutants.

The advantages of this method are that (i) once the DNA to be mutagenized is cloned, no further cloning steps are required; (ii) no special purification or treatment is necessary; (iii) the method is useful for creating deletions in any duplex molecule, M13 RF or plasmid; in addition, sequencing DNA directly from plasmids makes deletion mutagenesis of plasmids particularly attractive; (iv) only two unique restriction sites are required and they may be of any type; (v) in some cases deletion mutants of M13 mp phages can be recognized directly by their ability to provide lacZα-complementing activity and this in turn provides a rapid way to prepare genetic fusions of lacZ to the target gene.

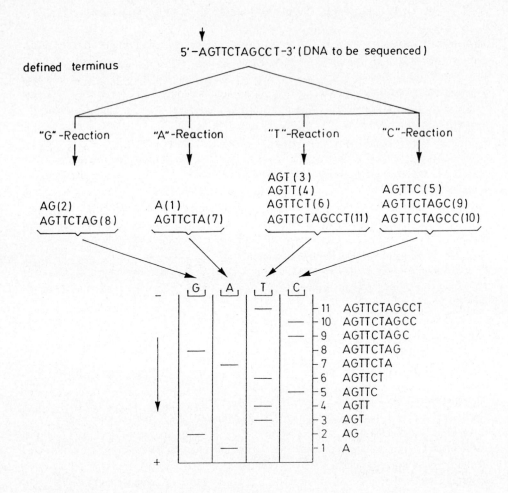

Fig. 5-2 General principle of DNA sequence analysis.
A mixture of labeled single-stranded oligodeoxynucleotides is generated in four separated reactions. In all four reactions the oligonucleotides have the same defined 5'-end; they differ, however, in their 3'-ends terminating either with a G, A, T, or C, respectively. The reaction mixtures are separated next to each other on a denaturing polyacrylamide gel. The DNA-sequence can be read directly from the autoradiogramm of the gel.

5.4 DNA Sequencing Techniques

Classical sequencing techniques are based on high-resolution denaturing polyacrylamide electrophoresis procedures which are capable of resolving single stranded oligodeoxynucleotides of up to 500 bases in length. In practice, for a given region to be sequenced, a set of radiolabeled single stranded oligodeoxynucleotides is generated, the members of which have one fixed end but differ in length at the other end by each successive deoxynucleotide. To determine the nucleotide sequence, in four separate chemical or enzymatic reactions all of the oligodeoxynucleotides have to be synthesized that terminate at the variable end in A, C, G, or T. The oligodeoxynucleotide products of the four reactions are resolved on adjacent lanes on a sequencing gel. The DNA sequence can be read directly from the four ladders of oligodeoxynucleotides on the autoradiograph, as illustrated in Fig. 5-2.

The chemical and enzymatic DNA sequencing methods differ primarily in the means by which the ladder of oligodeoxynucleotides is generated. The enzymatic DNA sequencing method has become used more frequently because of its simplicity, quickness and flexibility. Together, these two techniques have been employed to sequence almost 15 million bases of DNA (the current size of GenBank) of which about 1 million bases are from the human genome, and the data base is being enlarged by more than 30% every year.

Employing mass spectrometry or tunnelling electron microscopy for DNA sequencing might increase the rate at which DNA can be sequenced dramatically in the future [71].

5.4.1 Chemical DNA Sequencing

In the chemical method of DNA sequencing developed by Maxam and Gilbert [42, 72] the target DNA is radioactively labeled at one end (3'- or 5'-end). In four separate reactions the labeled DNA is cut with a base specific chemical reagent under limiting conditions and the reaction products are separated on a sequencing gel. Because only end-labeled fragments are observed following autoradiography of the sequencing gel, the DNA sequence can be read from the four DNA ladders as illustrated in Fig. 5-3.

The chemical sequencing method is based on the specificity of dimethyl sulfate and hydrazine to specifically modify purine- and pyrimidine-bases (Fig. 5-4) within the DNA molecule. Piperidine is then added to catalyze strand breakage at these modified molecules. The specificity of the method resides in the modification reaction (hydrazine, dimethyl sulfate, or formic acid), which reacts with only a few percent of the bases. The second reaction, piperidine strand cleavage, must be quantitative. The chemical mechanism can be explained as follows (see also Fig. 5-4):

- G-reaction: Dimethyl sulfate methylates the nitrogen 7 of guanine, which then opens between carbon 8 and nitrogen 9. Piperidine catalyzes the displacement of the modified guanine from its sugar.
- G+A-reaction: Formic acid weakens the N-glycosidic bonds of the adenine and guanine residues by protonating purine ring nitrogens. The purines can be displaced by piperidine.
- T+C-reaction: Hydrazine splits the ring of thymine or cytosine. The fragments of the bases can be displaced by piperidine.
- C-reaction: In the presence of NaCl, only cytosine reacts with hydrazine. The modified cytosine can be displaced with piperidine.

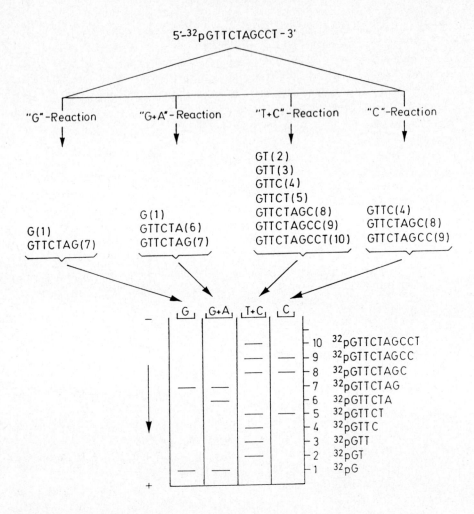

Fig. 5-3 General scheme of chemical DNA-sequencing.
At the top of the Figure the sequence of a 5'-end labeled DNA fragment is shown. The numbers in brackets indicate the lengths of the base specific degradation products. The oligonucleotides which are generated by the base specific chemical cleavage reactions are separated by polyacrylamide gel electrophoresis in four adjacent lanes. Bands are detected by autoradiography.

In all four reactions, the piperidine also catalyzes the phosphodiester bond cleavage at the position where the modified base has been displaced by piperidine. According to the four reactions the resulting products are separated in the lanes G, G+A, T+C and C on a sequencing gel (see Fig.5-3).

The major disadvantage of the chemical sequencing method is that preparation of the DNA prior to sequencing is very time-consuming, involving several gel fractionations and DNA extractions. The development of sequencing vectors such as pUR250 [73], pGV451 [74] and pSP65LS [75] eliminates the time-consuming steps otherwise required, because they allow the direct subcloning of the DNA fragment of interest and end-labeling of a single predetermined terminus adjacent to the sequence of interest.

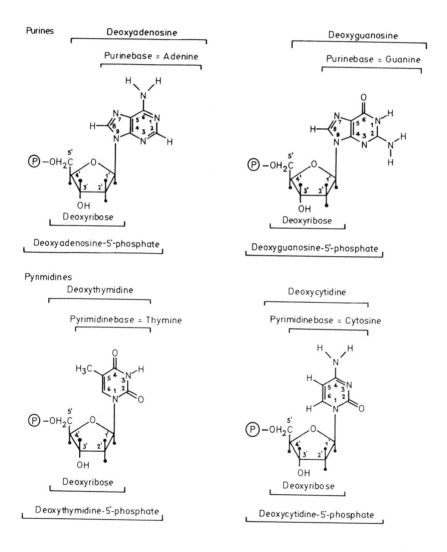

Fig. 5-4 Structure of the four deoxyribonucleotides.

66

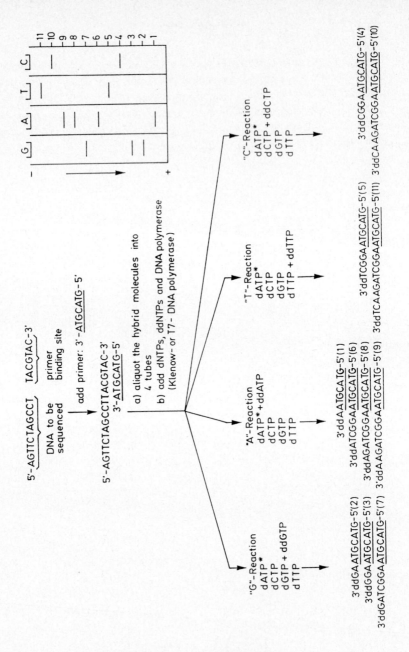

Fig. 5-5 General scheme for enzymatic DNA-sequencing.
First an oligodeoxynucleotide primer is annealed to the single-stranded DNA to be sequenced. The annealed product is divided into 4 aliquots. The four dNTPs (one out of the four carries a radioactive label, see *), one of the four ddNTPs and DNA-polymerase are added. A mixture of labeled oligodeoxynucleotides is gerated which terminate at their 3'-ends with the added specific dideoxynucleotides. The oligonucleotides are separated by polyacrylamide gel electrophoresis in four adjacent lanes. Bands are detected by autoradiography.

5.4.2 Enzymatic DNA Sequencing

The dideoxy, enzymatic or chain termination method was introduced in 1977 by F. Sanger [43] and uses *E. coli* DNA polymerase I to synthesize a complementary copy of a single stranded DNA template. DNA polymerases cannot initiate DNA chains. Therefore, chain elongation occurs at the 3'-end of a short complementary primer which is annealed adjacent to the DNA segment to be sequenced, as illustrated in Fig. 5-5. Chain growth involves the formation of a phophodiester bridge between the 3'-hydroxyl group at the growing end of the primer and the 5'-phosphate group of the incorporated deoxynucleotide. Thus, overall chain growth is in the 5'-3'-direction. Since intact *E. coli* DNA polymerase I also has 5'-3'-exonuclease activity, the large fragment (Klenow-fragment) of *E. coli* DNA polymerase I is employed, which can still carry out the elongation reaction. Alternatively, reverse transcriptase (either from Moloney-Murine Leukemia virus or from avian myeloblastosis virus) can also be used. In addition, the use of T7 bacteriophage DNA polymerase [76] has improved and simplified DNA sequence analysis. The use of the Taq DNA polymerase from the thermophilic bacterium *Thermus aquaticus*, which has a temperature optimum for polymerization of 75 °C to 80 °C should be particularly advantageous for sequencing DNA templates that exhibit strong secondary structures at lower temperatures [77], since a higher reaction temperature will disrupt DNA secondary structures and inhibit reannealing of denatured, double-stranded templates.

The enzymatic sequencing method is based on the ability of DNA polymerases to use both 2'-deoxynucleotides and 2',3'-dideoxynucleotides as substrates. When a dideoxynucleotide is incorporated at the 3'-end of the growing primer chain, the elongation is terminated selectively at A, C, G or T due to the missing 3'-hydroxyl group of the primer chain. In practice, the enzymatic DNA sequencing involves the following steps (see Fig. 5-5):

- Preparation of the DNA to be sequenced as single stranded molecule.
- Annealing of a short, chemically synthesized oligonucleotide primer to the 3'-end of the region to be sequenced. The annealed oligonucleotide serves as primer for DNA polymerase.
- Dividing of the hybrid molecules into four aliquots. Each of it contains all four dNTPs, one of which is either ^{32}P- or ^{35}S-labeled, and one of the four 2', 3'- ddNTPs. The dNTP/ddNTP concentration is adjusted such, that termination of the elongation primer chain occurs at each base in the template resulting in a population of radiolabeled extended primer chains, which have a fixed 5'-end determined by the annealed primer and a variable 3'-end terminating at a specific base.
- The radiolabeled reaction products are denatured by heating and then separated on a sequencing gel in adjacent lanes, and the DNA sequence can be read directly from the autoradiograph of the gel.

Most enzymatically based sequencing reports have used as cloning vector bacteriophage M13 DNA, which replicates as a double stranded DNA molecule but is packaged as single stranded DNA in the virus [50]. This method, however, usually requires the subcloning of DNA fragments from plasmids into the double stranded M13 replicating form, in which the stability of some DNA inserts has, at times, been problematic [51]. To eliminate subcloning steps, Wallace et al. [78] introduced a procedure using linearized plasmid DNA as template for sequencing. In this method, sequencing primers were synthesized to align chain elongation next to the vector cloning site and were hybridized to heat-denatured DNA. Chen and Seeburg [51] could demonstrate that alkali denaturation of supercoiled plasmid DNA is a more efficient procedure to carry out primer-template hybridization than heat denaturation. DNA sequencing of alkali-denatured supercoiled DNA can produce sequencing gels with the same quality as those employing M13 vectors. The past several years have witnessed several efforts to make the direct sequencing of plasmid DNA feasible by the chain termination method [78, 79, 80, 81, 82, 83, 84].

The most important advantage of the plasmid sequencing method is that the production of unidirectional overlapping deletions (see above) and the DNA sequencing can be performed on a single vector, avoiding subdividing steps into M13 or the use of special vectors producing single stranded DNA.

5.4.3   Choosing Between Chemical and Enzymatic DNA Sequencing

There is no doubt that dideoxy sequencing is more simple and more rapid than chemical sequencing and therefore the tool of choice for DNA sequencing. But there are also disadvantages, which relate to the property of the Klenow polymerase to terminate sometimes prematurely the synthesis due to DNA secondary structure, especially in GC rich templates. Although the use of reverse transcriptase, T7 DNA polymerase and Taq DNA polymerase, and various modified nucleotides may help in some of these problems, there often remain DNA regions which are poorly sequenced by the chain termination method. Problems associated with polymerase synthesis can be eliminated by utilizing the chemical method, when difficult stretches of DNA have to be determined. For sequencing short oligonucleotides for the mapping of the DNA binding sites of DNA binding proteins (footprinting) the chemical method has to be employed. The genomic sequencing method, developed for studies of DNA methylation and DNA-protein interactions in living cells [85] is also based on the chemical approach. But for routine DNA sequencing and for large sequencing projects the dideoxy method is more efficient.

5.4.4   Sequencing Gels

The radiolabeled single-stranded oligodeoxynucleotides produced in the sequencing reactions are fractionated on a high-resolution denaturing polyacrylamide gel on the basis of their sizes. Considerable efforts have been made to improve the electrophoretic methods. Very thin and temperature-controlled gels improve the resolution [86]. Field-strength gradients with a gradient [87] or the use of wedge-shaped gels [50] result in an increased number of bases which can be read in a single gel run. In the enzymatic method ^{35}S-labeled oligodeoxynucleotides produced by the use of ^{35}S-dATP, gives considerably better resolution than ^{32}P-labeled material, because there is less scattering of the weaker ß particles generated by ^{35}S decay. The use of this compound leads to much sharper bands, thus the sequencing ladders can be read farther.

The ambitious goals to sequence entire genomes require the automation of the complete DNA sequencing process. A first step towards automation of the electrophoresis is the direct transfer electrophoresis technique, described by Pohl and Beck [88]. In this method, the radiolabeled sequencing products are loaded onto a short polyacrylamide slab gel. A blotting membrane is moved with a constant speed across the bottom of the gel and collects the molecules as they leave the sequencing gel. Employing this technique the full separating capacity of the gel can be exploited and the gel may be used several times.

5.4.5 Direct Enzymatic Sequencing of Genomic DNA

The identification of mutations and polymorphisms in human genes by DNA sequencing contributes substantially to an understanding of the molecular nature of the disease and enables a variety of practical approaches for diagnosis. Sequence information is usually obtained by cloning the genes from the individual to be analyzed. The cloning is labor-intensive and will take several weeks to carry out. Methods for direct DNA sequencing are desirable. A method for direct chemical sequencing of mammalian genomic DNA has been reported [85], but has not been widely adopted. Direct dideoxy sequencing of eukaryotic DNA has only recently been applied successfully in yeast [89]. The development of the polymerase chain reaction [24] to selectively amplify a short segment of DNA more than 10^6-fold make the direct analysis of a single-copy DNA sequence possible for identifying mutations or sequence variants in genomic DNA or expressed mRNA. Similar to conventual sequencing, both strands of amplified DNA can be sequenced with AMV reverse transcriptase [33]. PCR amplification, originally developed for diagnostic applications, has been used to rapidly clone highly polymorphic genes such as those of the major histocompatibility region [23, 31]. An alternative method to the direct sequencing of amplified DNA segments is the transcript sequencing of amplified DNA [90]. In this method, a phage RNA polymerase promoter is incorporated into at least one of the PCR primers. The PCR amplified DNA is then transcribed and the resulting template RNA is sequenced with reverse transcriptase.

5.4.6 Dideoxy Sequencing of Native RNA and Uncloned cDNA

The use of reverse transcriptase with appropriate oligodeoxynucleotide primers and chain terminating dideoxynucleotides has been valuable for obtaining sequence information directly from RNA templates [91, 92]. RNA can be sequenced with a specific ^{32}P-labeled oligonucleotide which is annealed to the enriched poly(A)$^+$-RNA and dideoxynucleotide chain terminators plus reverse transcriptase.

Uncloned cDNA can be sequenced by a primer extension reaction [93]. In this procedure unlabeled DNA primer homologous to the poly (A)$^+$-RNA is hybridized and the complementary strand is synthesized using reverse transcriptase and deoxynucleotides. The RNA-cDNA-hybrids are heat-denatured, and specific ^{32}P-labeled oligonucleotide primers are annealed to the cDNA followed by the conventional chain termination sequencing reaction using Klenow polymerase in the presence of dNTPs and ddNTPs.

This method can be used to extend data obtained from incomplete cDNA clones. It is possible to walk along an mRNA transcript by sequencing mRNA and then synthesizing an oligonucleotide primer to the newly sequenced region, followed by further sequencing. This method is also applicable for determining splice junctions and the size and sequence of 5'- untranslated regions of an mRNA. In addition, this approach can be used for sequencing RNA transcripts produced in vitro from plasmid templates.

5.4.7 Automation in DNA-Sequencing

Four methods have been described until to-date for automated DNA-sequencing [94, 95, 96, 97].

All of them apply non-radioactive labeling techniques, i. e., fluorescent dyes are used to label the DNA. These fluorescent dyes are then excited during electrophoresis by an Argon laser-beam. All described methods are based on the Sanger DNA chain termination protocol with dideoxynucleoside triphosphates as terminators. The reaction mixtures are electrophoretically separated on usual sequencing gels and the bands migrating through the gel are analyzed on defined positions during the run. The obtained data are saved and processed to yield the final DNA sequence of the analyzed piece of DNA.

All four described methods share several common features, but differ in a number of details: i) The fluorescent dyes are either attached to the sequencing primer or they are incorporated during the reaction in the form of labeled dideoxynucleotide triphosphates. ii) Different types of fluorescent dyes are used. iii) Either four different dyes are used which makes it possible to load the four base specific reaction products in a single lane, or a single dye is used to label the DNA fragments with the consequence that the reaction products have to be analyzed in four adjacent but separate lanes. iv) The collected data are processed and visualized in different ways.

Because of these differences the four techniques are described in the following separately, one of them in more detail.

In the method developed by Prober et al. [96] a primer is hybridized to a defined position of the DNA and the statistical termination of the following polymerase reaction is achieved by adding ddNTPs as chain terminators. The four different ddNTPS are labelled with four slightly different fluorescent dyes. In this way the fluorescent label is attached to the 3'-end of a DNA-fragment. The dyes are four very similar succinyl fluoresceins which differ only in their R1 or R2 groups as shown in Fig. 5-6a. The slightly different absorption spectra of the four succinyl fluoresceins SF-505, SF-512, SF-519, and SF-526 (Fig. 5-6a) are shown in Fig. 5-6b.

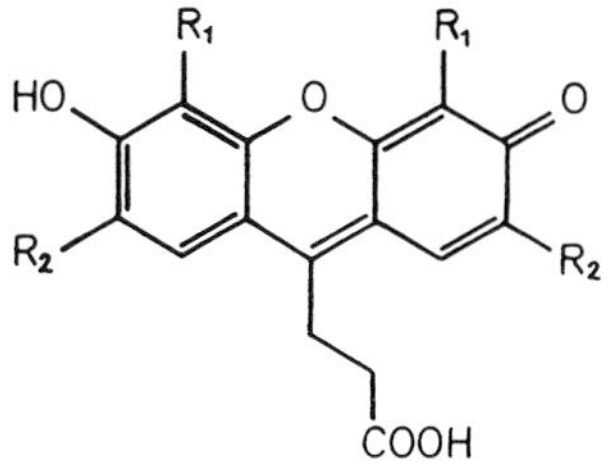

Fig. 5-6a Chemical structure of the four dyes used to label dideoxynucleoside triphosphates for use as chain-terminators in the automated DNA sequence analysis according to Prober et al. (1987). SF, succinyl fluoresceins; 505, 512, 519, 526, wavelength maxima for emission of the four dyes.

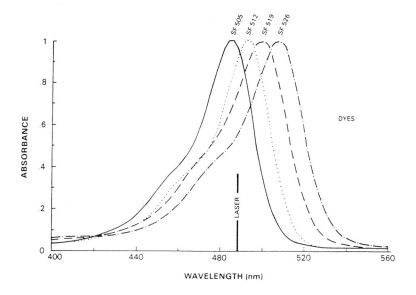

Fig. 5-6b Normalized absorption spectra of the four dyes, measured in 50 mM Tris-HCl, pH 8.2. The vertical bar (laser) indicates the position of the argon laser line at 488 nm used for the fluorescence excitation.

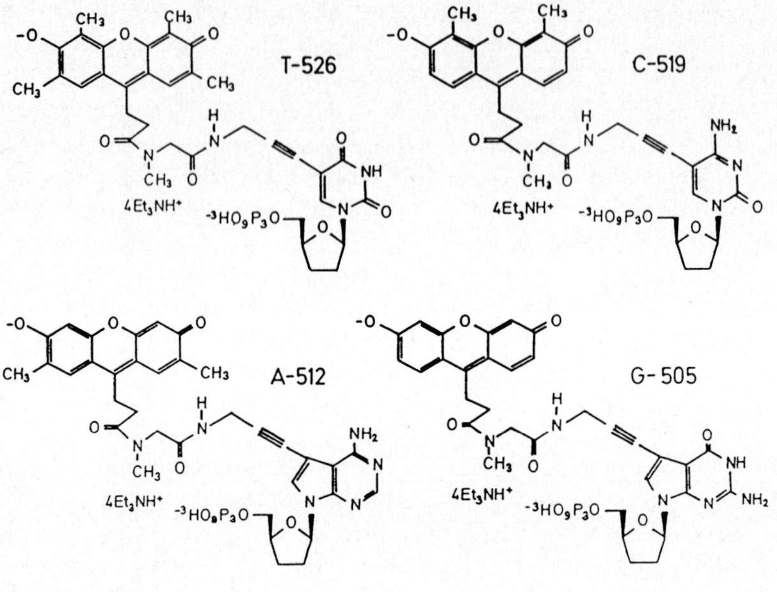

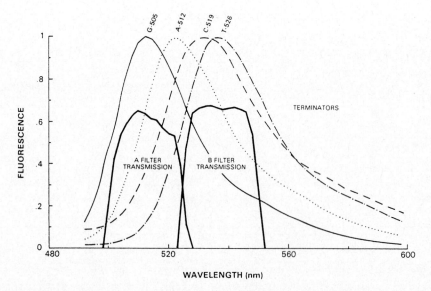

Fig. 5-7a   Chemical structure of the succinyl-fluorescein labeled dideoxynucleoside triphosphates.

Fig. 5-7b   Fluorescence emission spectra of the dye labeled dideoxynucleotides, measured in 8% polyacrylamide gels. The absolute emission intensity values of the four compounds vary relative to each other by less than a factor of 2.

Superimposed on the emission spectra are transmission functions of the interference filters used in the fluorescence detection systems.

Fig. 5-7 shows the molecular structure (a) and the normalized fluorescence emission spectra (b) of the four dye labeled dideoxynucleotides G-505, A-512, C-519, and T-526. The absolute emission intensity values of the four compounds vary relative to each other by less than a factor of 2. Through the nucleotide specific labeling of the DNA-chain at its 3'-end, one of the major problems in enzymatic DNA sequencing which is the unspecific termination of the polymerase reaction caused by secondary structure of the DNA-template does not exist any more. A further remarkable characteristic of the sequencing reaction is that it is carried out in a single tube containing all four dideoxynucleotides. During the electrophoretic separation of the labelled DNA-fragments an argon laser beam is directed sequentially to each of the sequencing lanes of the gel in a defined horizontal position. Upon entering the gel the beam excites fluorescence in the terminator labeled DNA fragments. Fluorescence is detected by two elongated stationary photomultiplier tubes which span the width of the gel. In front of each photomultiplier a filter stack is placed with one of the complementary transmission functions. Baseline corrected ratios of signals in the two photomultipliers are used to identify the labeled DNA-fragments currently in the excitation region. Around one nucleotide can be analyzed per minute and lane, and since 12 lanes can be used per gel it is possible to determine the sequence of around 12 x 300 nucleotides within 5 hours. The obtained data are saved and can be further processed and finally transferred to a data library.

In the automated DNA-sequencing method developed by Ansorge et al. [95] a 5'-endlabeled primer is used (Fig. 5-8). With this primer and the usual unmodified ddNTPs four sets of DNA-fragments with base specific termini are generated and electrophoretically separated in four adjacent lanes. The Argon-laser beam goes through the entire width of the gel and excitates the fluorescent dye which is attached to the DNA-fragments via the primer. Through light channels and different filters the fluorescence reaches the detector with the photomultipliers. The collected data are stored and processed.

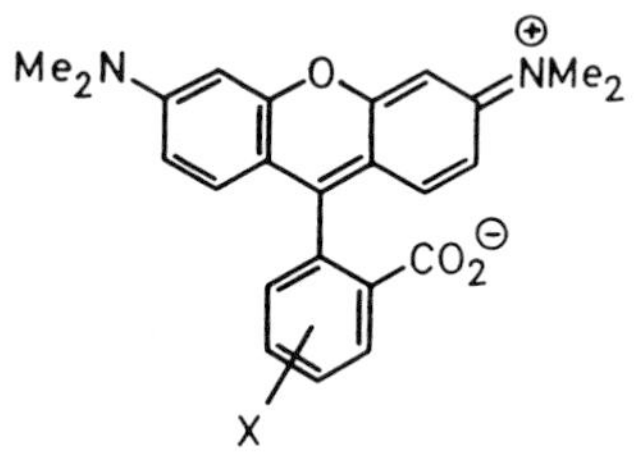

Fig. 5-8 Structure of the fluorescent dye (tetramethylrhodamine) which is used to label the primer for the automated DNA sequence analysis according to Ansorge et al. (1987).

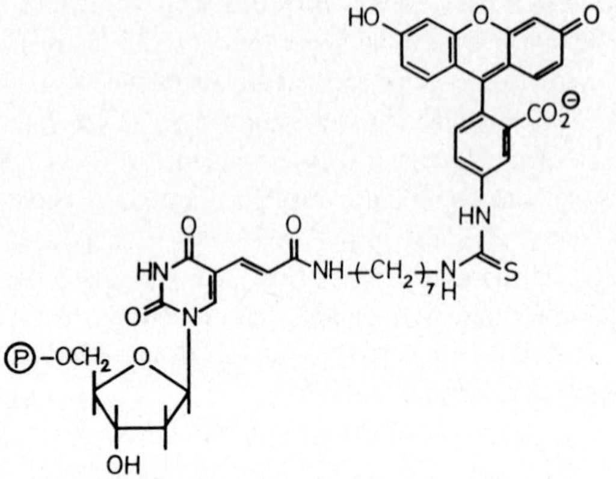

Fig. 5-9 Structure of the fluorescein-isothiocyanate coupled uracil analogue which is incorporated (instead of T) in the sequencing primers that are used for the automated DNA sequence analysis according to Brumbaugh et al. (1988).

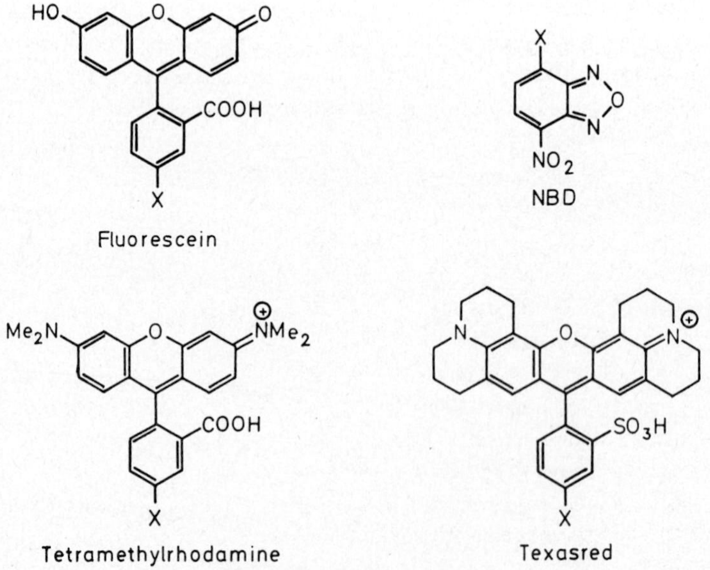

Fig. 5-10 Structure of the four different fluorescent dyes used in the automated DNA sequence analysis according to Smith et al. (1986). X, linker to the primer oligonucleotide.

The technique developed by Brumbaugh et al. [97] is rather similar to the Ansorge system. A slightly different fluorescent dye is used (Fig. 5-9) which is not necessarily attached to the 5'-end of the primer, but instead is fixed at the thymidine positions of the primer. Therefore more than a single dye can be attached to the primer increasing in this way the intensity of the signal. After reflection in a movable mirror the laser beam reaches the labeled DNA from the back of the gel. The collected and processed data are visualized two-dimensionally as bands migrating through the gel and the obtained picture therefore strongly resembles a typical classical DNA-sequence analysis autoradiogram.

The method described by Smith et al. [94] also uses fluorescence labeled primers, but here different primers each of them sharing the same sequence but differing in the attached fluorescent dye are used for the four base specific sequencing reactions (Fig. 5-10). This base specific labeling procedure results in the possibility to load and analyze all four sequencing reactions in a single lane. In order to determine which fluorescent dye was excited the signals are channelled sequentially through four different filters.

5.4.8 Multiplex-Sequencing

Using the methods described above it seems realistic to determine the sequence of 5,000 to 10,000 nucleotides per day and person. A technique named multiplex-sequencing which was developed by Church and Kieffer-Higgins [49] opens the possibility to obtain at least 10 times more DNA-sequence information with the same input of man-power. The method is based on the chemical sequencing strategy described in chapter 5.4.1. Yet in contrast to the classical method it is not only one DNA-fragment which is chemically treated per reaction but instead a mixture of more than 20 different DNA-fragments is processed at the same time. These fragments, however, do not only differ by their general sequence content, but also in their different ends which are individual and specific oligonucleotide sequences. These different terminal oligonucleotide sequences are obtained by cloning the DNA-fragments to be analyzed into different vectors in which the cloning site is bordered by these different oligonucleotide sequences, each of them terminating with a NotI recognition site (Fig. 5-11). Recombinant plasmids which now deviate not only in their inserted DNA-fragments but also in the different insert-flanking oligonucleotide linkers are combined and the inserts plus the flanking oligonucleotide linkers are excised from the plasmids by a digest with the restriction enzyme NotI. The excised - and in contrast to the classical sequencing reaction still unlabeled - DNA-fragment mixtures are subjected to the base-specific chemical sequencing reactions as described in chapter 5.4.1. The reaction mixtures are electrophoretically separated in the usual way on denaturing polyacrylamide gels. After electrophoresis the separated DNA-fragments are transferred onto nylon-filters by electroblotting. These filters are then hybridized sequentially with different radioactively labeled oligonucleotide probes. Each of these specific probes is complementary to one of the different terminal oligonucleotide sequences attached to the insert-DNA. Autoradiography of the hybridized filters then uncovers the DNA-sequencing ladders of the individual inserts. After washing off the radioactive probe, hybridization with the next probe which is complementary to another terminal linker sequence, and autoradiography are repeated several times. In this way more than 40 hybridizations can be completed from each filter and therefore more than 40 autoradiogramms can be obtained. Since 10-50 filters can be hybridized at the same time the sequence information of more than 100,000 nucleotides can be obtained per day.

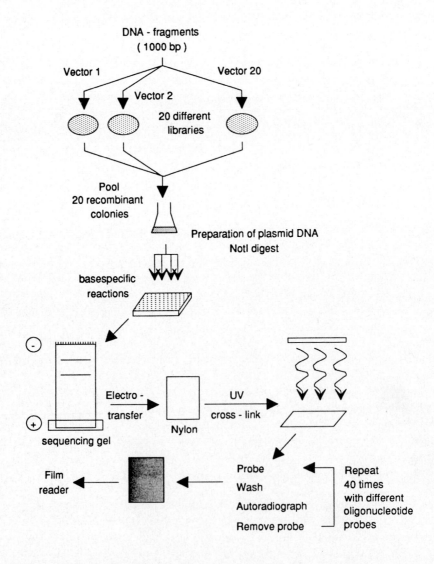

Fig. 5-11 General scheme of multiplex DNA sequencing (see text for details, adopted from Church and Kieffer-Higgins, 1988).

5.5 References

[1] Clarke, L., and Carbon, J., *Cell 1976, 9,* 91-99.

[2] Seed, B., Parker, R.C., and Davidson, N., *Gene 1982, 19,* 201-209

[3] Leach, D.R.F., and Stahl, F.W., *Nature 1983, 305,* 448-451.

[4] Bell, G.I., Selby, M.J., and Rutter, W.J., *Nature 1982, 295,* 31-35.

[5] Boissy, R., and Astell, C.R., *Gene 1985, 35,* 179-185.

[6] Dunn, I.S., and Blattner, F.R., *Nucleic Acids Res. 1987, 15,* 2677-2698.

[7] Frischauf, A.-M., Lehrach, H., Poustka, A., and Murray, N., *J. Mol. Biol. 1983, 170,* 827-842.

[8] Rimm, D.L., Horness, D., Kucera, J., and Blattner, F.R., *Gene 1980, 12,* 301-309.

[9] Collins, F., and Hohn, B., *Proc. Natl. Acad. Sci. USA 1979, 75,* 4242-4246.

[10] Okayama, H., and Berg, P., *Mol. Cell. Biol. 1982, 2,* 161-170.

[11] Gubler, U., and Hoffman, B., *Gene 1983, 25,* 263-269.

[12] Williams, J.G., in: *Genetic Engineering*: Williamson, R. (ed.) New York: Academic Press, *1981*; Vol. 1, pp 1-59.

[13] Huynh, T.V., Young, R.A., and Davis, R.W., in: *DNA Cloning*: Glover, D.M. (ed.), Washington, D.C.: IRL Press, *1985*; Vol. 1, pp 49-78.

[14] Helfman, D.M., Feramisco, J.R., Fiddes, J.C., Thomas, G.P., and Hughes, S.H., *Proc. Natl. Acad. Sci. USA. 1983, 80,* 31-35.

[15] Young, R.A., and Davis, R.W., *Proc. Natl. Acad. Sci. USA 1983, 80,* 1194-1198.

[16] Short, J.M., Fernandez, J.M., Sorge, J.M., and Huse, W.D., *Nucleic Acids Res. 1988, 16,* 7583-7600.

[17] Hedrick, S.M., Cohen, D.I., Nielsen, E.A., and Davis, M.M., *Nature 1984, 308,* 149-153.

[18] Mosley, B., Beckman, M.P., March, I.J., Idzerda, R.L., Gimpel, S.D., Van den Bos, T., Friend, D., Alpert, A., Anderson, D., Jackson, J., Wignall, J.M., Smith, C., Gallis, B., Sims, J.E., Urdal, D., Widmer, M.B., Cosman, D., and Park, L.S., *Cell 1989, 59,* 335-348.

[19] Tedder, T.F., Strueli, M., Schlossman, S.F., and Saito, H., *Proc. Natl. Acad. Sci. USA 1988, 85,* 208-212.

[20] Nevins, J.R., *Methods Enzymol. 1987, 152,* 234-241.

[21] Mechler, B.M., *Methods Enzymol. 1987, 152,* 241-248.

[22] Lynch, D.C., *Methods Enzymol. 1987, 152,* 248-253.

[23] Saiki, R.K., Gelfand, D.H., Stoffel, S., Scharf, S.J., Higuchi, R., Horn, G.T., Mullis, K.B., and Erlich, H.A., *Science 1988, 239,* 487-491.

[24] Saiki, R.K., Scharf, S., Faloona, F., Mullis, K.B., Horn, G.T., Erlich, H.A., and Arnheim, N., *Science 1985, 230,* 1350-1354.

[25] Saiki, R.K., Bugawan, T.L., Horn, G.T., Mullis, K.B., and Erlich, H.A., *Nature 1986, 324,* 163-166.

[26] Embury, S.H., Scharf, S.J., Saiki, R.K., Gholson, M.A., Golbus, M.A., Arnheim, N., and Erlich, H.A., *N. Engl. J. Med. 1987, 316,* 656-661.

[27] Lüdecke, H.J., Senger, G., Claussen, U., and Horsthemke, B., *Nature 1989, 338,* 348-350.

[28] Ou, L.-Y., Kwok, S., Mitchell, S.W., Mack, D.H., Sninsky, J.J., Krebs, J.W., Feorino, P., Warfield, D., and Schochetman, G., *Science 1988, 239,* 295-297.

[29] Bos, J.L., Fearon, E.R., Hamilton, S.R., Verlaan de Vries, M., van Boom, J.H., van der Eb, A.J., and Vogelstein, B., *Nature 1987, 327,* 293-297.

[30] Scharf, S.J., Friedmann, A.F., Brautbar, C., Szafer, F., Steinmang, L., Horn, G., Gyllenstein, U., and Erlich, H.A., *Proc. Natl. Acad. Sci. USA 1988, 85,* 3504-3508.

[31] Chiang, Y.L., Sheng-Dong, R., Brow, M.A., and Larrick, J.W., *Biotechniques, 1989, 7,* 360.

[32] Higuchi, R., von Beroldingen, C.H., Sensabaugh, G.F., and Erlich, H.A., *Nature 1988, 332,* 543-546.

[33] Engelke, D.R., Hoener, P.A., and Collins, F.S., *Proc. Natl. Acad. Sci. USA 1988, 85,* 544-548.

[34] Gyllenstein, U.B., *Biotechniques 1989, 7,* 700-708.

[35] Brandhorst, B.P., and McConkey, E.H., *J. Mol. Biol. 1974, 85,* 451-463.

[36] Latham, H., and Darnell, J.E., *J. Mol. Biol. 1965, 14,* 1-12.

[37] Brown, A.J.P., Leibold, E.A., and Munro, H.N., *Proc. Natl. Acad. Sci. USA 1983, 80,* 1265-1269.

[38] Harpold, M.M., Dobner, P.R., Evans, R.M., and Bancroft, F.C., *Nucleic Acids Res. 1978, 5,* 2039-2053.

[39] Lemke, G., and Axel, R., *Cell 1985, 40,* 501-508.

[40] Hastie, N.D., and Held, W.A., *Proc. Natl. Acad. Sci. USA 1978, 75,* 1217-1221.

[41] March, C.J., Mosley, B., Larsen, A., Cerret, D.P., Braedt, G., Prince, V., Gillis, S., Henney, L.S., Krunheim, S.P., Grabstein, K., Canlon, P.J., Hopp, P., and Cosman, D., *Nature 1985, 315,* 641-647.

[42] Maxam, A., and Gilbert, W., *Proc. Natl. Acad. Sci. USA 1977, 74,* 560-564.

[43] Sanger, F., Nicklen, S., and Coulson, A.R., *Proc. Natl. Acad. Sci. USA 1977, 74,* 5463-5467.

[44] Anderson, S., Bankier, A.T., Barrell, B.G., de Bruijn, H.M.L., Coulson, A.R., Drouin, J., Eperon, I.C., Nierlich, D.P., Roe, B.A., Sanger, F., Schreier, P.H., Smith, A.J.H., Staden, R., and Young, I.G., *Nature 1981, 290,* 457-465.

[45] Sanger, F., Coulson, A.R., Hong, G.F., Hill, D.F., and Petersen, G.B., *J. Mol. Biol. 1982, 162,* 729-773.

[46] Baer, R., Bankier, A.T., Biggin, M.D., Deininger, P.L., Farrell, P.J., Gibson, T.J., Hatfull G., Hudson, G.S., Satchwell, S.C., Seguin, C., Tuffnell, P.S., and Barrell, B.G., *Nature 1984, 310,* 207-211.

[47] Smith, L., and Hood, L., *Biotechnology 1987, 5,* 933-939.

[48] Van Dilla, M.A., Deaven, L.L., Albright, K.L., Allen, N.A., Aubuchon, M.R., Bartholdi, M.F., Brown, N.C., Campbell, E.W., Carrano, A.V., Clark, L.M., Cram, L.S., Crawford, B.D., Fuscoe, J.C., Gray, J.W., Hildebrand, L.E., Jackson, P.J., Jett, J.H., Longmire, J.L., Lozes, C.R., Luedemann, M.L., Martin, J.C., McNinch, J.S., Meincke, L.J., Mendelsohn, M.L., Meyne, J., Moyzis, R.K., Munk, A.C., Perlman, J., Peters, D.C., Silva, A.J., and Task, B.J., *Biotechnology 1986, 4,* 537-552.

[49] Church, G.M., and Kieffer-Higgins, S., *Science* **1988**, *240*, 185-188.

[50] Messing, J., *Methods Enzymol.* **1983**, *101*, 20-78.

[51] Chen, E.Y., and Seeburg, P.H., *DNA* **1985**, *4*, 165-170.

[52] Bankier, A.T., Weston, K.M., and Barrell, B.G., *Methods Enzymol.* **1987**, *155*, 51-93.

[53] Gardner, R.C., Howarth, A.J., Hahn, P., Brown-Luedi, M., Shepherd, R.J., and Messing, J., *Nucleic Acids Res.* **1981**, *9*, 2871-2888.

[54] Anderson, S., *Nucleic Acids Res.* **1981**, *9*, 3015-3027.

[55] Hong, G.F., *J. Mol. Biol.* **1982**, *158*, 539-550.

[56] Hong, G.F., *Methods Enzymol.* **1987**, *155*, 93-110.

[57] Lin, H.-C., Lei, S.-P., and Wilcox, G., *Anal. Biochem.* **1985**, *147*, 114-119.

[58] Poncz, M., Solowiejczyk, D., Ballantine, M., Schwartz, E., and Surrey, S., *Proc. Natl. Acad. Sci. USA* **1982**, *79*, 4298-4302.

[59] Misra, T.K., *Gene* **1985**, *34*, 263-268.

[60] Misra, T.K., *Methods Enzymol.* **1987**, *155*, 119-139.

[61] Henikoff, S., *Gene* **1984**, *28*, 351-359.

[62] Henikoff, S., *Methods Enzymol.* **1987**, *155*, 156-165.

[63] Reif, H., and Saedler, H., *Mol. Gen. Genet.* **1975**, *137*, 17-28.

[64] Ahmed, A., and Johansen, E., *Mol. Gen. Genet.* **1975**, *142*, 263-275.

[65] Ahmed, A., *J. Mol. Biol.* **1984**, *178*, 941-948.

[66] Ahmed, A., *Gene* **1985**, *39*, 305-310.

[67] Ahmed, A., *Methods Enzymol.* **1987**, *155*, 177-204.

[68] Dale, R., McClure, B., and Houchins, J., *Plasmid* **1985**, *13*, 31-40.

[69] Dale, R., and Arrow, A., *Methods Enzymol.* **1987**, *155*, 204-214.

[70] Barcak, G.J., and Wolf, R.E., *Gene* **1986**, *49*, 119-128.

[71] Lee, G., Arscott, P.G., Bloomfield, V.A., and Evans, D.F., *Science* **1989**, *244*, 475-477.

[72] Maxam, A., and Gilbert, W., *Methods Enzymol.* **1980**, *65*, 499-559.

[73] Rüther, U., *Nucleic Acids Res.* **1982**, *10*, 5765-5772.

[74] Volckaert, G., *Methods Enzymol.* **1987**, *155*, 231-250.

[75] Eckert, R., *Gene* **1987**, *51*, 247-254.

[76] Tabor, S., and Richardson, C.C., *Proc. Natl. Acad. Sci. USA* **1987**, *84*, 4767-4771.

[77] Erlich, H.A., Gelfand, D.H., and Saiki, R.S., *Nature* **1988**, *331*, 461-462.

[78] Wallace, R.B., Johnson, M.J., Suggs, S.V., Miyoshi, K., Bhatt, B., and Itakura, K., *Gene* **1981**, *16*, 21-26.

[79] Vieira, J., and Messing, J., *Gene* **1982**, *19*, 259-268.

[80] Guo, L.-H., Yang, R.C.A., and Wu, R., *Nucleic Acids Res.* **1983**, *11*, 5521-5540.

[81] Hattori, M., and Sakaki, J., *Anal. Biochem.* **1986**, *152*, 232-238.

[82] Heinrich, P., Rosenstein, R., Böhmer, M., Sonner, P., and Götz, F., *Mol. Gen. Genet.*, **1987**, *209*, 563-569.

[83] Stallard, R.L. Certa, U., and Bannworth, W., *Anal. Biochem.* **1987**, *162*, 197-201.

[84] Lim, H.M., and Pene, J.J., *Gene Anal. Techn.* **1988**, *5*, 32-39.

[85] Church, G.M., and Gilbert, W., *Proc. Natl. Acad. Sci. USA* **1984**, *81*, 1991-1995.

[86] Garoff, H., and Ansorge, W., *Anal. Biochem.* **1981**, *115*, 450-457.

[87] Biggin, M.D., Gibson, T.J., and Hong, G.F., *Proc. Natl. Acad. Sci. USA* **1983**, *80*, 3963-3965.

[88] Pohl, F.M., and Beck, S., *Methods Enzymol.* **1987**, *155,* 250-259.

[89] Huibregtse, J.M., and Engelke, D.R., *Gene* **1986**, *44,* 151-158.

[90] Stoflet, E.S., Koeber, L.D.D., Sarkar, G., and Sommer, S.S., *Science* **1988**, *239,* 491-494.

[91] Mierendorf, R.C., and Pfeffer, D., *Methods Enzymol.* **1987**, *152,* 563-566.

[92] Köhrer, K., Kutchan, T.M., and Domdey, H., *DNA* **1989**, *8,* 143-147.

[93] Geliebter, J., Zeff, R.A., Melvold, R.W., and Nathenson, S.G., *Proc. Natl. Acad. Sci. USA,* **1986** *83,* 3371-3375.

[94] Smith, L.M., Sanders, J.Z., Kaiser, R.J., Hughes, P., Dodd, C., Connell, C.R., Heiner, C., Kent, S.B.H., and Hood, L.E., *Nature* **1986**, *321,* 674-679.

[95] Ansorge, W., Sproat, B., Stegemann, J., Schwager, C., and Zenke, M., *Nucleic Acids Res.* **1987**, *15,* 4593-4602.

[96] Prober, J.M., Trainor, G.L., Dam, R.J., Hobbs, F.W., Robertson, C.W., Zagursky, R.J., Cocuzza, A.J., Jensen, M.A., and Baumeister, K., *Science* **1987**, *238,* 336-341.

[97] Brumbaugh, J.A., Middendorf, L.R., Grone, D.L., and Ruth, J.L., *Proc. Natl. Acad. Sci. USA* **1988**, *85,* 5610-5614.

6. Computer-Aided Analysis of Biomolecular Sequence and Structure in Genome Research

S. Suhai

6.1 Introduction

Of all of the information in the biosciences the chemical and spatial structures (i.e., sequence and geometry) of nucleic acids and proteins are likely to be among the most fundamental for our understanding of the key processes of life. The explosive growth of DNA and protein sequence data, together with the steady accumulation of newly determined protein structures, has an increasing impact, beyond basic biological research, on many areas of medical and agricultural sciences. About 25-30 million nucleotides from several hundred organisms have by now been sequenced (Fig. 6-1), and their number will double about every two years.

Not only the ever increasing amount of genomic data, but especially their complexity makes the use of computer-based methods inevitable in genome research. We had to realize in the past few years, however, that the development of biological data-generating procedures is about to outstrip the efficiency of methods for their handling, analysis and assimilation. Though computer technology is a dynamically growing field of its own, comparison of the increase of computer speed with the production rate of sequence information over the past two decades reveals (Fig. 6-2) that hardware development alone will not solve the problem. Projections based on near-term trends indicate, furthermore, that the problem could rapidly grow worse: large-scale projects to sequence the genomes of higher organisms are likely to produce in the near future a data flow of a million nucleotides per day, i.e., two orders of magnitude faster than the current rate. The only possibility to cope with this situation seems to be to develop completely new methods and techniques of database management, analysis, etc., based on algorithms which make optimal use of modern computer architectures (parallel and vector processors, networks, etc).

In order to understand the genomic information in higher organisms we will, of course, not only have to make computations more and more efficient. We will also have to develop qualitatively new databases and new analysis methods to represent, for instance, how such genomes encode in the order of 100,000 gene products, and how, during development, these products interact with each other and with the genome itself to produce

S. Suhai

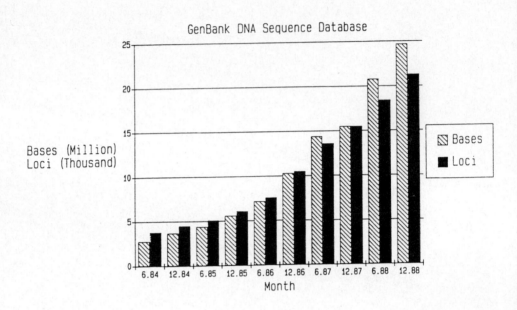

Fig. 6-1 Development of the GenBank nucleic acid sequence database containing 24,690,876 bases from 21,248 loci (Release 58.0, Dec. 1988). Collected and distributed by IntelliGenetics, Inc., in cooperation with the Los Alamos National Laboratory.

the adult organism. Until now, molecular biomedicine has been mainly aimed at decoding single genes or small gene families and tried to analyze control mechanisms involving only a few genes and gene products at a time. A tremendous effort will, therefore, be needed to be able to simulate the control of major cellular subsystems of the genome by computers, which is clearly a necessary step if we are to properly understand the molecular logic of life. The new sophisticated database structures, truly reflecting the complexity of the genomic information, will increase the computational problems certainly much greater than the expected thousandfold increase in the quantity of the data itself.

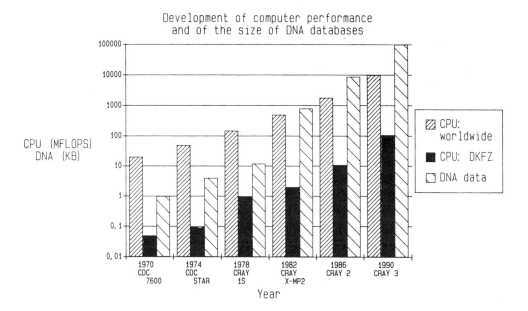

Fig. 6-2 Comparison of the increase in computer power with the development of the amount of DNA sequence information in kilobases, kb) over the past two decades. Computer speed is measured by the number of million floating point operations (MFLOPs) per second the central processing unit (CPU) could perform on the most advanced general purpose scientific computer at a given time. CPU: DKFZ stands for the speed of the central computer at the German Cancer Research Center, Heidelberg, to characterize the computer power available for scientists at a typical biomedical research institute. All three groups demonstrate an exponential increase (observe the logarithmic scale), but DNA data exhibit the largest exponent. The estimate for these data in 1990 (about 100 megabases) is rather conservative in view of ongoing major genome sequencing projects.

6.2   Overview of Genomic Information Available in Computerized   Databases

6.2.1   Databases for Physical and Genetic Linkage Maps

Though gathering and analyzing biomolecular sequences is one of the focal points of molecular genetics, we should still not forget that the order of the nucleotide building blocks provides us with only one of several possible structural descriptions of a given genome. In fact, the sequence level, as a genome map of medium resolution, is imbedded in the hierarchy of several further "maps" with lower and higher resolutions, respectively.

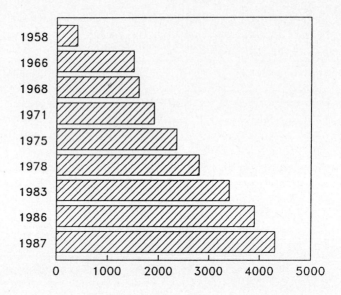

Fig. 6-3 Number of human gene loci identified from 1958 to 1987 and compiled by V.A.McKusick, The Johns Hopkins University Medical School, Baltimore. The data are available free of charge as the Online Mendelian Inheritance in Man (OMIM) database and cross-referenced in the Human Gene Mapping Library (HGML, see below) so that information on expressed genes can be linked to map data.

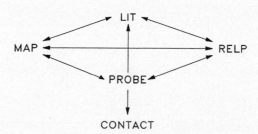

Fig. 6-4 Interrelationships of the five databases unified in the Human Gene Mapping Library HGML, Howard Hughes Medical Institute, New Haven). LIT: about 6,000 literature citations and abstracts; MAP: all information on about 1,600 mapped genes or loci (HGM Workshop symbol, chromosomal localization, status of assignment, etc); PROBE: information on 2,000 probes (mapped and unmapped genes, DNA segments, RFLPs, symbol of the corresponding locus, etc); RFLP: all information on nearly 1,400 RFLPs (one entry per mapped gene locus), and, finally, CONTACT shows how the 1,400 scientists possessing probes can be contacted.

In the most general sense, mapping means measuring the positions of certain landmarks on the genome. At the highest level (lowest resolution), about 4,300 genes are represented, for instance, in the encyclopedia of expressed human genes (Fig. 6-3), more than 1,200 of which have now been assigned, e.g., to positions on the human cytogenetic map [1] by the techniques of somatic cell genetics [2] and are collected and maintained (through computer networks) as a worldwide available database [3].

In contrast, genetic linkage maps describe the arrangement of genes and DNA markers on the basis of the pattern of their inheritance. Since the advent of recombinant DNA methods [4], restriction fragment length polymorphisms (RFLPs) have been widely used to map genomes of several organisms. In the case of the human genome, e.g., an average resolution of about 10 cM (approximately 10 million bases) was recently reported [5], and the data are again available online as part of the Human Gene Mapping Library (Fig. 6-4).

HGML also supplies software to search the databases, supports remote access via TELENET and relevant subsets of the database(s) can be captured, e.g., on a personal computer. Importantly, its five constituent databases are not only properly structured and interrelated, but they also contain cross-references to other relevant databases like OMIM, ATCC (American Type Cell Culture), and GenBank. It is, however, a rather tedious task to increase the resolution of RFLP maps by including more and more informative three-generation families; an average resolution of 4-5 cM can be expected for the human genome in about three years [6]. Fig. 6-5 shows the distribution of RFLPs for cloned genes and anonymous ones on the human genome.

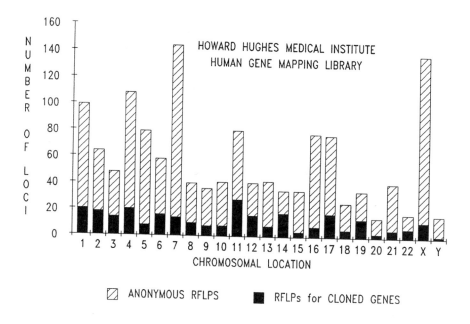

Fig. 6-5 Distribution of 1367 identified RFLP loci (as of June 1988) on the human genome (HGML, New Haven).

Physical maps specify the distances between landmarks along a chromosome using (in the ideal case) nucleotides as units. The most important landmarks are cleavage sites of restriction enzymes, and the maps are calibrated by measuring the sizes of DNA fragments. A special type of physical map that provides information on the approximate location of expressed genes is a complementary DNA (cDNA) map obtained by localizing cDNAs on genomic maps using reverse transcription of mRNAs. cDNAs also permit the localization of genes of unknown functions, including genes that are expressed only in differentiated tissues and at particular stages of development and differentiation. Physical maps are gaining in importance relative to genetic linkage maps, in general because they describe the arrangement of genes at the most fundamental level. They will never displace genetic linkage maps, however, which are distinctive in their ability to map traits that can be recognized only in whole organisms. As physical markers that can be followed genetically, RFLPs are the key to linking the genetic and physical maps and to exactly aligning them at a large number of genome sites. This will greatly facilitate finding the actual DNA sequences that correspond to a gene once such a gene is localized on the genetic linkage map.

Although research is, of course, proceeding from lower to higher resolutions, disclosing more and more fine details of genetic, molecular, atomic, and electronic structure, it would certainly be a mistake to regard the databases corresponding to higher mapping levels as intermediate results of historical value. In fact, all types of mapping presuppose an inherent trade-off between the level of detail (resolution) in the map and the extent to which the map provides a convenient overview of the mapped object (connectivity). At the "high end" of physical mapping, e.g., the banding patterns observed by light microscopy during cytological mapping allow an average chromosome to be subdivided into 10 to 20 regions (each of a length of 10-12 million nucleotides). This "bird's view" of the genome has to be contrasted with the resolution of "low-end" physical maps obtained by restriction enzymes (about 10,000 nucleotides). Prospects of bridging this 1,000-fold gap in resolution are, however, good. During the past years, enzymes have been discovered that cleave DNA into fragments of sizes ranging from 100,000 to 1 million nucleotides. Pulsed-field gel electrophoresis allows, on the other hand, to separate fragments as large as 10 million nucleotides [7] (the limit of previous methods was about 20,000). The order of the resulting 50 to 500 chromosome fragments could be determined either by using two distinct sets (obtained with two enzymes) or linking probes [8] generated by selectively cloning short DNA segments that surround each of the cleavage sites.

Several further databases for physical mapping data for less complicated genomes are in progress (but not yet publicly available) [9]. *Escherichia coli:* genome size 4.7 Mb, 580 kb sequenced (12%), complete 'Not I' map available, 'Eco R1 + Hind III + Bam H1' map in preparation; *Saccharomyces cerevisiae:* genome size 14-15 Mb, 350 kb sequenced (2.4%), 95% of the genome is cloned and available in about 400 contigs correlated with a complete Sfi/Not restriction map; *C. elegans:* genome size 80 Mb, construction of an overlapping clone library by 'fingerprinting' is in progress; *Drosophila melanogaster:* genome size 165 Mb, 270 kb sequenced (0.16%), about 10% of the genome is available in contigs, data collection of about 1,000 clones in various vectors, cross-referenced with genetic and cytological mapping information, is being maintained at UCLA.

6.2.2 DNA and Protein Sequence Databases

The collection of biopolymer sequence information has a relatively short history. It was a polypeptide of 30 residues, the B-chain of bovine insulin, whose primary sequence was first identified by Sanger and Tuppy in 1951. The first complete nucleotide sequence (of an alanine transfer RNA from yeast) was derived by Holley and co-workers in 1965. Sequencing remained tedious and slow work until the real methodological breakthrough came with the new techniques of Maxam and Gilbert and of Sanger and co-workers in 1977. While about 12,000 nucleotides were identified during the 14 years between 1965 and 1978, their number increased to one million in 1982 and to 25 millions by the end of 1988.

Until quite recently, there have been three major organizations, the DNA Data Bank of Japan (DDBJ, Mishima), the EMBL Data Library (Heidelberg), and the GenBank (Los Alamos and Mountain View), which independently collected and distributed DNA/RNA data. Though all of them used tables of sites and features to describe the roles and locations of higher order sequence domains and elements within the genome of an organism (a sample file from GenBank is shown in Fig. 6-6), the independent development of feature table formats and annotation standards created significant difficulties for the users.

```
//
LOCUS        ACARR58S      162 bp      rRNA            RNA         15-FEB-1984
DEFINITION   a.castellanii (amoeba) 5.8s ribosomal RNA.
ACCESSION    K00471
KEYWORDS     5.8S ribosomal RNA; ribosomal RNA.
SOURCE       amoeba (strain atcc 30010) rrna.
  ORGANISM   Acanthamoeba castellanii
             Eukaryota; Protozoa; Sarcodina; Rhizopoda.
REFERENCE    1   (bases 1 to 162)
  AUTHORS    Mackay, R.M. and Doolittle, W.F.
  TITLE      Nucleotide sequences of acanthamoeba castellanii 5s and 5.8s
             ribosomal ribonucleic acids: phylogenetic and comparative
             structural analyses
  JOURNAL    Nucl Acid Res 9, 3321-3334 (1981)
  STANDARD   simple staff_review
COMMENT      [1] also sequenced a.castellanii 5s rrna <acarr5s>.
FEATURES        from  to/span     description
   rRNA           1       162     5.8s rrna
BASE COUNT     40 a      39 c       44 g      39 t
ORIGIN       5' end of mature rrna
        1 aactcctaac aacggatatc ttggttctcg cgaggatgaa gaacgcagcg aaatgcgata
       61 cgtagtgtga atcgcaggga tcagtgaatc atcgaatctt tgaacgcaag ttgcgctctc
      121 gtggtttaac cccccgggag cacgttcgct tgagtgccgc tt
//
```

Fig. 6-6 Sample sequence data file from the GenBank DNA/RNA database (Los Alamos National Laboratory and IntelliGenetics, Inc).

The above named three data collectors agreed in 1988 to merge their databases and to join their efforts to cover the world sequence literature. Furthermore, they also designed a common Feature Table Format [10] to provide a flexible framework to describe sequence regions which perform a biological function, affect or are the result of the expression of a biological function, interact with other molecules, affect replication or recombination, have a secondary or tertiary structure, etc. Implementation of specific database search procedures on the basis of this definition (which is in progress among others also at the DKFZ) will tremendously increase the utility of this databank for biomedical sciences. Table 6-1 gives a short overview of the major internationally (almost free of charge) available sequence collections implemented, e.g., on the central computer of the DKFZ. Some local databases (for oncogenes, keratins, tubulins, papilloma viruses, etc.) complement them in our institute to serve site-specific research interests.

Table 6-1 Internationally available DNA and protein sequence databases implemented in the program package HUSAR [11] on the central computer of the DKFZ.

Database	Type	Source	Bases/AA[+]	Loci
GenBank	DNA,RNA	IntelliGenetics, Inc.	24,691,207	21,252
EMBL	DNA,RNA	EMBL, Heidelberg	24,208,366	20,698
NBRF/PIR	protein	Natl.Biomed.Res.Found.	2,378,146	9,138
SwissProt	protein	University of Geneva	2,498,816	8,702
VecBase	DNA	MIPS, Martinsried	580,138	140
HIV-Base	DNA	Los Alamos Natl Lab.	166,529	26
HIV-Prot	protein	Los Alamos Natl.Lab.	61,659	198
ENZYME	enzymes	Cold Spring Harbor		1,018

[+] Data from Dec. 1988

DNA sequencing still entailed a lot of manual laboratory work until now, and its output was, therefore, limited, even for hard working biologists, to a few hundred nucleotides per day. The situation may rapidly change these days. DNA sequencing automats with a performance of 6,000-8,000 nucleotides per day have just appeared on the market and other ones with a reported performance of several hundred thousand nucleotides per day are just around the corner. This technical revolution may substantially change the biologist's attitude toward collecting sequence information. As the rather uniform distribution of DNA data shows (Fig. 6-7), different taxonomic groups have been investigated until now with nearly equal weight, depending on the research interests of thousands of research groups all over the world. The 24 million nucleotides of GenBank originate, for instance, from more than 21,000 different *loci*, and only 14 of them have a length of over 30,000 nucleotides (the longest sequence being the genome of the Epstein-Barr virus of 172,282 nucleotides).

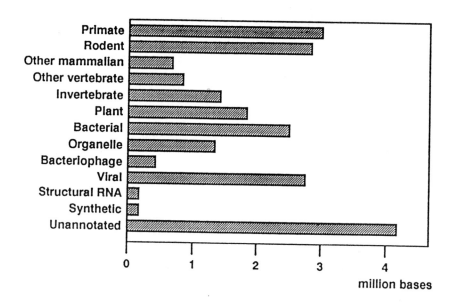

Fig. 6-7 Distribution of sequence data in the GenBank collection over different taxonomic categories. The largest group in terms of bases are the primate sequences, comprising 14.2% of GenBank. The largest group in terms of *loci* are the rodent sequences, which have 14.6% of all the *loci*.

High-speed sequencing will now make it possible to work out in the near future the complete genome for simpler organisms like the bacterium *Escherichia coli* (with an estimated total length of 4 million nucleotides) or *Saccharomyces cerevisiae* (yeast, 13.5 million nucleotides). It also seems realistic to think about concentrated efforts to uncover the genomic information of even higher organisms (of the order of magnitude of a few billion bases). In fact, three such large-scale international projects are just in the phase of maturation (in Japan, in the United States, and in Europe), and it does not seem improbable that genomes of some domestic animals will also be known about the end of our century.

6.2.3 Databases for Atomic and Electronic Structure of Biomolecules

For a number of biological phenomena, the above (strictly molecular) approach would be too crude. To be able to study, for instance, the interaction of regulatory proteins with DNA or to establish the role of structural features in chemical carcinogenesis, one will have to regard the above mentioned subunits as real chemical molecules instead of purely mathematical symbols. Also the investigation of their conformational properties and capabilities to form complicated but well defined three-dimensional structures requires the performance of molecular mechanical and molecular dynamical calculations solving the simultaneous equations of motion for several thousand atoms over macroscopic time scales in these molecules. These calculations will have to take into account the molecular structure of nucleotides or amino acids on the atomic scale calculating interatomic interactions with the aid of empirical potential functions but neglecting the fine (electronic) structure of the constituent atoms themselves.

We have to go even one step further if we are to study biochemical or biophysical processes related to the polarization, excitation or redistribution of the electronic system of biomolecules itself. It must be remembered that atomic nuclei just provide a framework on which to hang varying electron densities. When a substrate molecule approaches, for instance, the active site of an enzyme, it is the varying charge density of the two species which will interact. In a certain sense, both molecules see a structured cloud of electron density, not just a set of pointlike atoms. Such subatomic properties can be investigated only by quantum-mechanical methods solving the appropriate Schrödinger equation. The instruments of molecular quantum mechanics will have to be combined, furthermore, with those of other physical disciplines (like solid state theory, quantum electrodynamics, statistical mechanics, etc.) if the research involves cooperative properties of biopolymers in which electron delocalization over several subunits, formation of well ordered supramolecular structures over large distances or interactions with external fields play a dominant role.

The databases at these deeper levels of genome structure are much less mature than those for sequence information though their size (and importance) steadily increases. At the atomic resolution, there are two publicly available data collections containing crystallographically determined structures for biologically relevant molecules. The Protein Data Bank (Brookhaven National Laboratory) supplies 3D geometrical data for 461 macromolecules, mostly proteins (371 entries), oligonucleotides and sugars. (It is well known, on the other hand, that unfortunately a large number of refined macromolecular structures will be stored in publicly not available form.) Though some of these data can also be lexically analyzed, their real value is shown only by using powerful molecular graphics equipments. The Cambridge Crystallographic Database contains about 40,000 molecular data partly relevant for biomedical research. At the electronic level, most of the information for biomolecules is still stored in private databases, but the Gaussian, Inc. (Carnegie Mellon University, Pittsburgh) is developing a collection of molecular electronic indices (Central Gaussian Archive) and provides such data for about 20,000 molecules, many of which are biologically active compounds.

6.2.4 Future Trends for Genomic Databases

Though the amount of data collected until now both at the level of biomolecular sequence and structure is very impressive, the reorganization of this information from data collections to well structured and efficiently accessible real databases is still only under development.

The major problems with existing genomic databases and subjects of intensive research at different institutions are, in summary, the following:

a) The missing link between high-level (low-resolution) genetic databases and low-level (high-resolution) sequence data: database(s) for ordered clones, contigs, etc. are needed with cross-references in both directions (up and down).

b) Data collections are being distributed as 'flat files'; each institution is developing its own (in most cases incompatible) software to structure and analyze them. Definition of common protocols and data structures should help to make software units compatible and transferable. The GenBank-EMBL-DDBJ proposal for a common Feature Table for nucleotide databases is just a first step in this direction. Similar agreements with protein sequence, structure, etc. databases are needed.

c) In the somewhat more remote future, the basic philosophy of data entry and maintenance will have to be changed. Scientists in the lab, who understand the full meaning of the new data, should be enabled (through appropriate software and computer network) to place it into the proper context within the database and establish its relation with existing knowledge. Beyond software supporting basic annotation, automatic error checking and consistency proves, they will also have to possess more advanced analytic tools, for instance, to be able to optimize homologies to related sequences during submission (which should then become part of the database), to define pointers for highly informative sites (e.g., protein-binding sites, structural motifs, regulatory sequence patterns, etc.). Such information could also form the basis to cross-link different databanks (sequence, structure, enzyme, vector, etc.).

d) Parallel to our proceeding with major genome projects, we will have to pass from centralized to distributed database management systems handling data distributed among a number of interconnected machines that may be physically remote from each other. Major data generation centers should serve as local nodes to this network updating the primary data collection only at regular intervals.

e) The present day organization of genomic databases reflects the fact that genome research has, up until now, concentrated on the structure and function of single genes or small groups of genes. Our existing databanks are nothing more than an additive collection of information about a few thousand nearly independent genes. As molecular genetics will pass on to searching for the overall logical structure of larger genomes and asking questions about how, during development or function, different parts of these genomes interact with each other, we will need far more sophisticated database structures and retrieval tools than are now in use. Principally, the organization of and search for such highly branched structures will be possible only by using completely new software tools like artificial intelligence programming techniques.

f) To be able to build comprehensive pictures of complicated genomes, databases should support the need to view the data in different contexts at the same time, i.e., from a physical point of view (mapping, taxonomy) or from a functional one (encoded gene products, their structure and interactions, etc.). Supplementary information on organization and coordination (who is doing what?) will also become increasingly important since genome research is imaginable only as an (international) collaborative effort. To provide such a flexibility, the databases will have to be richly structured and profoundly linked so that one will be able to take data stored primarily under one organizational scheme and reorganize them according to the structure of some other data set to which they are to be linked.

6.3   From Sequence Information to Biological Knowledge

Molecular biological techniques centered around DNA manipulations are, both in their experimental and theoretical phases, so intimately connected with computer-aided sequence analysis that we can pick out only a few examples to demonstrate how this alliance works (the program package BSA contains alone, e.g., about 300 different functions to support research in this field [13]). The major applications of computer-aided methods are: direct support of experiments (planning the sequencing project, preparation of cloning vectors, restriction enzyme mapping, etc.), prediction of coding regions in DNA, database search for homology, multiple sequence alignments to localize common functional groups and to construct evolutionary trees, prediction of RNA or protein secondary structures, and, finally, methods for statistical and informational analyses of sequences [17].

Several methods support the determination and analysis of restriction maps of sequences cloned in lambda or cosmid vectors, their correlation with digest - and hybridization patterns in total digestions and Southern blot experiments (Fig. 6-8). Using a digitizer tablet, gel patterns from partial and complete digestion experiments can be directly entered and the molecular weights of the DNA bands can be calculated from their mobilities relative to appropriate size marker fragments. For the currently fastest technique, the shotgun strategy, long DNA molecules are broken (using the prepared enzymes) into fragments of 300-400 nucleotides in length, which are cloned and then selected randomly and their sequences determined. The relationship (overlap) between any pair of fragments is investigated by computer and a special mathematical algorithm (based on graph theory) orders them to find the unique arrangements of all fragments which reproduce the original DNA molecule [17].

The projected daily output of the planned "sequencing factories" - one megabase or so - will likely contain 40 to 50 genes together with large regions of noncoding DNA that are present both between and within genes. Neither the identity of the genes nor their boundaries will become apparent by merely looking at the sequence. A combination of existing mathematical algorithms (Fig. 6-9) with artificial intelligence techniques, that are now being developed, will help to distinguish genes from intervening sequences and protein coding exons from noncoding introns. This technique will not only be much faster than looking for and translating the corresponding messenger RNAs, but it will also contribute to the classification of the cellular location and of the function of the corresponding gene products. Combined with methods also predicting the structure of the deduced proteins, site-specifically binding ligands could be engineered for the identification of the proteins in their respective cellular environment.

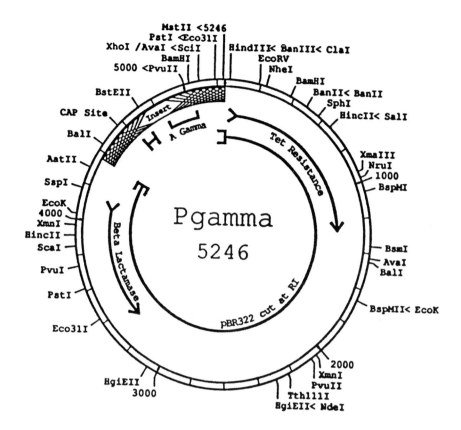

Fig. 6-8 Circular plot of a plasmid construct displaying restriction patterns, inserts and known genetic elements (output from the program PlasmidMap, which has been integrated in HUSAR from the UWGCG package [12]).

The ability to translate DNA stimulated the development of new methods to rapidly gain insight also into the function and cellular location of the coded protein. The most direct computational approach to obtaining such hints is to search databases for homologous sequences of known function. A number of important discoveries in the past years were the results of such a strategy, e.g., the similarity between the cAMP-dependent kinase from cattle and the *src* proteins from the Rous avian and Moloney sarcoma viruses [18] confirming the hypothesis that the src genes originate from the host genome (this was the first evidence for the homology of an oncogenic product with a cellular protein of known function). The degree of the homology between the *v-sis* oncogen and the growth factor PDGF [19, 20] leads, on the other hand, to the conclusion that the chromosomal *sis* gene codes for a growth factor.

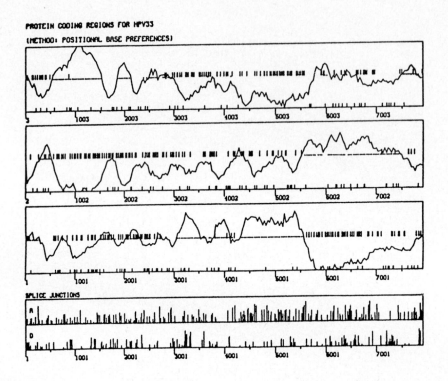

Fig. 6-9 Putative protein coding regions of the virus HPV33 as predicted by the program BasePref in HUSAR (integrated from the BSA package [11]). The heights of the three curves (representing the three reading frames of the DNA) are proportional to the probability that a given DNA region codes for a protein. The mathematical procedure underlying this program takes advantage of the uneven use of amino acids by proteins and the structure of the genetic code table [16]. Possible splice junctions are shown in the lower part of the figure.

Database search has to be efficient enough to remain an interactive research tool. Its speed depends, besides the size of the database itself, on the hardware performance and on the quality of the algorithms used by the programs. Fig. 6-10 demonstrates on the example of the DKFZ environment that though the size of the DNA databases increased by a factor of twenty in the past six years, the search time was reduced from one and a half hour to one minute. This acceleration by more than three orders of magnitude was caused only partly (a factor of 50-60) by an improvement of hardware; a factor of comparable size originates from the use of much more advanced algorithms in recent times. Even if the ambitious sequencing projects will succeed and the size of the DNA library will exceed the 100 Mb mark in 1990, there still remains hope to reasonably work with it if methods will be developed that will make more efficient use of innovative computer architecture (e.g., vector and parallel processors), new mathematical algorithms and software tools (parallel programming, methods of artificial intelligence, etc.).

At a methodologically more advanced level, a larger group of functionally related proteins from different organisms will be simultaneously aligned (Fig. 6-11) to find those parts of them which are best conserved in evolution. These highly conserved sequence patterns may serve as templates to identify functional regions of new proteins.

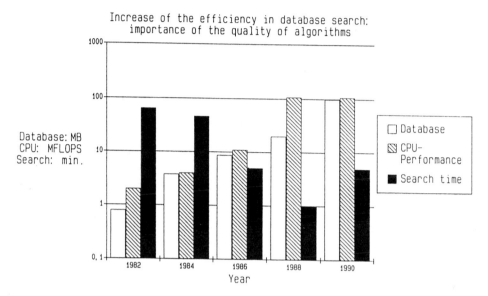

Fig. 6-10 The computer time required to search the GenBank database with a DNA sequence of the length 1,000 bases in the past years on the central computer of the DKFZ (observe the logarithmic scale) as compared to database and hardware development (values for 1990 are estimates).

```
                    -- PROGRESSIVE (similarity) ALIGNMENT --

Sequences are entered in the following order:
1) HBGH  HEMOGLOBIN  GAMMA   CHAIN, HUMAN
2) HBBH  HEMOGLOBIN  BETA    CHAIN, HUMAN
3) HBAH  HEMOGLOBIN  ALPHA   CHAIN, HUMAN
4) MYOH  MYOGLOBIN,  HUMAN
5) MYCR  MYOGLOBIN,  GASTROPOD, CERITHIDEA RHIZOPHORARUM
6) HBRL  HEMOGLOBIN, RIVER LAMPREY (LAMPETRA FLUVIATILIS)
7) HEHA  HEMOGLOBIN, HAGFISH (MYXINE GLUTINOSA)

                        ---- Alignment ----
                        (gap penalty = 8)

HBGH                     G H F T E E D K A T I T S L W_      G K V       N V E D
HBBH                     V H L T P E E K S A V T A L W_      G K V       N V D E
HBAH                     V L S P A D K T N V K A A W_        G K V G A H A G E
MYOH                     G L S D G E W Q L V L N V W_        G K V E A D I P G
HBRL  P I V D S G S V A P L S A A E K T K I R S A W_        A P V Y S N Y E T
HEHA  P I T D H G Q P P T L S E G D K K A I R E S W_        P Q I Y K N F E Q
MYCR                     S L Q P A S K S A L A S S W_K T L A K D A A T I Q N

HBGH  A G G E T L G R L L V V Y P_W T Q R F F_D S F_G N L S S A S A I M G N
HBBH  V G G E A L G R L L V V Y P_W T Q R F F_E S F_G D L S T P D A V M G N
HBAH  Y G A E A L E R M F L S F P_T T K T Y F_P H F_  D L S H           G S
MYOH  H G Q E V L I R L F K G H P_E T L E K F_D K F_K H L K S E D E M K A S
HBRL  S G V D I L V K F F T S T P_A A Q E F F_P K F_K G M T S A D Q L K K S
HEHA  N S L A V L L E F L K K F P_K A Q D S F_P K F_S A K K S     H L E Q D
MYCR  N G A T L F S L L F K Q F P_D T R N Y F_T H F_G N M S D A     E M K T T

HBGH  P K V K A H G K K V L T S L G D A I K H L D D       L K G T F A Q L_S
HBBH  P K V K A H G K K V L G A F S D G L A H L D N       L K G T F A T L_S
HBAH  A Q V K G H G K K V A D A L T N A V A H V D D       M P N A L S A L_S
MYOH  E D L K K H G A T V L T A L G G I L K K K G H       H E A E I K P L_A
HBRL  A D V R W H A E R I I N A V N D A V A S M D D T E K M S M K L R D L_S
HEHA  P A V K L Q A E V I I N A V N H T I G L M D K E A A M K K Y L K D L_S
MYCR  G V G K A H S M A V F A G I G S M I D S M D D A D C M N G L A L K L_S

HBGH  E L H_C D K L H V D P E N F K L L G N V L V T V L A I H F G K E F T P
HBBH  E L H_C D K L H V D P E N F R L L G N V L V C V L A H H F G K E F T P
HBAH  D L H_A H K L R V D P V N F K L L S H C L L V T L A A H L P A E F T P
MYOH  Q S H_A T K H K I P V K Y L E F I S E C I I Q V L Q S K H P G D F G A
HBRL  G K H_A K S F Q V D P Q Y F K V L A A V I A D T V                   A
HEHA  T K H_S T E F Q V N P D M F K E L S A V F V S T M
MYCR  R N H_I Q R   K I G A S R F G E M R Q V F P N F L D E A L G G G A S G

HBGH  E V Q A S W Q K M V T G V A S A L S S R Y H
HBBH  P V Q A A Y Q K V V A G V A N A L A H K Y H
HBAH  A V H A S L D K F L A S V S T V L T S K Y R
MYOH  D A Q G A M N K A L E L F R K D M A S N Y K E L G F Q G
HBRL  A G D A G F E K L M S M I C I L L R S A Y
HEHA  G G K A A Y E K L F S I I A T L L R S T Y D A
MYCR  D V K G A W D A L L A Y L Q D N K Q A Q A L
                        -- Results --

Length of sequences with gaps = 168
Actual sequence lengths: 146, 146, 141, 153, 149, 148, 151,

Similarity Scores:
        HBGH     HBBH     HBAH     MYOH     HBRL     HEHA     MYCR
HBGH    0.00   176.60   145.40   135.30   129.40   125.00   130.00
HBBH  176.60     0.00   146.40   133.20   128.40   124.20   127.30
HBAH  145.40   146.40     0.00   132.90   127.10   121.30   126.00
MYOH  135.30   133.20   132.90     0.00   127.50   120.70   128.40
HBRL  129.40   128.40   127.10   127.50     0.00   152.50   122.10
HEHA  125.00   124.20   121.30   120.70   152.50     0.00   122.00
MYCR  130.00   127.30   126.00   128.40   122.10   122.00     0.00
              *** Sreal (total) = 2781.70 ***
```

Fig. 6-11 Automatic multiple alignment of hemo- and myoglobin sequences by the TREE program of HUSAR [11] using the method of Feng and Doolittle [14]. The similarity scores obtained may be used to construct a phylogenetic tree as well.

6.4 Biopolymers at Atomic Resolution: Computer Simulation of Structure and Function

Though biopolymer sequence analysis is an important first step, the function of biological molecules depends in the end on their conformation in three dimensions. We must determine, visualize and analyze such three-dimensional structures in order to understand mechanisms of enzyme catalysis, recognition of nucleic acids by proteins, interactions between cell surface molecules and their ligands, antibody-antigen binding, and many other dynamic events central to cell biology. Conformational energy calculations facilitate not only our understanding as to how interatomic interactions lead to the three-dimensional structures of biomolecules, but also provide theoretical models as to how these molecules interact with other molecules. Beyond chemical *sequence specificity,* conformational *structural specificity* is a most important source of biological information and plays a central role in biopolymer function [21].

For a sufficiently detailed calculation of the conformation and dynamics of biological molecules, one has to know how the potential energy of the molecular system varies with the position of its constituent atoms. Since quantum calculations (which could provide this potential from first principles) are, in general, not feasible for such systems, a number of model potential energy functions have been developed which approximate the exact energy of a molecule (or assembly of molecules) in the form of a sum of terms involving internal coordinates of the molecule (like bond lengths, bond angles, interatomic distances, etc.). These terms represent the basic physical contributions to interatomic interactions (covalent, electrostatic, dispersion, etc.) in a parametrized, analytically simple form [22]. Since all the results of computer simulations critically depend on the quality of these approximate functions, extensive further work remains to be done to develop more and more reliable potential functions that properly describe all functional groups of interest in biomolecules [23].

Though the simulation methods to predict (static) structure and those to analyze dynamic properties (motion) apply the same abstract model of the molecule, namely an assembly of atoms held together by forces which are calculated (as derivatives) from the above potential functions, they are rather different in their further steps and purposes. To optimize the conformation of a molecule, energy minimization methods make use of the fact that the potential energy is a simple analytic function of the atomic coordinates and calculate its gradient for any set of atomic positions. This information will be used to generate a new set of coordinates in order to reduce the energy and, by repeating this process over and over again, to optimize the molecular structure.

Molecular dynamics studies take a somewhat different route [24]. In a sense, they animate the molecule by assigning initial velocities to the atoms corresponding to the distribution of the kinetic energy at a given temperature. From the force acting on each atom, they calculate its acceleration (by Newton's second law) and, advancing time by a small step (usually 10-15 seconds or less), solve the equations of motion to calculate the new atomic positions and velocities, respectively. At the new set of coordinates the forces are recalculated and the process is repeated several thousand times.

In addition to static biomolecular conformation, increasing attention also has been focused during the past years on the dynamic aspects of structure and function [25]. It has long been inferred from a variety of experimental studies that substantial structural fluctuations occur in biomolecules which are essential to their activity. The recent interest in biomolecular dynamics has largely been stimulated by theoretical studies that have provided a detailed picture of the atomic motion in proteins and nucleic acids [26]. As a result of computer simulations, it is now recognized that the atoms, of which biopolymers are composed, are in a state of constant motion at ordinary temperatures. The x-ray structure of a protein provides the average atomic positions, but the atoms exhibit fluid-like motions of sizable amplitudes about these averages. Use of the average positions still allows discussion of many aspects of biomolecule function in the language of structural chemistry. However, the recognition of the importance of fluctuations opens the way for more sophisticated and accurate interpretations of a number of biological phenomena. At the present stage of the molecular dynamics of biomolecules, there is a general understanding of the motion that occurs on a subnanosecond time scale. Simulations have shown that the structural fluctuations are sizable; particularly large fluctuations are found where steric constraints due to molecular packing are small. For motions on a longer time scale, our understanding is more limited. When the motion of interest can be described in terms of a reaction path (e.g., hinge-bending, local activated events), methods exist for examining the nature and the rate of the process. However, for motions that are slow due to their complexity and involve large-scale structural changes, extensions of the available approaches are required.

Computer simulations of molecular structure have two main fields of application in biosciences: as tools to predict new structures and as instruments to improve the accuracy of models constructed from x-ray scattering, nuclear magnetic resonance (NMR) or other experimental data. In the first case, they help to propose possible three-dimensional structures for new molecules on the basis of sequence homologies to other molecules whose structures are established. A related problem is to predict and analyze possible structural changes as a consequence of mutations (insertion, deletion or substitution of residues) in molecules of known structure to design, for instance, enzymes and other proteins with altered characteristics (protein engineering) [27].

The degree of success in predicting the structure of a new or altered protein from its sequence adopting the known structure of an homologous protein as starting point (template) of the computer simulation, depends upon the portion of identical residues in both structures. At a minimum, about 50 % identical residues are required for present day simulation methods to have a good chance of success. It is interesting to note, however, that active sites of proteins with much less overall homology show substantially greater similarities in their geometrical arrangements than other molecular regions. This is certainly the consequence of an evolutionary pressure to preserve the relative spatial orientation of functionally important groups in these molecules. It provides, on the other hand, a possibility to successfully model active sites even for more distantly related proteins.

Structural simulations became indispensable tools also for x-ray crystallography due to the fact that, in the case of biopolymers, diffraction studies provide fewer data than are required for a complete determination of all atomic positions in the crystal. In using these computer-based methods, x-ray data are first synthetically generated from atomic position distributions observed in theoretical molecular dynamics simulation. The data obtained will be processed by refinement algorithms and the resulting atomic mean positions and temperature factors compared with results obtained directly from the simulation. The assumptions built into the refinement are then varied to find optimal agreement with the simulation results. Computer simulations are most useful also in providing atomic position distribution functions that can be used to model the fluctuations around the observed atomic positions.

Molecular dynamics simulations may profoundly contribute to the refinement of NMR determinations of protein and nucleic acid structures, as well. In a pioneering study [28], for example, a set of 159 internuclear distances obtained from nuclear crossrelaxation experiments on the headpiece of the DNA binding lac repressor protein was used to deduce an initial model of the protein molecule, comprising three alpha helices whose locations and relative orientation approximately confined the measured distances. Subject to the NMR constraints, the model was successfully refined by dynamics calculations to remove the conflicts between model and measurement. In applying molecular dynamics to physical studies of biomolecules, comparisons between the results of calculations and the experimental data are, in general, very important. Even more interesting is the possibility of extending the interpretation of experiments. Also, experimental data can be generated by simulations and analyzed as if they were real data to test the method used. Finally, new effects can be predicted from the simulation as a stimulus for additional experimental investigations.

Recent experimental methods for the specific replacement of amino acid residues in enzymes by genetic manipulations open new perspectives for enzymology. They not only allow to study the role of individual residues in substrate binding and catalysis but also to modify the active site of the enzymes and tailor them for specific applications. Complementing such experiments, simulation methods are useful to predict or rationalize the effects of such modifications in the molecular structure [27]. The mechanical properties (e.g., flexibility) of DNA also play a critical role in gene regulation and in packing genetic material in viruses and other assemblies [29]. When cyclic adenosine monophosphate (cAMP) is attached, for instance, to the catabolite gene activator protein (CAP), CAP binds to DNA at a number of sites where it stimulates binding of RNA polymerase and gene transcription. For a variety of reasons, it appears that CAP does not interact directly with RNA polymerase, but rather distorts DNA so as to facilitate the binding of the polymerase. Conceivably, simulation studies, under way also in our institute, would be helpful to characterize the detailed nature of the DNA distortion involved in this important molecular complex.

Simulations are likely to play an important role also in studying thermal mobility of proteins in relation with their antigenicity. It was observed that antibodies to peptide fragments of a protein were relatively likely to bind to the native protein if the corresponding protein segment was highly mobile. Molecular dynamics could thus help as a design tool to indicate the mobile regions of protein antigens in planning peptide vaccines or affinity elements for chromatographic fractionation of antisera, respectively. In more fundamental terms, the correlation of mobility and antigenicity is of interest with respect to the immune response. The flexibility of antigens suggests that different antigens can adopt structures complementary to a given antibody; the immunologic repertoire corresponding to a given amount of genetic information would thereby be increased.

Intermolecular complexes of some DNA and protein molecules involve interesting hinge bending motions in one of the partners. The lactose repressor protein of *E. coli* has been shown, for instance, to undergo such a motion upon binding the sugar lactose; in the closed conformation, the protein dissociates from the operator region of the lac operon, allowing transcription of the associated genes. A recent study of a mutant repressor [29] shed additional light on the importance of hinge bending in this system. The rate constant for binding lactose is reduced by a factor of 150-300, probably due to infrequent opening of the gate to the binding site. Even in the absence of lactose, the repressor has a low affinity for the operator because it is locked into the liganded conformation. Studies of other repressor molecules show that hinge bending motions are also involved in binding these molecules to DNA. Again, molecular dynamical simulations, in progress in our laboratory as well, are the ideal tools for the study of such time-dependent phenomena.

Though until the 1970s molecular quantum mechanics has been limited to the investigation of small, biologically less relevant molecules, in the past fifteen years there is a rapid qualitative and quantitative change in this field brought about not so much by theoretical advances but by the advent of increasingly powerful computers. Molecular biophysicists are now enabled by them to solve an approximate form of the Schrödinger equation to an arbitrary level of accuracy for larger and larger molecules (even for biopolymers) and there are quite a few examples where these theoretical calculations produce more precise details of molecular properties than experiment. For a number of molecules such calculations are the only source of information since corresponding experiments cannot be performed at all (because the species cannot be synthesized or are unstable and short-lived under experimental conditions, etc.).

There may be also factors of biological importance which cannot be discovered in any other way than by such calculations. There are good reasons to suppose, for instance, that during intermolecular interactions (DNA-protein, enzyme-substrate, etc.) induced changes of conformation on binding of one molecular species to another are an essential feature. Theoretical calculations may indicate non-equilibrium structures for the partners during interaction (not shown by crystallography or NMR spectroscopy) and a whole range of possible molecular shapes within a given energy above the most stable form. In this way, they are a complementary and important addition to experimental structural studies [31]. The wavefunction of the molecule, the result of the quantum mechanical computation, contains also information about a wide range of further sub-molecular properties beyond the energy, which provide, for instance, an attractive research tool for structure-activity studies and for quantum pharmacology in general [32]. The biological activity of molecules is ultimately explicable in terms of the behavior of their electrons. Therefore, the most widely useful result of quantum mechanical calculations are the electron distribution and its change during intermolecular interactions and the potential field which electrons and nuclei generate. Especially efficient may be the use of these properties in designing new drugs or investigating biochemical phenomena if the results are converted from numbers to pictures with the help of computer graphical equipments.

Detailed knowledge of the electronic structure of biomolecules is indispensable, finally, for the proper investigation of their optical activities. Such information seems essential to understanding photomutations in the genetic material at the earth's surface, for instance, a process confined to electronic excitations in a specific energy region due to the ozone layer. Optical rotatory dispersion (ORD) and circular dichroism (CD) represent, on the other hand, two of the most sensitive and versatile probes of biopolymer conformation. ORD is proportional to the difference in the refractive index for left- and rightcircularly polarized light, respectively, CD is the difference in the extinction coefficients for these two forms of polarized light. Both phenomena are manifestations of molecular dissymmetry, a fundamental property of all biopolymers. Quantum mechanical calculations to elucidate these important molecular properties from first principles for even quite complex systems are feasible today, using the ideas of exciton theory from solid-state physics for stereoregular polymers [33, 34].

6.5 Concluding Remarks

We hope the above examples provide a sound basis for the assumption that a number of mathematical and physical methods combined with powerful computational facilities may substantially extend the horizon of problems genome research can successfully attack, and that thus they may contribute to the solution of important biological problems as well. If there is a further cause to contemplate upon the present situation in this field, it is that theoretical models rationalizing the experimental findings of molecular biology fulfil their promises more slowly than desired. There also seems to be a kind of uncertainty concerning the real position of a "theoretical molecular biology" in the network of established disciplines. It would certainly not be wise to ask that it should become a branch of mathematics; molecular physics or physical chemistry would be a sufficiently exacting demand. Its fulfillment is as probable as that of early dreams of quantum physicists in the 1920s, predicting a speedy absorption of experimental chemistry as a branch of applied mathematics. There is, however, an obvious parallel between the present situation of molecular biology and that of atomic physics at the end of the first quarter of our century. Advanced experimental techniques made it possible at that time to fill huge data catalogues with the frequencies of spectral lines; yet all attempts at empirical systematization or classically founded explanation remained unsuccessful until a completely new conceptual framework emerged (quantum mechanics) which made sense of all the data.

The fear is not unjustified at the moment that data accumulation in molecular biology will get so far ahead of its assimilation into an appropriate theoretical framework that the data itself might eventually prove an encumbrance in developing such new concepts. Though the emergence of such a unifying biological theory (if it exists at all) is admittedly only wishful thinking in our days, we would certainly need a better balance between data acquisition and analysis in this field. This sobering reflection should not, however, obscure the more important fact that the collection of molecular data and the willingness to analyze it with established methods of exact sciences has profoundly changed biology not only into a successful and fashionable science but also into a real intellectual adventure.

6.6 References

[1] McKusick, V. A., Mendelian Inheritance in Man: *Catalogs of Autosomal Dominant, Autosomal Recessive, and X-Linked Phenotypes*, 7th ed., Baltimore: Johns Hopkins University Press, *1986*.

[2] Weiss, M. C., and Green, H., *Proc. Natl. Acad. Sci. U.S.A. 1967, 58*, 1104-1111.

[3] OMIM, *Online Mendelian Inheritance in Man*, William H. Welch Medical Library, Johns Hopkins University, Baltimore.

[4] Botstein, D., White, R.L., Skolnick, M., and Davis, R.W., *Am. J. Hum. Genet. 1980, 32*, 314-331.

[5] Donis-Keller, H. et al., *Cell 1987, 51*, 319-337.

[6] Cohen, D., Centre d'Etudes du Polymorphisme Humain, Paris (personal communication).

[7] Schwartz, D. C., and Cantor, C. R., *Cell 1984, 37*, 67-75.

[8] Poustka, A., and Lehrach, H., *Trends Genet. 1986, 2*, 174-179. [9] *Mapping Our Genes*, Congress of the United States, Office of Technology Assessment, Washington, DC, *1988*.

[10] The DDBJ/EMBL/GenBank Feature Table: Definition, Version 1.01, September 10, *1988*.

[11] HUSAR (Heidelberg Unix Sequence Analysis Resources) is a program package providing an integrated environment for molecular biological sequence and structure manipulations on the CONVEX 210 scientific supercomputer at DKFZ. It incorporates, besides the most recent versions of all internationally available databases, major parts of the program packages UWGCG [12], BSA [13], UCSD [14], REMA [15] and several other ones.

[12] Genetics Computer Group Sequence Analysis Software Package, *Nucleic Acids Res. 1984, 12*, 387-395.

[13] *Biological Sequence Analysis (BSA)*, DKFZ, Heidelberg, *1988*.

[14] Feng, D.F., and Doolittle, R.F., *J. Mol. Evol. 1987, 25*, 351-360.

[15] Zehetner, G., and Lehrach, H., *Nucleic Acids Res. 1986, 14*, 335-349.

[16] Staden, R., *Nucleic Acids Res. 1984, 12*, 551-567.

[17] Suhai, S., in: *Molekular- und Zellbiologie*, Blin, N., Trendelenburg, M.F., and Schmidt, E.R. (eds.) Berlin: Springer Verlag, *1985*; pp. 52-64.

[18] Barker, W., and Dayhoff, M., *Proc. Natl. Acad. Sci. U.S.A. 1982, 79*, 2836.

[19] Doolittle, R.F. et al., *Science 1983, 221*, 275.

[20] Waterfield, M.D. et al., *Nature 1983, 304*, 35.

[21] Wong, C.F., and McCammon, J.A., *J. Am. Chem. Soc. 1986, 108*, 3830.

[22] Burkert, U., and Allinger, N.L., *Molecular Mechanics, ACS Nonograph 177,* Washington, DC, *1982.*

[23] Suhai, S., in: *Molecules in Physics, Chemistry, and Biology, Vol. 4: Molecular Phenomena in the Biological Sciences,* Maruani, J. (ed.) Dordrecht: Kluwer Acad. Publ., *1989*; pp. 133-192.

[24] McCammon, J.A., and Harvey, S.C., *Dynamics of Proteins and Nucleic Acids,* Cambridge: The University Press, *1987.*

[25] Ehrenberg, A. et al. (eds.), *Structure, Dynamics and Function of Biomolecules,* Berlin: Springer Verlag, *1987.*

[26] Nicolini, C. (ed.) *Structure and Dynamics of Biopolymers,* Dordrecht: Nijhoff Publ., *1987.*

[27] Oxender, D.L., and Fox, C.F. (eds.) *Protein Engineering,* New York: Alan R. Liss, Inc., *1987.*

[28] Kaptein, R., Zuiderweg, E.R.P., Scheek, R.M., Boelens, R., and van Gunsteren, W.F., *J. Mol. Biol. 1985, 182,* 179.

[29] Widom, J., *BioEssays, 1985, 2,* 11.

[30] Chakerian, A.F. et al., *J. Mol. Biol. 1985, 43,* 183.

[31] Austin, R. et al. (eds.) *Protein Structure: Molecular and Electronic Reactivity,* Berlin: Springer Verlag, *1987.*

[32] Richards, W.G., *Quantun Pharmacology,* London: Butterworth, *1983.*

[33] Suhai, S., *Int. J. Quant. Chem.: Quant. Biol. Symp. 1984, 11,* 223-235.

[34] Suhai, S., *J. Mol. Structure 1985, 123,* 97-108.

7. Linkage Studies and Mapping

F. Clerget-Darpoux and M.P. Baur

7.1 Introduction

The concept of linkage and the methods for testing linkage are not new but the recent advances in DNA techniques, allowing the detection of a large number of genetic markers, have revived the interest of linkage analysis in human genetics as well as in the genome analysis in domestic animals. In parallel to the development of molecular genetics, there is a rapid evolution in linkage analysis methods and in related computer programs. This paper provides a short review on this topic. Interested readers may refer to the books of Bailey [1] and of Ott [2].

7.2 Definition of Linkage

Two loci are linked if one can observe non independent segregation of alleles at these loci. This means of course that these loci are polymorphic and they are referred to as genetic markers. Until recently the number of such markers was very small but the new DNA techniques allow identification of many allelic variations which provide an increasing number of genetic markers on the whole genome [3].

Let us consider two markers A and B with two co-dominant alleles A_1, A_2 for A and B_1, B_2 for B. An individual $A_1A_2B_1B_2$ may produce four kinds of gametes A_1B_1, A_2B_2, A_2B_1 and A_1B_2. If the loci A and B are on different chromosome pairs and if there is no gametic selection, the proportion of each kind of gametes is 1/4. If the loci A and B are on the same chromosome pair, with A_1 and B_1 on one chromosome and A_2, B_2 on the other, the individual $A_1A_2B_1B_2$ will produce gametes A_1B_1 and A_2B_2 called in this situation "parental gametes" and also gametes A_1B_2 and A_2B_1 called "recombined gametes" (see Fig. 7-1).

Indeed during the meiotic process crossover events may occur between the chromatids and recombined gametes correspond to an odd number of crossovers between loci A and B. The closer the loci A and B, the lower the probability of a crossover event and the smaller the proportion of recombined gametes. The proportion of recombined gametes among those which are produced is called the recombination fraction and denoted Θ. If the frequencies of parental and recombined gametes are the same, then $\Theta = 1/2$. This means that the allele transmitted at locus B is independent of the one at locus A. Contrariwise, if Θ is smaller than 1/2, the loci A and B are linked. Testing linkage is equivalent to test $\Theta = 1/2$ against the alternatives $\Theta \leq \Theta < 1/2$.

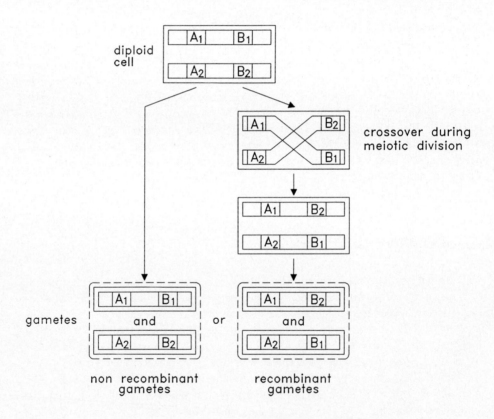

Fig. 7-1 Two loci A and B on the same chromosome. A double heterozygous individual with known phase A_1B_1/A_2B_2 produces "parental gametes" A_1B_1 and A_2B_2 and "recombinant gametes" A_1B_2 and A_2B_1 with a proportion Θ for recombinant gametes and $1-\Theta$ for the parental, non recombinant gametes.

7.3 Methods for Testing Linkage

Linkage analysis requires observations on the segregation of alleles or of related traits. Such observations are possible only through parent to offspring transmission. Unrelated individuals cannot give information on linkage.

Many methods for testing linkage have been proposed. The most simple and natural one is, when possible, to count recombinants and non recombinants among offspring [4]. Doing so is possible only if the genotype can be unequivocally deduced from the phenotype and if the parental phases are known (see Fig. 7-2). Such a situation is rare in practice. Credit for the earliest method to apply independently of the parental phase may be given to Bernstein in 1931 [5]. The statistic he proposed is however poorly efficient and applies only to two generation families. The likelihood approach is much more efficient. This approach and its statistic properties were introduced by Fisher [6]. Most of the genetic analyses in human genetics are presently based on its principles. In addition to testing hypothesis, this approach allows the estimation of parameters. For a set of data, under a specific hypothesis the parameter estimates are the values which maximize the likelihood. Under very general conditions, these estimates have a number of optimum properties: they are consistent, efficient and asymptotically normally distributed.

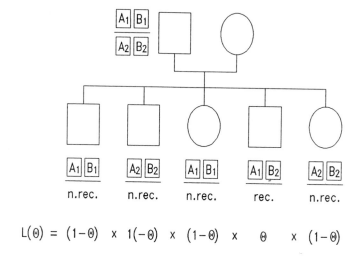

$$L(\theta) = (1-\theta) \times 1(-\theta) \times (1-\theta) \times \theta \times (1-\theta)$$

Fig. 7-2 Pedigree with known phase in the father and all paternal haplotypes known in five offspring. Likelihood L of this observation is a function of the recombination fraction θ.

The likelihood of a model M when data R are observed, is computed as the probability of observing these data under the model M. Such a computation may become difficult following the complexity of the model and of the data. From Fisher to now, many improvements have been brought to the likelihood computation by Haldane and Smith [7], Smith [8] and particularly in 1971 by Elston and Stewart [9] who proposed a very general algorithm allowing computation for any extended pedigree and for qualitative as well as quantitative data.

In 1955, another approach was proposed by Morton [10]: the lod score method. This method streamlines the application of an exact maximum likelihood in the sense that the likelihood of linkage is compared to the likelihood of no linkage but the test is based on a sequential procedure. This approach will be developed in the next section. Several computer programs are available for computing the likelihood of recombination fractions or lod scores: LIPED (Ott [11], [12]), PAP (Hasstedt and Cartwright [13]), FAP (Neugebauer et al. [14]), LINKAGE (Lathrop and Lalouel [15]). Maximization of the likelihood or lod score function may be obtained through the GEMINI procedure (Lalouel [16]) which is implemented in particular in the LINKAGE programs as a subroutine.

Other approaches such as the sib pair method (Penrose [17]), repeat statistics (Green and Woodrow [18], Weeks and Lange [19]) may be used for testing linkage. Such approaches may be particularly interesting when testing linkage between a marker locus and a disease locus for which genetic parameters are uncertain. Such methods are indeed independent of the genetic model underlying the disease.

7.4   The Lod Score Method

Let us consider a family sample $S = \{F_1, ..., F_n\}$. The phenotypes at the marker loci A and B are known for every member of each family F_i.

Under the following assumptions: random mating, no inbreeding, no gametic selection, known probabilities of genotypes given phenotypes and absence of epistasis, the likelihood of different recombination fractions may be computed for the family F_i. Let $L_i(\Theta_1)$ be the likelihood for $\Theta = \Theta_1 < 1/2$ and $L_i(1/2)$ be the likelihood for $\Theta = 1/2$. The lod score of the family F_i at Θ_1 is defined as:

$$Z_i(\Theta_1) = \log_{10} \frac{L_i(\Theta_1)}{L_i(1/2)}$$

The lod score of the sample at Θ_1 is

$$Z(\Theta_1) = \Sigma_{i=1, ..., n} \, Z_i(\Theta_1).$$

The recombination fraction Θ_1 may take any value in the interval [0,1/2] and $Z(\Theta)$ can be considered as a function defined on this interval (see Fig. 7-3).

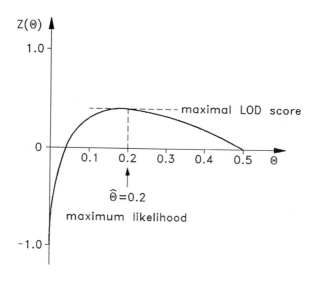

Fig. 7-3 The lod score function $Z(\Theta)$ for the given example of Fig. 7-2. Function $Z(\Theta)$ reaches its maximal value at $\widehat{\Theta} = 0.2$.

A sequential test procedure has been proposed by Morton [10]. The principle and properties of this test had been developed by Wald [20]. Two alternative hypotheses H_0 and H_1 are considered:

H_0 : no linkage $\quad \Theta = 0.5$

H_1 : linkage at $\quad \Theta = \Theta_1 < 0.5$

If the lod score $Z(\Theta_1)$ is greater than or equal to 3, which means that the likelihood of H_1 is at least one thousand fold greater than the likelihood of H_0, then H_0 is rejected and one concludes to linkage. If the lod score at Θ_1 is smaller than or equal to -2, which means that the likelihood of H_0 is at least one hundred fold greater than the likelihood of H_1, then linkage at Θ_1 is rejected. Between the two values -2 and 3 no decision can be made between H_0 and H_1 and data must be accumulated until reaching these stopping criterion values. Morton [10] showed that for these values, the reliability of the test is greater than 95 % for autosomal loci and the power for detecting linkage greater than

80 %, when the true recombination fraction is smaller than or equal to 0.10. The advantage of a sequential procedure is that the number of observations required to detect linkage at a given value Θ_1 is minimum.

In practice, however, the test is not carried out in a sequential way. First the maximum $Z_{max}(\Theta)$ of the lod score function over the interval [0, 1/2] is determined and the critical value of 3 is used for rejecting H_0. The limit -2 is used to determine the Θ values to exclude. It may happen that there exist some Θ values for which $Z_{max}(\Theta) > 3$ and some other for which $Z(\Theta) < -2$. So the test is more similar to a classical test procedure based on a fixed sample size [21]. However, the value of 3 has been chosen in order to make the test very conservative, even with a low prior probability of linkage.

When linkage has been established, the estimate of the recombination fraction is the value which maximizes the lod score function (see Fig. 7-3). Different rules have been given for the confidence interval. Some are directly related to the criteria of the test. The confidence interval Θ_1,Θ_2 may be chosen such that

$$Z(\Theta_1) = Z(\Theta_2) = Z_{max}(\Theta) - 3.$$

It can be also taken as the set of values where $Z(\Theta) > -2$. Other rules are based on the Θ estimate, assuming that Θ follows a normal distribution and that the likelihood has a normal distribution density. Such an assumption is very disputable, particularly for the boundaries $\Theta = 0$ and $\Theta = 1/2$. A consensus has been reached at the Human Gene Mapping 8 [22] for adopting as boundaries of the confidence interval Θ_1 and Θ_2 such that

$$Z(\Theta_1) = Z(\Theta_2) = Z_{max}(\Theta) - 1.$$

7.5 Linkage Between a Disease Locus and a Marker Locus

One goal of linkage analysis is to localize disease genes relatively to marker genes. The same method may be applied as long as the disease is monogenic and the parameters at the disease locus, gene frequencies and penetrance values, are known. Indeed when computing the lod score for a pedigree and thus the likelihood of the different recombination fraction values, we must be able to infer from phenotypes the probability of each genotype. Using wrong parameter values at the disease locus may induce a large bias on the recombination fraction estimate and decreases the power of detecting linkage. It can lead to falsely rejecting linkage but conversely does not lead to spurious evidence of linkage [23].

7.6 Mapping Function and Gene Order

The closer the genetic markers to each other and the more the genome is covered, the easier the localization of disease genes will be. Consequently it is not enough to detect linkage between genetic markers, one must also get their order and define a practical measurement, a map distance. The recombination fraction is not an additive measurement. If we consider three loci A, B, C in this order on a same chromosome pair and if Θ_{AB}, Θ_{BC}, Θ_{AC} are the recombination fractions respectively between A and B, B and C and A and C, one generally has:

$$\Theta_{AC} \neq \Theta_{AB} + \Theta_{BC}$$

Thus, it is useful to define a mapping function which is an additive function of the recombination fraction and depends on assumptions about the occurrence of crossovers [24]. If we assume that crossovers occur independently from each other (no interference), then:

$$\Theta_{AC} = \Theta_{AB} (1 - \Theta_{BC}) + \Theta_{BC} (1 - \Theta_{AB})$$
or
$$\Theta_{AC} = \Theta_{AB} + \Theta_{BC} - 2 \Theta_{AB} \Theta_{BC} \qquad (1)$$

The function $x(\Theta) = -1/2 \, Ln(1 - 2\,\Theta)$, called the Haldane mapping function, is additive:

$$x(\Theta_{AC}) = x(\Theta_{AB}) + x(\Theta_{BC})$$

x measures the genetic distance and the unit is the morgan. When there is no independence of crossover events (interference), the relation (1) does not hold, and:

$$\Theta_{AC} = \Theta_{AB} + \Theta_{BC} - 2C \, \Theta_{AB} \, \Theta_{BC}$$

where C is called the coincidence coefficient. If $C > 1$ there is negative interference and if $C < 1$ there is positive interference. Crossovers in the segment AB respectively decrease and increase the probability of crossovers in segment BC. A test of interference may be performed by a likelihood ratio test. For given observations at loci A, B and C, the likelihood of Θ_{AB}, Θ_{BC}, Θ_{AC} with the constraint given by the relation (1) is compared to the likelihood of Θ_{AB}, Θ_{BC}, Θ_{AC} without any constraint. If the data give a significant evidence for interference, the coefficient C may be estimated simultaneously with Θ_{AB}, Θ_{BC}, Θ_{AC}. In practice, because the power to detect interference is low, the mapping function is chosen prior to the analysis. In addition to the Haldane function corresponding to the absence of interference, several different mapping functions which correspond to different modelisations of C have been proposed by Kosambi [25], Carter and Falconer [26], Rao et al. [27], Lalouel [28] and Felsenstein [29]. For a given mapping function, the likelihood of different possible orders for the loci which are studied can be computed and one may compare the relative probabilities of each order [30]. A goodness of fit test for each order may also be performed as proposed by Ott and Lathrop [31], retaining as possible orders only those which fit the data. The number of orders to consider increases quickly with the number of studied loci. Several algorithms have been recently proposed for ordering loci in a safe computer time [32, 33].

Generally, the recombination fraction estimates appear to be higher in females than in males. To test such a difference, we still apply a likelihood ratio test where the likelihood of $\Theta_m = \Theta_f$ is compared to the likelihood of Θ_m, Θ_f without constraint. As for the interference, it is usually more efficient to use an a priori relation between the two recombination fractions than to estimate them simultaneously. A linear relation is generally proposed between the genetic distances x_m in males and x_f in females: $x_m = 2\ x_f$.

The recombination fraction may depend on different variables such as sex and age, but it may also vary between families. Two different tests have been proposed to search for linkage heterogeneity in a family sample. The first one has been developed by Morton [34] and the other one by Smith [35]. They have been called respectively the "redivided sample test" and the "admixture test". In the redivided sample test, the family sample is subdivided into m subsamples according to a specific criterion. Let $Z_i(\Theta)$ be the lod score of the ith subsample, r_i the value of Θ which maximizes this lod score function and r the value of Θ which maximizes the lod score of the total sample. To test the homogeneity of the recombination fractions between all the subsamples, we may use the quantity:

$$2LN(10)[\Sigma_i\ Z_i(r_i) - Z(r)]$$

which approximately follows a chi-square distribution, with m-1 degrees of freedom. In the admixture test, it is assumed that the sample contains two kinds of families: those where the disease is determined at a locus linked to the marker with a recombination fraction Θ, in a proportion α, and the other 1-α families where the disease segregates independently of the marker. Let $L(\alpha,\Theta)$ be the likelihood of α,Θ. Under the null hypothesis of homogeneity ($\alpha=1$), the quantity:

$$-2\ [\max\ Ln(\alpha=1,\Theta) - \max\ Ln\ (\alpha,\Theta)]$$

follows asymptotically a chi-square distribution with one degree of freedom. In other terms, if this quantity is larger than 3.84, we will conclude to the existence of heterogeneity with a type one error of 5 %. Misspecifying the genetic parameter values at the disease locus will not lead to false conclusion of heterogeneity (at least if we do not constrain the Θ value) but may induce a large bias in the estimate of α [36].

7.7   Conclusion

Linkage analysis is now a very powerful tool in genetics. The polymorphisms detected by the new DNA techniques may be mapped and will provide numerous genetic markers available on the genome. Such genetic markers will allow to localize genes of monogenic diseases. After localization, one may hope to clone the gene itself and to study its biological activity. Linkage analysis with genetic markers may be also useful for the analysis of complex diseases in showing the role of a specific (but not necessarily unique) factor and, if linkage heterogeneity has been assessed, in distinguishing subgroups with different risk factors for the disease.

7.9 References

[1] Bailey, N.T.J., *Introduction to the mathematical theory of genetic linkage,* Oxford: Clarendon Press, *1961*.

[2] Ott, J., *Analysis of human genetic linkage,* Baltimore and London: John Hopkins University Press, *1985*.

[3] Bostein, D., White, R.L., Skolnick, M.H., and Davis, R.W., *Am. J. Hum. Genet. 1980, 32,* 314-331.

[4] Haldane, J.B.S., *J. Genet. 1919, 8,* 299-309.

[5] Bernstein, F., *Z. Abst. Vererb. 1931, 57,* 113-138.

[6] Fisher, R.A., *Ann. Eugen. 1935, 6,* 187-201.

[7] Haldane, J.B.S., and Smith, C.A.B., *Ann. Eugen. 1947, 14,* 10-31.

[8] Smith, C.A.B., *J. Roy. Statist. Soc. 1953, 15B,* 153-184.

[9] Elston, R.C., and Stewart, J. *Hum. Hered. 1971, 21,* 523-542.

[10] Morton, N.E., *Am. J. Hum. Genet. 1955, 7,* 277-318.

[11] Ott, J., *Am. J. Hum. Genet. 1974, 26,* 588-597.

[12] Ott, J., *Am. J. Hum. Genet. 1976, 28,* 528-529.

[13] Hasstedt, S.J., and Cartwright, P.E., "PAP-Pedigree Analysis Package." Technical Report Number 13, Department of Medical Biophysics and Computing, University of Utah, Utah, USA, *1981*.

[14] Neugebauer, M., Willems, J., and Baur, M.P., Analysis of multilocus pedigree data by computer. In: *Histocompatibility Testing 1984:* Albert, E.D., Baur, M.P., Mayr, W.R. (eds.) Berlin, Heidelberg, New York: Springer- Verlag, *1984*; pp. 333-341.

[15] Lathrop, G.M., and Lalouel, J.M., *Am. J. Hum. Genet. 1984, 36,* 460-465.

[16] Lalouel, J.M., GEMINI - a computer program for optimization of general nonlinear functions. Technical Report No. 14, University of Utah, Department of Medical Biophysics and Computing, Salt Lake City, Utah, *1979*.

[17] Penrose, L.S., *Ann. Eugen. 1935, 6,* 133-138.

[18] Green, J.R., and Woodrow, J.C., *Tissue Antigens 1977, 9,* 31-35.

[19] Weeks, D.E., and Lange, K., *Am. J. Hum. Genet. 1988, 42,* 315-326.

[20] Wald, A., *Sequential analysis,* New York: Wiley, *1947*.

[21] Chotai, J., *Ann. Hum. Genet. 1984, 48,* 359-378.

[22] Coneally, P.M., Edwards, J.H., Kidd, K.K., Lalouel, J.M. Morton, N.E., Ott, J., and White, R., Report of the comittee on methods of linkage analysis and reporting (HGM8). *Cytogenet. Cell. Genet. 1985*, 40, 356-359.

[23] Clerget-Darpoux, F., Bonaiti-Pellie, C., and Hochez, J., *Biometrics 1986, 42,* 393-399.

[24] Sturt, E., and Smith, C.A.B., *Cytogenet. Cell. Genet. 1976, 17,* 212-220.

[25] Kosambi, D.D., *Ann. Eugen. 1944, 12,* 172-175.

[26] Carter, T.C., and Falconer, D.S., *J. Genet. 1951, 50,* 307-323.

[27] Rao, D.C., Morton, N.E., Lindsten, J., Hulten, M., and Yee, S., *Hum. Hered. 1977,* 27, 99-104.

[28] Lalouel, J.M., *Heredity 1977, 38,* 61-77.

[29] Felsenstein, J., *Genetics 1979, 91,* 769-775.

[30] Lathrop, G.M., Lalouel, J.M., Julier, C., and Ott, J., *Proc. Natl. Acad. Sci. USA 1987,* 81, 3443-3446.

[31] Ott, J., and Lathrop, G.M., *Genet. Epidemiol. 1987, 4,* 51-57.

[32] Lathrop, G.M., Chotai, J., Ott, J., and Lalouel, J.M., *Ann. Hum. Genet. 1984, 51,* 235-249.

[33] Lander, E.S., and Green, P., *Proc. Natl. Acad. Sci USA 1987,* 84, 2363-2367.

[34] Morton, N.E., *Am. J. Hum. Genet. 1956, 8,* 80-96.

[35] Smith, C.A.B., *Ann. Hum. Genet. 1963, 27,* 175-182.

[36] Clerget-Darpoux, F., Babron, M.C., and Bonaiti-Pellie, C., *J. Psychiatr. Res. 1987, 21,* 625-630.

8. Gene and Chromosome Homologies in Different Species

G. F. Stranzinger

8.1 Introduction

The diversity of living organisms has been created during evolution by genetic mechanisms in combination with natural selective forces of the environment [1]. During civilization and domestication man has started to alter the genome of captured animals. Artificial selective interactions with animal breeding strategies developed many breeds and genetic variants. A wide variation of genetic traits in animals was created and this also includes negative, mostly recessively inherited aberrations and phenotypic malformations. These traits cause problems and are a waste of energy, since they are of little general use for animal and food production. Cytogenetic and molecular genetic methods allow analysis of the genome on every level of genetic expression [2]. It is difficult to imagine that there exist equal interest and possibilities for the study of the genome for all life systems. Comparative considerations and extrapolations of genetic information will therefore become more important in the future. Comparative gene mapping is an efficient way to analyze the genome of domestic animals.

The chemical construction of the hereditary information on life is quite similar in all life systems and has become accessible through different methods and techniques [3]. Differences occur in the organizational construction of genetic information in higher forms of life, e.g. as chromosomes. These genetic units are selfduplicating and can change the genetic content by recombination and mutation. Expressed are those changes in the gene product and consequently also in the phenotypic picture of each organism. If they contain valuable genetic information in a given environment and are able to multiply, a new variation is created. To investigate this continuously ongoing alteration of life a variety of techniques and methods were developed which can be summarized via genome analysis by gene mapping. The information gained consists of morphological and functional characters of genetic units. A fast expansion of the knowledge on the genome of different species can be gained by comparative aspects of homologies, keeping in mind that conservation of functional units of genetic information might have an advantage over dysfunctional units. Completely independent from the general importance of the increasing knowledge on genes and genomes of species and their interactions with the environment is the special knowledge necessary to interfere with the genome by genetic engineering. Genetic counseling for man, new breeding techniques for farm animals, and the use of transgenic experimental animals for cancer research e.g. will use the information produced by gene mapping and molecular genetic techniques. Accepting the fact that the human being is the center of interest, and microorganisms and small laboratory animals are investigated for reasons of finance and information, one should also consider the importance of farm animals for nutrition, clothing, work and pleasure in human society [4]. Therefore, one should not only use, but also know more about these species. In future it will be important to make efficient use of natural resources, such as animals and their products. Indirect energy from the sun over grassland, which can not be used other than via bovides, will be an alternative energy resource. Increase of knowledge on the qualitative genetics of farm animals will allow for much more efficient and productive breeding strategies and will therefore save energy and reduce waste [5]. A comparative approach in genetics will therefore become all the more important to solve the existing problems.

8.2   Survey of Methods

For a better understanding of gene mapping and comparative genetics a few notions are necessary and methods on cytogenetics and molecular genetics have to be explained.

8.2.1 Dimensions and Genetic Measurements

There is a fundamental difference in the identification of physical or genetic distance of two loci in the genome.

The physical assignment directly identifies a clear morphological site of a chromosome region. This implies that the localization of a gene can be confirmed by a visible mutation on a chromosome which causes a variation of traits connected to this event or e.g. by a hybridization process of a preidentified base sequence within a special chromosome region using a known gene probe.

Under special conditions the genetic distance of two loci can be measured by the recombination event, which allows to estimate the recombination fraction in consecutive generations [6]. In other words, the linkage of two loci can be confirmed if they are located on one chromosome region in a distance, that recombination can be measured and cosegregation exist.

The combination of cytogenetic and molecular genetic methods enables the geneticist to demonstrate and visualize chemical units such as base pairs. This allows to construct a physical distance which can be measured in base pairs (bp) and kilobases (kb) and to combine it with genetic distances within recombining systems, expressed in centimorgan (cM) units. Assuming a constant distribution of bases in the construction of a chromosome, one cM contains approximately 1000 kb. Conversions of genetic and physical dimensions are shown in Fig. 8-1.

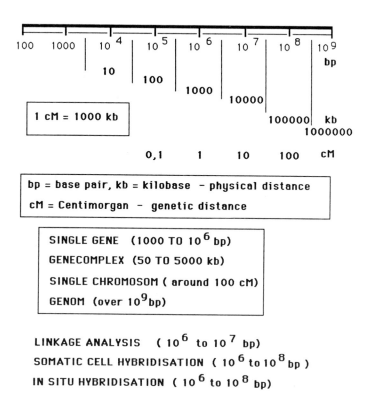

Fig. 8-1 Diagram of genetic and physical distances and their possible comparison in different methods of mapping and in gene complexes.

Linkage and synteny of special genes are the primary cause of a complex genetic expression. In some cases functional needs provided reasons for keeping many genes together in one chromosome region. Very little is known of this, since the comparative approach to different species and their chromosomes relative to function of defined chromosome regions is quite new. Precisely described chromosomes or chromosome regions from one species can now be used as candidates to select appropriate genes or gene products for the mapping approach of complex chromosome regions in other species.

Without a discussion of the evolutionary development of the existing genetic complexity of life, one may assume that formation of variability of genetic information occurs by mutations in the genome. Single point mutations and those related to chromosome mutations produce, within a given chromosome region, a base variation and cause genetic diversity [7]. Such a variation within a given locus is called an allele. For most genetic traits known polymorphic conditions prevail. Some loci have in excess of 1000 alleles, as demonstrated with the B blood group locus in cattle.

Rather different considerations are necessary for genome mutations, since varying levels of gene products are produced due to a variation in the number of existing loci within chromosomes (monosomy, trisomy) or the whole genome (polyploidy). Chromosome mutations also change the linear order of special loci in a given chromosome region (inversion) or between different chromosomes (reciprocal translocation). Such may have implications on the recombination behavior and segregation in meiosis, on the maturation and fertilization process of gametes, embryo development, viability and reproduction of organisms.

8.2.2 Methods and Techniques Used for Mapping Purposes

In general one can divide methods used for gene mapping into two groups: First one may use living individuals (*in vivo* studies) within large families that express genetic variation in combination with chromosome markers. This approach is an old one, commonly used in humans, but also applicable to farm animals (recombination studies and mutation mapping).

Secondly one may use *in vitro* systems such as cell cultures or desoxyribonucleic acid (DNA) material from cells. Using such material one may study the morphological feature of chromosome construction, gene products or the genes themselves. It is not yet possible to describe the recombination behavior of these genes within a complex organism. The combination of the different approaches is an ultimate goal for the correct use of genetic information *in vivo* and *in vitro*.

8.2.3 Cytogenetic Techniques

The precise definition and identification of chromosomes within metaphases and individuals is important for further mapping efforts. Cell culture and chromosome preparation techniques are largely routine and can be applied to most species and tissues following existing protocols [8]. It should be mentioned that differences in the success rate of preparations for useable metaphases do occur between species. Cytogenetic experience and sound laboratory work is always important. The precise description of chromosomes is possible by chromosome banding techniques summarized for farm animals by Gustavsson

[9]. The different staining behavior of chromosomes after various treatments during culture and on slides creates banding structures on chromosomes which indicate different base configurations, and histone and non-histone protein structures around the DNA. Fluorescing and normal staining dyes make differences of chromosome morphology visible. The general principles of international agreements for the standardization of banded chromosomes (ISCN [10]) are shown in Fig. 8-2.

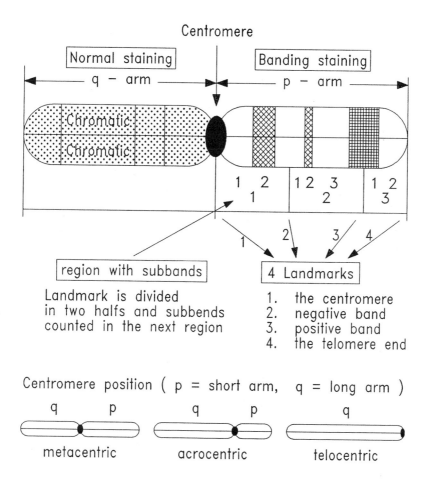

Fig. 8-2 ISCN Standard description of banded chromosomes.

This standardization was made for human chromosomes, and reflects worldwide agreement. Further agreements for farm animal karyotypes have to be worked out and published. A first step was made in 1976 during the Reading Conference [11]. An agreement was reached recently for the pig G banded karyotype [12] which is shown as an idiogram in Fig.8-3.

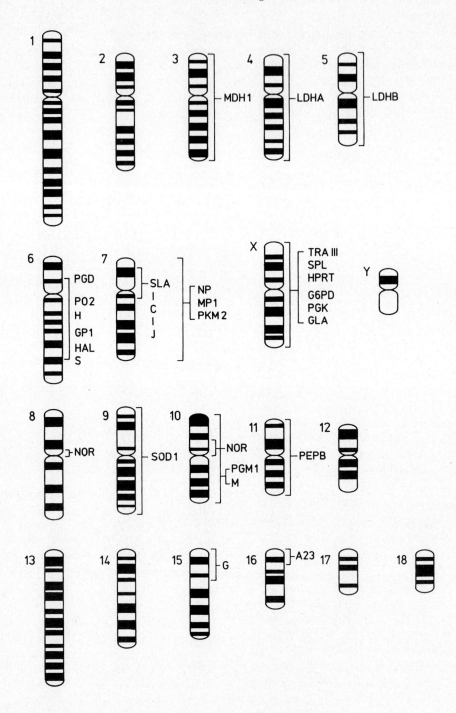

Fig. 8-3 Pig G-band idiogram standard with gene assignments.

With more than 50 different farm animal species globally and each of them different in either the number or structure of chromosomes, one can imagine the need for intensified engagement of cytogeneticists in this field.

8.2.4 Mapping Techniques

As mentioned above, the study of phenotypic variants in combination with chromosome markers within families allows to link genes to special chromosomes and also to define linkage groups by recombination studies. Numerous assignments using these methods contributed to the present knowledge of the human genome. In farm animals very little data have been produced by these methods, if only assignments to chromosomes are considered [13]). General linkage studies without chromosomal assignments are quite frequently published.

Historically the somatic cellhybridization technique followed family studies. The significant difference to the previously described technique is the use of an in vitro cell system, and therefore recombination of genes does not take place. In this situation the alignment of two or more loci within one chromosome region can be studied. Without knowing the distance between two loci, but only their existence on one chromosome, a synteny is formed. The distance can only be defined if in addition to the identification of the chromosome involved a chromosomal mutation can be induced in these special chromosomes, creating a new situation between the two genes. Since during and after the cell hybridization process an exchange of chromosomal material between the two species involved can take place (somatic reciprocal translocations and insertions), some discordances are created, which have resulted in difficulties in the interpretation of the results. With the help of a genomic DNA hybridization approach, these chromosomal rearrangements can be studied and the cell clones eliminated for further investigations. A great number of assignments and syntenies has been published in pigs [14, 15], cattle [16]) and carnivora [17].

More recently, molecular genetic approaches have been added to mapping procedures [18]. Some of these techniques are carried out in combination with cytogenetic methods. One is the *in situ* hybridization technique. Results in farm animals are summarized by Stranzinger [4] and contain assignments for the genes of the histocompatibility complex in horse, cattle, sheep and pig, the assignment of the betaglobin gene, the parathyroidhormon gene and the follicle stimulating hormone gene in cattle, the keratin genes A and B in cattle and sheep and the glucose phosphate isomerase gene (GPI) in pig. For the saturation of the genome with marker loci restriction fragment length polymorphisms (RFLPs) and variable number of tandem repeat studies (VNTR) are quite suitable and have been summarized by Lalley et al. [19] in the HGM9 [20] report and discussed for farm animals by Soller and Beckmann [21] and Prokop et al. [22]. These techniques use the direct approach toward the construction of genes on the molecular level and use the possibility to isolate DNA from cells, cutting the DNA by restriction enzymes, separating the fragments by electrophoretic techniques and hybridizing some of the fragments with known DNA sequences which have been labelled to visualize those fragments on gels or films. If hybridization takes place between DNA and DNA, the procedure used is called "Southern Blot", since E.M. Southern developed and described this method. The "finger printing" or minisatellite technique, which was first used for paternity testing by Jeffreys et al. [23] is very useful for demonstrating the high polymorphic situation in the genome, but could contribute very little to gene mapping information. It seems, that many new molecular techniques are on the way and it becomes more and more complicated to follow the developments in this field. The need for further development of gene libraries such as the one for human gene mapping results of the Howard Hughes Foundation at Yale should be emphasized. In addition, statistical and mathematical methods for linkage studies, DNA and chromosome measurements, automatization of technical procedures and other methods will become more and more important.

8.3 Sex Chromosome Homologies

From the earliest documents in man, it is known that traits can segregate with sex, and specially inherited disorders were known to be linked to sex . Hemophilia was already treated before Christ to the degree that if circumcision resulted in problems the third son was excluded from this ritual. Not only was the problem apparent, but selection against this sex linked inherited disorder took also place. In addition, the mother was marked as carrier of that trait, and was excluded from further motherhood with the husband set free to take another wife. Though very empirical the treatment was effective and also practiced in animals. Castration and tail cutting in sheep resulted in similar problems and stimulated selection against this disorder. On the other hand in some sheep breeds, the tails are cut only in female lambs. Males for some reason were allowed to keep the tail, possibly thus saving the males without knowledge of the long-term effect.

Mendels discovery on the laws of inheritance, Landsteiners work to elucidate the blood group factors and their genetic background, and recent sequencing of the gene for factor VIII that causes hemophilia A in human, are all important in order to understand these genetic pathways. The correct number of human chromosomes was published around 1956 even though as early as 1888 Waldeyer described the term "chromosome". The existence of sex chromosomes and their homologies were widely known, but it was up to Ohno [24] to develop theories of their evolutionary and functional role in sex determination. It can be generalized that in mammals males carry an X and a Y chromosome of different size and morphology along with the diploid number of autosomes. Therefore they produce two different gametes determining the sex of the zygote. Females have two X chromosomes and the gamete can only be of X determination. In birds it is just the other way around: females produce eggs with Z (X) and W (Y) chromosomes and males have only sperms carrying the Z apart of the haploid number of autosomes. It is also known that a pseudoautosomal recombination between the X and Y chromosome takes place, indicating that homologous areas still occur between these two morphologically different chromosomes. Even more interesting are those genes that are located on the nonhomologous area of the sex chromosomes, since they contribute in a very special way to the knowledge of sex chromosome homologies in different species. For many species it is accepted that the X chromosome carries the loci for many enzymes such as glucose-6-phosphate dehydrogenase (G-6-pd) on hypoxanthine-guanine-phosphoribosyltransferase (Hprt). Between human and mouse some 42 homologous genes were assigned to the X chromosome [19] and for the human alone over 300 genes are known to be on the X chromosome. For the Y chromosome only 5 homologous genes are known between human and mouse, but the testis determining factor (Tdf) with a high conservation of base pairs between X and Y represents an interesting case for further studies. There is still a controversial discussion on the role of the HY antigen regulating some parts of the sex determination. It is documented for goats to be linked to the gene for polledness [25] exhibiting pseudomales and intersexes in the homozygote state of polledness. These two genes may be located on an autosome or on the homologous part of the X and Y chromosome. The answer is yet to be given. The hypothetical assumption of Ohno et al. [26] seems to be correct namely that during evolution of the last 300 million years the two sex chromosomes were differentiated from a homologous pair by point and chromosome mutations. In this connection the Lyons-hypothesis [27] must be mentioned, since under normal circumstances a random inactivation of one of the two X chromosomes in female zygotes contributes to the gene compensation effect, which is necessary to allow a normal sex differentiation and further expression of genes located on the sex chromosomes. Gene probes specific for loci on the sex chromosomes might solve some of the questions related to sex differentiation and aberrations. In contrast to widely conserved linkage groups on the X chromosome, the banding patterns between X chromosomes of different species are not generally comparable. As will be outlined in the next chapter, it is interesting that between human and pig the X chromosome shows quite a unique banding pattern, but between the more closely related bovidae species relatively large differences occur in the sex chromosomes despite the very homologous autosomes. More information will be available after assignments of additional marker loci and genes to the X chromosome of different species. In addition, meiotic studies and the behavior of sex chromosomes in the synaptonemal complex will give more insight to these problems. Reproduction is a functional part of life and sex chromosomes play an important role in this context.

8.4 Chromosome Banding Homologies

Within closely related species it was already recognized that the chromosome morphology gives some indication on common evolutionary developments. As outlined above, chromosomes can be prepared from dividing cells and stained to make them visible in size, morphology, arm length ratio and centromerposition. Some further signs of nucleolus organizer regions and replication patterns are made visible through special treatment and staining procedures. A precise identification became possible subsequent to the discoveries of Caspersson et al. [28], that chromosomes can show banding pattern. This was the

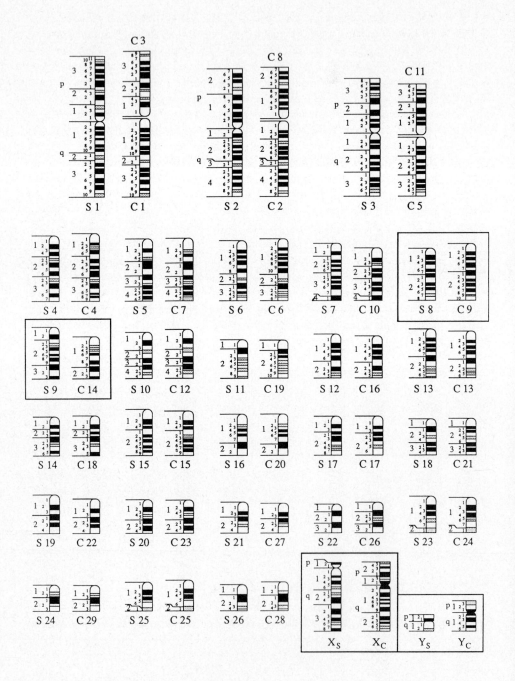

Fig. 8-4 Comparative G-banding idiogram of cattle and sheep (Hediger [32]).

starting point for further developments in cytogenetics, especially for gene and chromosome mapping. The precise identification of chromosomes alone is, however, not sufficient for the statement of banding and gene homologies, since some biological variation in the position of bands and their unique appearance does not allow the conclusion of similarity. Only the sequential combination of identification with banding patterns and the assignments of gene loci to these chromosomes in comparison to different species allowed a conclusion of homologies. These chromosome morphological comparisons were first demonstrated in the bovidae by Wurster and Benirschke [29], however banding homologous studies by Evans et al. [30] and Förster and Stranzinger [31] did not answer the problems. With *in situ* hybridization gene assignments between cattle and sheep Hediger [32] was the first to make conclusive comparisons.

It is also important to note the necessity for standardization agreements on chromosome banding karyotypes of different species. The Reading [11] standard alone is not sufficient, since many unsolved identifications are still open for discussion. Mainly the small telocentric chromosomes (Nr. 19 to 29) in cattle have to be described in more detail. For further discussions Hediger [32] proposed a standard G banding idiogram for both species of cattle and sheep, which is shown in Fig. 8-4.

Only two autosome chromosomes show significant banding differences between cattle and sheep. The rest is very similar despite the centric fusions. The banding structures and gene assignments of comparable chromosomes between cattle and sheep are shown in Fig. 8-5, which is taken from the thesis of Hediger [32].

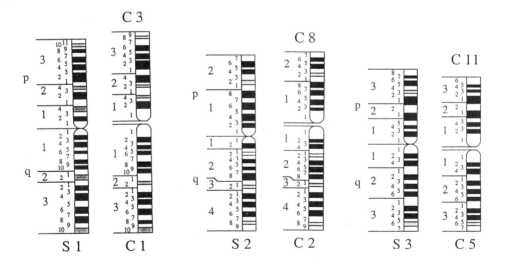

Fig. 8-5 Comparison of banded chromosomes in sheep and cattle demonstrating the identity of the three pairs of metacentric sheep chromosomes with the homologue telocentric cattle chromosomes.

The assignment for the keratin gene B in cattle and sheep is very impressive, since the sheep chromosome Nr. 3 consists of the fusion of cattle chromosomes 5 and 11. The assignment for the keratin gene B took place exactly on the same banding position of the comparable chromosome region of both chromosomes in cattle and sheep. This homology is the first exact demonstration of conserved regions between the bovidae and serves as a candidate position for further comparisons.

8.5   Synteny and Linkage Homologies

An attempt to systematically describe syntenies and linkage groups between farm animal species was made by Womack and Moll [33]) and Fries et al. [34]. In general terms these comparisons were made available for the human genome and other species in the proceedings of the human gene mapping reports (HGM 9,0 [20]). From this a very instructive example was constructed involving the linkage group for the parathyroidhormon gene (PTH), the hemoglobin beta gene (HBB) and the follicle stimulating hormone gene (FSH B) published for man, mouse, cat, chimpanzee and cattle (HGM [20]); Fries et al. [35], Nadeau and Taylor [36]. Large regions of chromosomes between the species must be homologous and conserved for a long time (Fig. 8-6). In the human the whole short arm of chromosome 11 must be identical in the linear construction of the genes with chromosome 15 from cattle as the flanking and intermediate genes for PTH, HBB and FSH and others show the same position. In the mouse a break between the loci for PTH and FSH is likely, since HBB and PTH are located on mouse chromosome 7 and FSH on mouse chromosome 2. Identical gene loci shown in human, cattle and mouse are also assigned to chimpanzee, cat and rabbit.

This example allows several conclusions. For the first time a definitive proof of conserved regions is established between chromosomes of very different species. Every candidate gene assignment and linkage group must be tested against other species, since evolutionary diversified situations do occur as demonstrated between mouse and man. If functional or epistatic relationships exist between such genes, this needs to be taken into account for experiments where segregation and function of participating genes are involved. In a survey PTH and FSH e.g. in man and mouse must be considered independently, and cannot be handled in the same way.

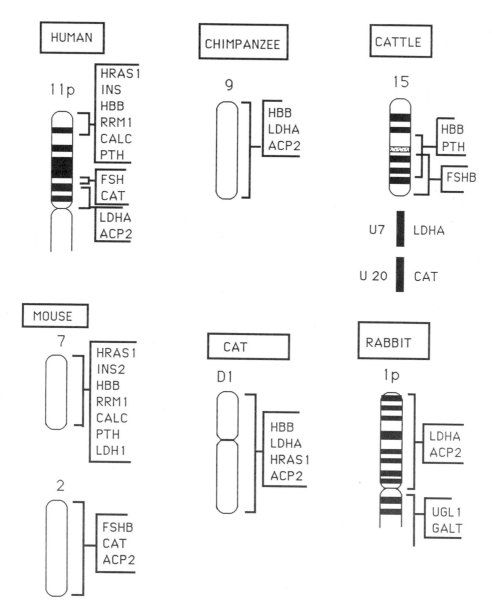

Fig. 8-6 Comparison of a known linkage group containing the loci for the parathyroid hormone (PTH), the beta globin (HBB) and the follicle stimulating hormone (FSH B) between different species and gene assignments. (Ref.: HGM 9 [20] and Fries et al. [34]).

8.6 *In situ* Hybridization Cross Homologies

The *in situ* hybridization method opens several options, since two major aspects are combined in this technique. Sequenced gene probes of known size, molecular construction and origin (function) are used to identify homologous areas on chromosomes irrespective of species and tissue used for chromosome preparation. The hybridization process only takes place when identical homologous regions are present in the major base sequences on both hybridization sites. Cross hybridization considers the fact that gene probes developed from different species can be used for hybridization on any chromosomal material. For example, we can demonstrate the situation on different species with the major histocompatibility complex (MHC). A histogram from cattle is shown in Fig. 8-7.

A probe was developed by Singer et al. [38] using a porcine genomic clone encoding a major histocompatibility antigen (SLA class I). This probe was used to study the expression in mouse L cells and thereafter used for *in situ* hybridization in pigs assigning the porcine major histocompatibility complex to chromosome 7 on the short arm [39].

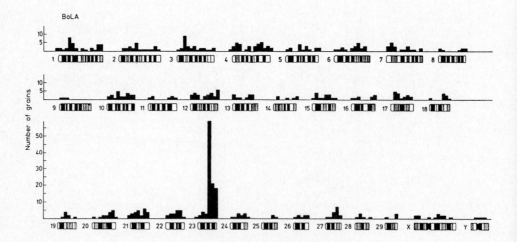

Fig. 8-7 Histogram of the BoLA assignment in cattle using a porcine MHC class I gene probe (Fries et al. [37]).

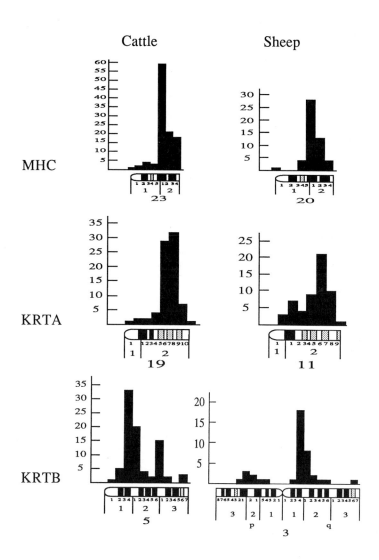

Fig. 8-8 Histograms of the grain distribution over the hybridization sites of the gene probes for swine lymphocyte antigens (SLA) and keratin genes (KRT A and B) for cattle and sheep chromosomes with identical banding pattern (Hediger [32]).

Geffrotin et al. [40] previously described the assignment of this locus to chromosome 7 in pigs with a human gene probe but the hybridization signal was not in agreement with the other publications that concluded the localization of the gene on the long arm. Echard et al. [41] confirmed the localization on the short arm using a different probe. Further studies of assignments of the MHC were made with the porcine probe class I to the horse ELA site on chromosome 20 [42], to the cattle BoLA site on chromosome 23 [37], and the OLA site in sheep on chromosome 20 [32]. These experiments proved, that a gene probe from one species can be used for hybridization in other species. These data have been confirmed using either the same probes or probes from other species [12]. In addition Hediger [32]) was able to use two probes for the keratin genes A and B for assignment in cattle and sheep, and received the confirmed results of the comparative mapping onto the same banded chromosomes for both probes, as keratin A maps to cattle chromosome 19 and sheep chromosome 11, and as keratin B maps to cattle chromosome 5 and sheep chromosome 3. The results are demonstrated in Fig. 8-8.

These cross hybridizations are the first positively demonstrated results in farm animals and in this respect will be used for further demonstration.

The detection of homologous sequences in normal equine males and sex-reversed females by in situ hybridization due to conserved repetitive DNA was published by Kent et al. [43]). This example is very important for further studies on sex reversal, that occurs in many species and under different circumstances.

Such a comparative approach will have important implications for the future, since the multitude of gene probes available from laboratory animals can be used for assignments in farm animals. Disease models in laboratory animals and humans can now be used for confirmation, and *vice versa*.

8.7 Gene and Gene Product Homologies

Construction of genes has been studied by sequencing the base pairs in lower organisms such as viruses and bacteria. This has also been done in higher organisms. The studies have yielded interesting results in that highly conserved genetic elements are present in almost all systems studied. It was concluded that quite unique sequences are present for every functional role in the transcriptional and translational process of a gene. A start codon for a transcriptional signal is called the CAT box, and a translational signal at the end of a promotor region is called the TATA box, indicating a precise array of nucleotides which are unique for all species. Special signals in the CAP site have protectional roles against degradation through enzymes. The end of a gene is, in most cases, encoded by a polyadenine tail (AUAAA). This information could also be called homologies, since they occur in many species with the same informational content. The genes also contain nucleotides which are called EXONS and are functional sequences that are eventually decoded to form the gene product. INTRONS are noncoding sequences that vary in size and number, interspersed between the coding sequences. Enhancer sequences at a certain site will enhance transcriptional signals and can also interact with the promotor region. Several examples are published to show homologies in genes expressing gene products for the same encoding region, such as the HOMEO box genes. Hart et al. [44] has given data on the homeo box homologies which are summarized in Table 8-1 for simplicity of this very complex situation.

Similar comparisons can be made at the level of amino acids. This was demonstrated by Wallace [45] for mitochondrial DNA mutations in relation to neuromuscular disease (Lebers mutation) for the NADH dehydrogenase subunit 4. The mutation, when compared to normal humans and to bovine, murine, xenopus, drosophila, aspergillus and leishmania has shown, that arginine at the position 340 is conserved throughout the animal and protist kingdom.

Table 8-1 Comparison of homeo box sequences and conserved regions between different species (data from Hart et al. [44]).

	DROSOPHILA Antp	XENOPUS MM3	MOUSE Hox−2.3	HUMAN Hu c 1
A				
M P	1 to 10	identical	identical	identical
I O	11 to 20	identical	identical	identical
N S				
O I	21 to 30	identical	Pos. 22 Y	Pos. 22 Y
A T	31 to 40	Pos. 37 V	Pos. 37 T	identical
C I				
I O	41 to 50	identical	identical	identical
D N	51 to 60	identical	identical	identical

On the level of gene products the homologies and differences are shown by electrophoretic analyses in case of proteins and enzymes. This approach is a very old one, since it was previously used to demonstrate protein variants between species. Some indications are also given by blood group factor comparison between species of common genetic background [46]. This can be expanded to milk proteins [47] and e.g. to interferons [48]. Most proteins and more complex gene products are a combination of several genes and therefore not easily analyzed in the genetic expression. Because this was not well understood and because the inheritance pattern cannot be reconstructed from the analysis of these complex traits, severe problems in animal breeding resulted. Reverse genetics might solve this. It will open the way to understand how such complex structures are built and which genes are involved. For these analyses the transgenic technique is very helpful, because isolated gene sequences can be tested very accurately on their gene products in complex systems . At present beginning of an understanding, how traits of interest for animal breeders are composed has just started. Polymorphisms, mutations, gene interactions and the expression of genes must be precisely understood at the molecular level in order to use this information effectively in animal breeding programs. There is no reason to discriminate any discipline in animal genetics. Every information and developed method in population genetics, biochemical genetics, statistics and cytogenetics will be very helpful to comprehend the ongoing change of the genome of our life systems.

8.8 Summary

The demonstration of genetic homologies on different levels of genetic expression needs to incorporate methods and techniques used for that purpose. First some fundamental aspects are discussed. Using the literature, the reader should be able to intensify his studies. Thereafter, the possibilities of homologies are discussed at different levels of genetic expression. Examples are given to demonstrate the present knowledge. Selected topics on sex chromosomes, banding chromosome homologies, synteny and linkage homologies, *in situ* hybridization homologies, gene and gene product homologies are given. In recent years information on farm animals has increased relative to these aspects. Animal breeders should draw conclusions from this information. The more effective selection in farm animals is to take place, the more precise information on the genome is necessary. Risk evaluation must be made and long term effects on single animals and generations have to be considered. There is no alternative to the molecular genetic approach for animal sciences in the future.

8.9 References

[1] Mayr, E.: *Artbegriff und Evolution*. Verlag Paul Parey. Hamburg und Berlin, *1967*.

[2] Fechheimer, N.S., *J. Dairy Sci. 1986, 69,* 1743 - 1751.

[3] Watson, J.D., Tooze, J., and Kurtz, D.T. In: *Rekombinierte DNA*. Eine Einführung. Spektrum der Wissenschaft Verlagsgesellschaft mbH und Co, Heidelberg, *1985*.

[4] Stranzinger, G., *J. Anim. Breed. Genet. 1988, 105,* 251 - 263.

[5] Fechheimer, N.S., Prospects for Genetic Engineering in Domestic Animals., pp 385 - 398 In: A.B. Chapman (Editor), *General and Quantitative Genetics*. Elsevier Science Publishers B.V., Amsterdam, *1985*.

[6] Ott, J. In: *Analysis of Human Genetic Linkage*. Johns Hopkins University Press, Baltimore, *1985*.

[7] Rieger, R., Michaelis, A., and Green, M.M. In: *A Glossary of Genetics and Cytogenetics*. Springer-Verlag New York Inc., *1968*.

[8] Nagl, W. In: *Zellkern und Zellzyklen*. Verlag Ulmer, Stuttgart, *1976*.

[9] Gustavsson, I. In: Banding Techniques in Chromosome Analysis of Domestic Animals. Academic Press, *Adv. in Vet. Sci. and Comp. Medicine 1980, 24,* 245 - 289.

[10] *ISCN Standard:* An International System for Human Cytogenetic Nomenclature. Report of the Standing Commitee on Human Cytogenetic Nomenclature. Karger-Basel, New York, *1985*.

[11] Reading Conference, *Hereditas 1980, 92,* 145 - 162.

[12] Gustavsson I. (Co-ordinator), *Hereditas 1988, 109,* 151 - 157.

[13] Fries, R., Stranzinger, G., and Vögeli, P., *J. of Heredity 1983, 74,* 426 - 430.

[14] Förster, M. In: *Ein Beitrag zur Zytogenetik und Zellgenetik mitotischer Chromosomen (Sus scrofa domestica)* Habilitationsschrift an der Technischen Universität München, Freising-Weihenstephan, *1984*.

[15] Dolf, G., and Stranzinger, G., *Genet. Sel. Evol. 1986, 18,* 375 - 384.

[16] Womack, J.E., *Developmental Genetics 1987, 8,* 281 - 293.

[17] Rubtsov, N., Graphodatsky, A., Matveeva, V.G., Radjabli, R.I., Nesterova, T.B., Kulbakina, N.A. and Zakian, S., *Cytogenet. Cell Genet. 1988, 48,* 95 - 98.

[18] Ruddle, F. H., *Nature 1981, 294,* 115 -120.

[19] Lalley, P.A., O' Brien, S.J., Créav-Goldberg, N., Davisson, M.T., Roderick, T.H., Echard, G., Womack, J.E., Graves, J.M., Doolittle, D.P., and Guidi, J.N., In: [20], pp 367 - 389.

[20] HGM 9 (Human Gene Mapping, Workshop 9): Paris Conference (1987), *Cytogenet. Cell Genet. 1988, 46,* 1 - 762.

[21] Soller, M. and Beckmann, J.S., *Reviews in Rural Science 1985, 6,* 10 - 18.

[22] Prokop, C.M., Plenz, G., Kratzberg, T. and Geldermann, H., *Journal of Animal Breeding Genetics 1988, 105,* 70 - 80.

[23] Jeffreys, A.J., Wilson, V., Thein, S.L., Weatherall, D.J., and Ponder, B.A.J., *American Journal of Human Genetics 1986, 39,* 11 - 24.

[24] Ohno, S., *Annual Review of Genetics 1969, 3,* 495 - 524.

[25] Elmiger, B.: *H-Y Antigen als Selektionskriterium für die Intersexualitätsprobleme bei hornlosen Ziegen.* Diss. ETH *1983 Nr. 7630*: 1 - 129.

[26] Ohno, S., Christian, L.C., Wachtel, S.S., and Koo, A.C., *Nature 1976, 261,* 597 - 599.

[27] Lyon, M.F., *Nature (London) 1961, 190,* 372 - 373.

[28] Caspersson, T., de la Chapelle, A., Schröder, J., and Zech, L. , *Exptl. Cell Res. 1972, 72,* 56 - 59.

[29] Wurster, D.H., and Benirschke, K., *Chromosoma 1968, 25,* 152 - 171.

[30] Evans, H.J., Buckland, R.A., and Sumner, A.T., *Chromosoma 1973, 42,* 383 - 402.

[31] Förster, M., and Stranzinger, G., *Zeitschrift für Tierzüchtung und Züchtungsbiologie 1975, 92,* 267 - 271.

[32] Hediger, R. In: *Die in situ Hybridisierung zur Genkartierung beim Rind und Schaf.* Diss. ETH *1988* Nr. 8725: 1 - 162.

[33] Womack, J.E. and Moll, Y.D., *Journal of Heredity 1986, 77,* 2 - 7.

[34] Fries, R., Beckmann, J.S., Georges, M., Soller, M., and Womack, J., *Animal Genetics 1989, 20,* 3 - 29.

[35] Fries, R., Hediger, R., and Stranzinger, G., *Genomics 1988, 3,* 302 - 307.

[36] Nadeau, J.H., and Taylor, B.A. In: *Proceedings of the National Academy of Sciences of United States of America 1984, 81,* 814 - 818.

[37] Fries, R., Hediger, R., and Stranzinger, G., *Animal Genetics 1986, 17,* 287 - 294.

[38] Singer, D.S., Camerini-Otero, R.D., Satz, M.L., Osborne, B., Sachs, D., and Rudikoff, S., *Proc. Natl. Acad. Sci., USA 1982, 79,* 1403 - 1407.

[39] Rabin, M., Fries, R., Singer, D., and Ruddle, F.H., *Cytogenet. Cell Genet. 1985, 39,* 206 - 209.

[40] Geffrotin, C., Popescu, C.P., Cribiu, E.P., Boscher, J., Renard, C., Chardon, P., and Vaiman, M., *Annales Genetiques 1984, 27,* 213 - 219.

[41] Echard, G., Yerle, M., Gellin, J., Dalens, M., and Gillois, M., *Cytogenet. Cell Genet. 1986, 41,* 126 - 128.

[42] Ansari, H.A., Hediger, R., Fries, R., and Stranzinger, G., *Immunogenetics 1988, 28,* 362 - 364.

[43] Kent, M.G., Elliston, K.O., Shroeder, W., Guise, K.S., and Wachtel, S.S., *Cytogenet. Cell Genet. 1988, 48,* 99 - 102.

[44] Hart, Ch. P., Fainsod, A., and Ruddle, F.H., *Genomics 1987, 1,* 182 - 195.

[45] Wallace, D.C., *Trends in Genetics 1989, 5 (1),* 9 - 13.

[46] Schmid, D.O., and Buschmann, H.G., *Blutgruppen bei Tieren.* Ferdinand Enke Verlag, Stuttgart, *1985.*

[47] Stewart, A.F., Bonsing, J., Beattie, C.W., Shah, F., Willis, I.M., and Mckinlay, A.G., *Mol. Biol. Evol. 1987, 4 (3),* 231 - 241.

[48] Romeo, G., Fiorucci, G., and Rossi, G. B., *Trends in Genetics 1989, 5 (1),* 19 - 24.

[49] Stranzinger, G., *Dtsch. tierärztl. Wschr. 1988, 95,* 257 - 312.

9. How to Demonstrate Genetic Individuality in Any Human and Animal Subject

H. Zischler, I. Nanda, C. Steeg, M. Schmid and J.T. Epplen

9.1 Introduction

DNA polymorphisms are playing an increasingly important role as genetic markers in individuality testing, linkage analysis and for many other purposes. Due to the low overall genetic variability of human and animal DNA [1] most of the hitherto described restriction fragment length polymorphisms (RFLPs) are only diallelic. Therefore and because of their limited degree of heterozygosity RFLPs are not highly informative with respect to individuality testing and linkage analysis. In contrast to this, loci containing stretches of simple repetitive DNA exhibit a considerable degree of polymorphism. Oligonucleotide probes specific for simple repetitive sequences hybridize simultaneously to many such hypervariable loci, yielding a complex DNA fingerprint pattern. For the probe $(CAC)_5$ e.g. the pattern is individual-specific in man [2].

The basic experimental procedure of oligonucleotide fingerprinting is outlined in Fig. 9-1. Genomic DNA is digested to short fragments with a frequently but sequence-specifically cutting restriction enzyme that removes most of the repeat flanking DNA in order to improve the electrophoretic resolution of repeat containing fragments. The restricted DNA is subsequently separated according to size by means of agarose gel electrophoresis and afterwards immobilized by drying the gels [3]. Since Southern blotting is omitted, no problems occur with respect to transfer efficiency and gel drying is less time consuming. After denaturation of double stranded DNA *in situ,* the ^{32}P-labelled oligonucleotide probe is hybridized in the gel, applying conditions that only perfectly matching target DNA/probe hybrids are stable. Dried gels can be rehybridized several times, the probes are easily removable (by denaturation or immersing gels in a saltless solution).

DNA-FINGER PRINT in the gel

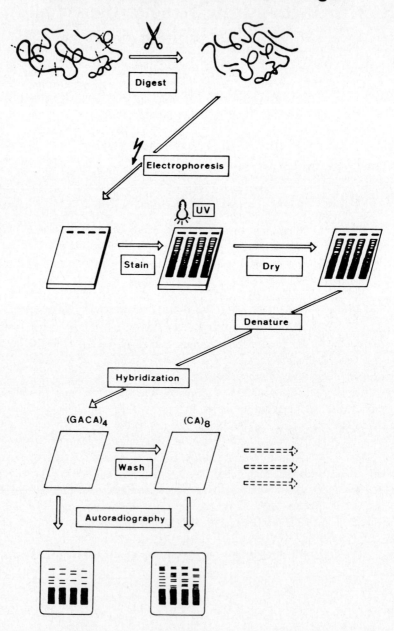

Fig. 9-1 Schematic representation of the basic oligonucleotide fingerprinting procedure.

9.2 Ubiquitous Dispersion of Simple Repetitive Sequences

Simple repetitive sequences make up a defined class of DNA with short motives (less than 10 bases) extensively repeated in a head to tail fashion. Such simple repetitive sequences could be detected in all classes of animals and plants investigated [3]. Many simple repeats are more or less evenly dispersed over the genome, as can be demonstrated by means of hybridization *in situ* [4]. Fig. 9-2 shows a human male metaphase probed with a ^{3}H-labelled, $(CAC)_5/(GTG)_5$ -containing fragment subcloned from a human genomic library.

(cac)$_n$

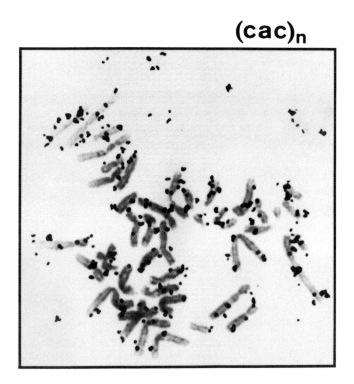

Fig. 9-2 Human male metaphase chromosomes probed with a ^{3}H-labelled $(CAC)_5/(CTG)_5$ -containing fragment at low stringency. Chromosomes were stained after hybridization with Giemsa's solution.

The low stringency conditions allow the probe to hybridize with many $(CAC)_5/(GTG)_5$ stretches of genomic DNA, as could be demonstrated in Southern blot experiments (data not shown). The silver grains are distributed all over the chromosomes with no obvious preferential localization. A better correlation of signal to chromosome morphology could be obtained after hybridization *in situ* with non-radioactively labelled oligonucleotide probes [4]: Hybridization *in situ* onto human metaphase chromosomes with the $(CAC)_5$ oligonucleotide yields signals coinciding with R-banding patterns. This means that a DNA fingerprint established by the $(CAC)_5$ oligonucleotide probe represents hypervariable loci statistically distributed all over the human genome. Such probes are therefore most promising tools to search for minute genome alterations for example in malignant tissues [5]. Furthermore single copy probes flanking hypervariable loci can be identified all over the chromosomal complement, making them suitable probes for linkage analysis.

9.3   Oligonucleotide Fingerprinting in Man

Various simple repetitive oligonucleotide probes were synthesized and tested in DNA fingerprinting experiments [3]. The number of principally different simple repeat sequences can be calculated as $4^{n-1}+2$ ($n \geq 2$) [3]; n is the length of the simple repeat in bases. All tested probes yield different patterns with respect to signal intensity (depending on the degree of repetitivity of the target simple sequences), to molecular weight range of hybridizing fragments and to information content. The information content of a DNA fingerprint is dependent on the average number of discernible polymorphic bands per individual and the respective band frequencies. The probe $(CAC)_5$ (or its complement $(GTG)_5$) is the hitherto most informative probe in man [2]. Fig. 9-3 depicts Mbo I-digested, electrophoretically separated DNA from a family with monozygous twins, a further pair of monozygous twins and nine unrelated individuals probed with $(CAC)_5$. In the family each offspring fragment can be traced back to either the father or the mother. Both twins of each pair show identical patterns of hybridizing fragments. As it is known from extensive family studies every band is transmitted according to the Mendelian laws. $(CAC)_5$ - fingerprints from all different human tissues tested show the same pattern, indicating somatic stability and excluding gross rearrangements in normal tissues [6].

The overall mutation rate of the bands in a fingerprint is substantially higher than that of an average DNA sequence [2]. Many hypervariable loci with potentially different genetic stabilities are recorded at the same time with one probe. Therefore the back tracing of offspring bands is only a rough estimate of the mutation rates for the individual DNA locus. Precise values can be calculated only if single copy probes for each locus are available and tested in family studies.

The ubiquitous dispersion of simple repetitive DNA containing loci, the easy handling and high specificity of oligonucleotide probes make this fingerprinting method an attractive technique for routine individualization studies. Moreover the recently described non-radioactive approach to oligonucleotide fingerprinting will facilitate the introduction of fingerprinting also in non-specialized laboratories. For this purpose a digoxigenin derivative is attached to a primary amino group of "Aminolink" (New Brunswick Scientific[TM]) which has been coupled like an additional "base" to the 5'-end of the oligonucleotide probe. Hybrids can be detected using a monospecific antiserum coupled to alkaline phosphatase

(Boehringer-Mannheim) in a color reaction. Because the degree of repetitivity of simple repeats is high, an excellent signal to background ratio can be achieved although each oligonucleotide has only one reporter group. Banding patterns for both the radioactive and the non-radioactive approaches are identical [4].

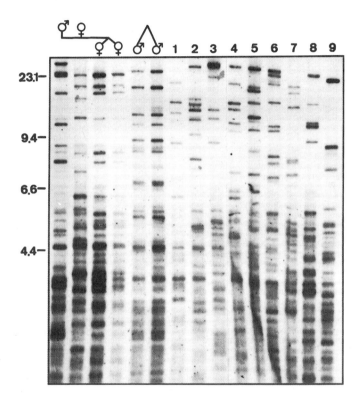

Fig. 9-3 In-gel hybridization of Mbo I digested and electrophoresed human DNA with the probe (CAC)$_5$. Samples from left to right: a family with monozygous twins, a further pair of monozygous twins and nine unrelated individuals. Molecular weight markers are given on the left in kilobases.

9.4 Fingerprinting Animal DNA

Simple repetitive DNA varies in sequence motives, quantity and genomic organization in all eukaryotic genomes tested so far. Nevertheless DNA fingerprinting with a panel of simple repetitive oligonucleotide probes is possible in all classes of vertebrates opening a broad field of applications: Individual identification, parentage testing, control of semen used in artificial insemination and genetic structure of animal populations with respect e.g. to inbreeding. Because of the fact, that in different species different oligonucleotide probes yield individual specific fingerprints (e.g. $(GTG)_5$ in Sapsaris, native dogs of Korea, unpublished results; $(GTG)_5$ and $(GT)_8$ in cattle (Buitkamp et al., in preparation); and $(GA)_8$ and $(GATA)_4$ in grasshoppers, *Poecillimon* (Achmann et al., in preparation); $(GGAT)_4$ in marmots (Rassmann et al., in preparation), a panel of different simple repetitive oligonucleotide probes is first evaluated with respect to informativity.

Apart from the above listed, immediate practical applications it is of interest to examine the possible role of simple repetitive DNA in sex determination. Simple quadruplet GATA/GACA repeats e.g. are in major component of the sex specific satellite of the female (heterogametic) snake *Elaphe radiata* [7]. The same simple quadruplet repeats are also sex specifically arranged in the male (heterogametic) mouse. The knowledge on sex chromosomal organization of simple repetitive DNA, its change in evolution and possible linkage to sex-determining loci (testis determining factor gene) may help to fully understand the mechanisms of sex determination in vertebrates.

Heretofore, as shown in Fig. 9-4, DNA's of several species from different vertebrate classes were Hinf I digested and hybridized in the gel with the oligonucleotide probes $(GTG)_5$, $(GACA)_4$, $(GT)_8$, $(TCC)_5$ and $(GATA)_4$. Each species is represented by the DNA of a male (left lane each) and of a female (right lane) individual. In order to demonstrate the effects of inbreeding on DNA fingerprints male/female pairs of two inbred strains of chicken (CC and CB) were examined in comparison with an outbred strain (CH). Even in this limited set of simple repetitive probes at least one probe is informative enough to allow to individualize each of the species in each vertebrate class. Variations in the information content of different oligoprobes for a certain species is exemplified in the inbred chickens. Whereas the male and female specimen of the CC and CB strains exhibit the identical fingerprint patterns with most simple repetitive probes tested, in the $(CAC)_5$ - fingerprint not all bands are shared in the CB pair of inbred chickens. This indicates a reduced inbreeding coefficient for the CB strain. Because male chickens are homogametic, the additional bands in the CB strain (marked by arrows) cannot be due to sex specific organization of $(CAC)_5$ loci. Sex specific arrangement of simple repetitive DNA can be seen in various examples: The male guppy (*Poecilia reticulata*) shows an additional $(GATA)_4$ -positive fragment in the high molecular weight range (e, short exposure). This probe (together with $(GACA)_4$) could be successfully used to identify guppy Y chromosome and confirm the XX/XY mechanism of sex determination. $(GATA)_4$ positive fragments can be also observed in the heterogametic female chicken (e', long exposure) and the heterogametic male mouse (e, short exposure). $(GATA)_4$ and $(GACA)_4$ containing sequences are enriched on the snake W chromosome, a phenomenon that correlates with the amount of heterochromatin on it [8]. Finally a high molecular weight fragment containing $(TCC)_5$ is sex specifically organized in chickens and mice (d). Hence we are able to correlate the sex specific organization of simple repetitive DNA with the evolution of vertebrate sex chromosomes [9].

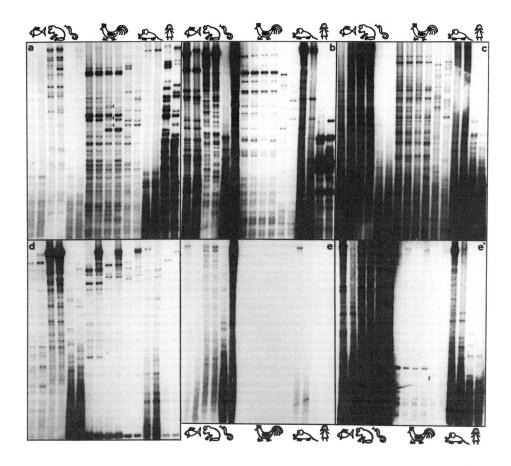

Fig. 9-4 In-gel hybridization patterns of Hinf I digested, genomic DNA with the probes $(CAC)_5$(a), $(GACA)_4$(b), $(GT)_8$(c), $(TCC)_5$(d) and $(GATA)_4$ (e, short exposure; e', long exposure). DNA samples from left to right: Male/female fish (*Poecilia reticulata*); male/female frog (*Gastrotheca riobambae*); male/female snake (*Elaphe taeniura*); male/female bird (*Gallus gallus domesticus*) inbred strains CC and CB, outbred individuals; male/female mammal (laboratory mouse strains C3H) and male/female human DNA.

9.5   Summary

Simple repetitive DNA of various sequence motives, amount and genomic organization can be detected in all eukaryotic genomes hitherto tested. These sequences frequently undergo mutations and are highly polymorphic in animal species. Depending on the species different oligonucleotide probes specific for certain simple repetitive DNA simultaneously hybridize to different hypervariable loci, thus yielding a complex DNA fingerprint pattern. The advantages of oligonucleotide probes are their easy handling, specificity, sensitivity, ubiquitous dispersion of target DNA and the fact, that they are chemically synthesized. In addition a non-radioactive approach to oligonucleotide fingerprinting facilitates the introduction of this technique into non-specialized laboratories. The implications of DNA fingerprinting in animal genetics are discussed for routine applications and the exploration of the organization of simple repetitive DNA in several animal species.

Acknowledgements

This work was supported by the Stiftung Volkswagenwerk. We thank Drs. Hala and Wick for the different chicken blood samples. The oligonucleotide probes are subject to patent applications and commercial enquiries should be directed to Fresenius AG, D-6370 Oberursel, FRG.

9.6   References

[1] Jeffreys, A.J., and Morton, D.B., *Anim. Genet. 1987, 18,* 1-15.
[2] Schäfer, R., Zischler, H., Birsner, U., Becker, A., and Epplen, J.T., *Electrophoresis 1988, 9,* 369-374.
[3] Epplen, J.T., *J. Hered. 1988, 79,* 409-417.
[4] Zischler, H., Nanda, I., Schmid, M., and Epplen, J.T., *Hum. Genet. 1989, 82,* 227-233.
[5] Lagoda, P.J.L., Seitz, G., Epplen, J.T., and Issinger, O.-G. *Hum. Genet.,* 84 (1).
[6] Nürnberg, P., Roewer, L., Neitzel, H., Pöpperl, A., Hundrieser, J., Epplen, C., Zischler, H., and Epplen, J.T., *Hum. Genet.,* 84 (1).
[7] Epplen, J.T., McCarrey, J.R., Sutou, S., and Ohno, S., *Proc. Natl. Acad. Sci. USA 1982, 79,* 3798-3802.
[8] Singh, L., Purdom, I.F., and Jones, K.W., *Cold Spring Harbor Symp. Quant. Biol. 1981, 45,* 805-814.
[9] Nanda, I., Feichtinger, W., Schmid, M., Zischler, H., and Epplen, J.T., *J. Mol. Evol.,* in press.

10. Information on RFLPs and VNTRs in Farm Animals

C.-M. Prokop

10.1 Introduction

The observation of hereditary molecular variability has wide applications in many fields of animal breeding, production and health. The application of protein polymorphisms in marker assisted selection programs is limited by their small number of variants and also by age or sex dependent appearance of protein polymorphisms. With the development of molecular genetic techniques it had become possible to establish a new class of genetic markers. These restriction fragment length polymorphisms (RFLPs) base upon variability at the DNA sequence level. RFLPs seem to be an almost unlimited source of genetic markers [1]. In human, RFLPs have found widespread application in diagnosis and linkage analysis of DNA loci. These techniques have also been used in anthropology and in studies of molecular evolution.

More recently, the techniques of RFLP analysis have been extended to domestic animals. The construction of detailed linkage maps required for the identification of monogenic and quantitative trait loci of economic importance could be assisted by these methods. However, the overall genetic variability of human and animal DNA is low. Jeffreys [2] has estimated that about 1 out of 100 base pairs is polymorphic in human genomes and 1% of these variants are detectable by RFLP technique. Searching for RFLPs produces mainly diallelic systems. So the use of these RFLPs in pedigree analysis and paternity control is limited by their low heterozygosity. For this purpose, a number of hypervariable regions have been found in man and in animals as well [3]. The loci have been characterized as variable numbers of tandem repeats (VNTRs) or as minisatellites. During the last few years several groups have been working in molecular genetics of farm animals. In this review the knowledge of RFLP analysis and the analysis of hypervariabel regions in farm animals is summarized.

In Fig. 10-1 the main steps for the identification of RFLPs with blot techniques by Southern [4] are shown. DNA may be isolated from various cell tissues, e.g. in most cases leucocytes. Then the DNA is digested by restriction enzymes. The DNA fragments are subjected to electrophoresis in agarose gels and transfered to nylon or nitrocellulose filters.

In this stage the hybridization with a labelled probe follows. The hybridized fragments can be shown by autoradiography. DNA from homologous or heterologous systems, cDNA, genomic DNA, or synthetic oligonucleotides are used as probes. Details may be found in the original papers.

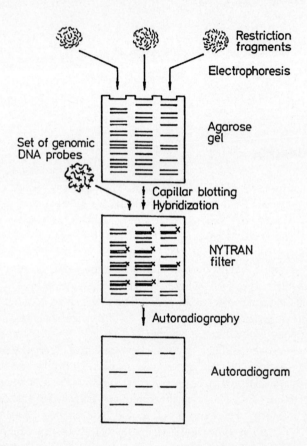

Fig. 10-1 Identification of Restriction Fragment Length Polymorphisms by Southern blot technique and hybridization [17].

10.2 Restriction Fragment Length Polymorphisms in Cattle

RFLPs in cattle are summarized in Table 10-1. It starts with the genes of the main histocompatibility complex (MHC), one of the best investigated genomic DNA regions. This region contains several genes, which are in different ways involved in the immunresponse [11]. For the identification of distinct genes in the MHC heterologous as well as

Table 10-1 Restriction Fragment Length Polymorphisms in cattle.

Gene loci	Probe	Restriction enzyme	Number of individuals	Authors
Major Histocompatibility Complex (MHC)	human cDNA DOß, DQα DQß, DRα, DRß, DZα, DYα, DYß	Bam HI Eco RI Taq I Pvu II	5 sire half sib families	Andersson et al. [6, 7, 8] Sigurdadottir et al. [9]
	bovine DRß genomic DNA	Eco RI Hind III	67	Muggli-Cockett & Stone [10]
Complement C4	human cDNA	Bam HI Eco RI Pvu II	5 sire half sib families	Sigurdadottir et al. [9]
T-complex	murine cDNA	Bam HI Eco RI Pvu II	84	Andersson et al. [8]
Growth hormone	bovine genomic DNA	Bgl II Bam HI	50	Beckmann et al. [12]
Growth hormone		Bgl II Bam HI Eco RI Hind III	78	Hallermann et al. [13]
ß-Actin	cDNA (rat)	Bgl II Pst I	50	Beckmann et al. [12]
Prolactin	bovine genomic DNA	Bam HI Eco RI Eco RV	50	Hallermann et al. [13]
Chymosin	bovine cDNA	Hind III Eco RI	10 10	Hallermann et al. [14]
Oxytocin-neurophysin I	bovine cDNA	Hind III	9	"
Lutropin ß subunit	bovine cDNA	Hind III	37	"

Table 10-1 Restriction Fragment Length Polymorphisms in cattle (continued).

Gene loci	Probe	Restriction enzyme	Number of individuals	Authors
Keratin III	bovine cDNA	Eco RV	10	Hallermann et al. [14]
Keratin VI	bovine cDNA	Bgl II	9	"
Keratin VII	bovine cDNA	Eco RI	41	"
Prolactin	bovine cDNA	Eco RV	18	"
Thyroglobin	bovine cDNA	Bam HI Eco RI Hind III Bgl II Eco RV Pvu II	12 ♂	Georges et al. [15]
Thyroglobin	bovine cDNA	Pst I Taq I	?	Ricketts et al. [16]
α_{s1}-Casein	bovine cDNA	Pst I	38 ♀	Prokop et al. [17]
	bovine genomic DNA	Pst I	53 ♀	Geldermann et al. [18]
k-Casein	bovine cDNA	Pst I Bam HI	44 ♀	Rando & Masina [19]
k-Casein	bovine cDNA	Taq I Hinf I Alu I	?	Damiani et al. [20]
k-Casein	bovine cDNA	Taq I Hind III	?	Kühn [21]
k-Casein	bovine cDNA	Hind III	?	Leveziel et al. [22]
α-Interferon	human cDNA	Eco RI	20	Adkinson et al. [23]
ß-Interferon	bovine cDNA	Hind III		"

Table 10-1 Restriction Fragment Length Polymorphisms in cattle (continued).

Gene loci	Probe	Restriction enzyme	Number of individuals	Authors
Fibronektin	human cDNA	Taq I	20	Adkinson et al. [24]
y-Crystallin	bovine cDNA	Eco RI		"
Protamin	bovine genomic DNA	Taq I	?	Krawetz et al. [25]
High Mobility Group I	bovine genomic DNA	Taq I	?	"

homologous probes were used. Andersson and his group studied the class II genes, their organization and linkage relationships [6, 7, 8]. They found series of RFLPs using a set of restriction enzymes. The material of animals was extended to 5 sire half-sib families, 48 dams and 50 offspring. 197 young sires were also analyzed in another study [9] dealing with the DQ and DR class II genes of the bovine MHC. Using a bovine genomic DRß probe, Muggli-Cockett and Stone [10] also found genetic variation in cattle. Apart from class II genes the group of Andersson studied the genes for the complement factor C4 and for the TCP of the T-cell complex [11]. For digestions of DNA the enzymes Bam HI, Eco RI and Pvu II were used. The authors tested 5 sire half-sib families. The hybridization was carried out with a human C4-cDNA and a murine TCP1-cDNA probe. Incidentally, the authors suggested a duplication in the TCP1 gene locus.

Beckmann et al. [12] described RFLPs for the gene loci of growth hormone (bGH) and ß-actin. For the bGH-locus polymorphisms exist after digestion with the enzymes Bgl II and Bam HI and hybridization with a bovine cDNA probe. The patterns for DNA digested by Bgl II or Pst I and hybridized with a rat muscle ß-actin cDNA probe also contain variable bands. In a later study [13] digestion with Hind III or Eco RI yielded informative patterns for the bGH locus as well. For the bGH loci the authors established a genomic location of the restriction sites in the flanking regions of the gene. In the same paper RFLPs for the prolactin locus are also presented. In all cases the examined animals were Israeli Holstein Friesian bulls and cows. The same group [14] screened the genomic DNA with 17 cloned DNA probes in order to demonstrate variability at several gene loci. RFLPs were observed at the chymosin, oxytocin-neurophysin I, lutropin ß, keratin III, IV, VII, prolactin and dihydrofolate reductase loci.

RFLPs for the thyroglobin locus (bTg) are published by Georges et al. [15]. The hybridization probe consisted of two fragments of a bTg-cDNA, called "right or left probe".

Using this probe they were able to differentiate variable patterns, a genomic map was constructed. The animals investigated were 12 bulls of the Belgian Blue breed. Associations between RFLP patterns for bTg loci and recessive hereditary goiter were found by Ricketts et al. [16]. From the results, a genetic lesion in the gene was suggested that causes aberrant splicing of the pre-mRNA in an intron /exon junction.

So far variable restriction sites in the casein loci were found for α_{s1}- and for k-casein. Prokop et al. [17] presented RFLPs for α_{s1}-casein loci using PstI and a α_{s1}-cDNA probe, and Geldermann et al. [18] showed polymorphic patterns using a genomic probe. In 1988 Rando et al. [19] found RFLPs for the k-casein after digestion with PstI and hybridization with a cDNA probe. This DNA polymorphisms were correlated with the protein types A and B. The switched codon was directly identified with the enzymes Taq I or Hind III [20, 21, 22].

RFLPs for α and ß interferon (IFNA, IFNB) were published [23]. When Adkinsons group tested a human cDNA for the IFNA and a bovine cDNA for IFNB they inferred for the IFNA a multiallelic polymorphism. The IFNB restriction pattern was interpreted as a result of 2 allelic variants. They also searched for RFLPs of fibronectin (FN1) and y-crystallin (CRYG) [24]. The results of hybridization with a human FN1-cDNA were interpreted as products of 3 alleles. Testing with the bovine CRYG-cDNA the authors found polymorphisms, which were based on a 2 allelic system. In addition they worked on somatic cell mapping in order to find synthetic groups for bovine genome mapping.

The analysis of the nuclear protein genes was published [25] with RFLPs for protamin and for the high mobility group I (HMG-1). Krawetz et al. could not detect a linkage to the sex chromosomes. From their result the authors concluded the presence of a single copy for the protamin gene and the existence of two copies for the HMG-1 gene. Further, they constructed a lineage of bovine species and breeds.

10.3   Restriction Fragment Length Polymorphisms in Pig

Analysis of RFLPs in pigs are summarized in Table 10-2. For the major histocompatibility complex (MHC) polymorphic patterns with 7 to 10 bands were performed using a human class I cDNA probe [26]. Polymorphic patterns for the DRß loci were obtained with a human cDNA probe. The 13 animals tested were also typed for their lymphocyte antigens (SLA). Some of the SLA types correlated with special restriction patterns. RFLPs could also be detected by human cDNA probes for the class I and the class II region of the MHC [27]. Sachs et al. [28] tested the DNA of the Göttingen Miniature pig. DNA from 4 families with 42 SLA typed individuals was screened for RFLPs of the complement C4. The linkage to the MHC was verified. Studying RFLPs and linkage for complement C4, C2 and factor B (Bf), Lie et al. [29] found polymorphisms for C4 and Bf but not for C2. Since they had an overlapping sequence of taq I fragment, in linkage analysis the genes were arranged C4-Bf-C2.

RFLPs in the ß globin locus of the pig were studied by a cDNA probe from rabbit and the enzyme Hind III [30]. From results of family studies 3 allelic forms of the gene were provided based on point mutations.

Polymorphisms for the glucose phosphate isomerase gene locus were presented by Davies et al. [31, 32]. Pvu II and Sac I were used along with a porcine cDNA probe. With 25 offspring from 4 families a linkage to the porcine chromosome 6 was verified by *in-situ*-hybridization. The glucose phosphate isomerase gene locus is of interest to animal breeding, because there is a linkage to the Halothan locus and correlation to the porcine stress syndrome.

With a human cDNA probe Bahou et al. [33] identified RFLPs for the von Willebrand factor in a breeding colony of pigs with a disease closely resembling the human bleeder syndrome. The authors also tested associations to clinical parameters.

Table 10-2 Restriction Fragment Length Polymorphisms in pig.

Gene loci	Probe	Restriction enzyme	Number of individuals	Authors
Major Histocompatibility Complex	human cDNA class I	Eco RI Hind III Bam HI	13	Chardon et al. [26]
	DRß	Taq I		
	human cDNA class II	Pvu II Taq I	130+67 offspring	Stolt et al. [27]
	human cDNA class I class II	Eco RI Bam HI Hind III Pvu II Taq I	4 families	Sachs et al. [28]
Complement C4	human cDNA	Eco RI	42	Kirzenbaum et al. [34]
Complement C4	human cDNA	Taq I	?	Lie et al. [29]
Factor B	human cDNA	Taq I		"
ß-Globin	cDNA (rabbit)	Hind III Eco RV	3 families	Rando and Masina [30]
Glucose-phosphate-isomerase	porcine cDNA	Pvu II Sac I	4 families	Davies et al. [31, 32]
von-Willebrand-factor	human cDNA	Hind III	35	Bahou et al. [33]

10.4   Restriction Fragment Length Polymorphisms in Horse

RFLP studies in the horse DNA are summarized in Table 10-3. For the major histocompatibility complex (MHC) Alexander et al. [35] described polymorphisms in class I as in class II genes of the MHC. Human cDNA probes were used for hybridization. In a collection of 3 families and 3 sire half-sib families Guerin et al. [36] also identified RFLPs in the MHC region. The authors concluded, that in the horse genome more than 1 gene is present for DRß and also for DQα.

Table 10-3 Restriction Fragment Length Polymorphisms in horse.

Gene loci	Probe	Restriction enzyme	Number of individuals	Authors
Major Histocompati- bility Complex	human cDNA	Hind III	?	Alexander et al. [35]
	Class I	Pvu II		
	Class II	Eco RI Taq I	6 families	Guerin et al. [36]
Complement C4	human cDNA	Eco RI	6	"
Complement C4	human cDNA	Eco RI Bam HI	?	Kay et al. [37]
Steroid-21- hydroxylase	murine cDNA	Taq I		"
Acetylcholine- receptor	murine cDNA	Pst I Taq I	?	Kay et al. [37]
		Mlu I Sfi I		Kay et al. [38]
ß-Globin	cDNA (rabbit)	Pst I Bam HI	?	Rando et al. [39]
z-Globin	equine cDNA	Hinf I	?	Flint et al. [40]

At the gene loci for complement C4 and the steroid-21-hydroxylase (21-OH), which are also clustered in the MHC Kay et al. [37] found RFLPs testing digested equine DNA with human cDNA probes. In addition the authors estimated the distance between C4 and 21-OH to be no more than 13 kbp. All screened animals were typed for their lymphocyte antigens (ELA). With the genes for acetylcholine receptor proteins (AChR) RFLPs were also found by conventional electrophoresis [38]. Later, pulsed field electrophoresis, rare cutting enzymes and murine cDNA probes for the subunits α1 and for the α2 were used. No cohybridization occurred with the components ß, y or d. Thus the authors suggested reasonable rearrangement or the existence of more than one gene.

RFLPs for globin genes in horse were described by Rando et al. [39] for the ß-globin locus and by Flint et al. [40] for the z-globin locus. Using a cDNA probe from rabbit, Rando et al. identified polymorphisms which based on two alleles. Flint et al. [40] detected a hypervariable region in the z-globin gene of the horse which generate a large number of variants.

10.5 Restriction Fragment Length Polymorphisms in Sheep and Goat

In Table 10-4 the knowledge on RFLPs in sheep and goat is summarized.

Table 10-4 Restriction Fragment Length Polymorphisms in sheep and goat.

Gene loci	Probe	Restriction enzyme	Number of individuals	Authors
α-Globin	ovine cDNA	Bam HI Pst I Bst EII	60 sheep 7 goats	Rando et al. [39]
Embryonic Globin	caprine cDNA	Bam HI Eco RI Hind III	30 sheep 10 goats	Rando et al. [42]
Hemoglobin A,B	caprine cDNA	Pst I Bam HI Hind III Eco RI	24 sheep 8 goats	Di Gregorio et al. [43]
Oncogene n-myc	human genomic DNA	Bgl II	?	Montgomery & Hill [44]

Rando et al. [41] used DNA from sheep and goats for RFLP studies of globin genes. After hybridization with an ovine cDNA, polymorphic patterns for α-globin were found in sheep and in goat DNA. In sheep and in goat they found identical patterns and suggested some variants of the gene loci for α-globin in sheep based on unregulated crossing over and amplification.

A caprine cDNA for embryonic globin identified a biallelic polymorphism in sheep and goat being associated with ßA and ßB [42]. The authors discussed the organization of the ß-globin cluster and assumed different recombination rates upstream or downstream of the adult ß-globin gene locus. Variable restriction patterns for the genes of hemoglobin A and B were detected by Digregori et al. [43] with a cDNA probe from rabbit.

In order to describe the variability of New Zealand sheep breeds (Romney and Merino) Montgomery and Hill [44] screened with a human n-myc-oncogene probe. They demonstrated a polymorphic restriction site with the help of Bgl II.

10.6   Restriction Fragment Length Polymorphisms in Chicken

In chicken there is only little information on RFLPs. Polymorphisms in the class II gene region were published by Andersson et al. [45]. They showed RFLPs by use of human cDNA probes. Kodoma et al. [46] investigated a W chromosome specific repeat unit from the Xho I family and found RFLPs after hybridization with cloned genomic probes. In addition an unusual electrophoretic behavior of the repeat units was described.

10.7   Reports on Variable Number of Tandem Repeats and Minisatellites in Farm Animals

Reports on highly polymorphic DNA regions, called variable number of tandem repeats (VNTRs) or minisatellites, are listed in table 10-5, as far as investigated in farm animals. Several cloned and subcloned sequences were tested [47, 48, 49, 50]. The following probes were used: the minisatellite 33.15 of Jeffrays, the per gene from Drosophila, the M13 protein III repeat, a anonymous probe EE 41.01 and a hypervariable region of α-globin. Synthetic oligonucleotides were tested as well [51]. Further discussion of the synthetic oligomers see Zischler et al. [3, this volume].

Table 10-5 Variable Number of Tandem Repeats and minisatellites in farm animals.

Probe	Species	Authors	
M13	cattle	Vassart et al.	[47]
Minisatellite 33.15 M13 EE 41.01	horse	Broad et al.	[48]
Minisatellite 33.15 M13 α-Globin-HVR PER gene (Drosophila)	cattle horse pig	Georges et al.	[49]
M13	cattle horse pig	Ryskov et al.	[50]
(CAC)$_5$	cattle	Buitkamp et al.	[51]

10.8 Conclusions

In recent years the screening for RFLPs and VNTRs in farm animals attracted considerable interest. As shown in table 10-6, in more than 90 % of the cases cDNA probes were used for hybridization. Heterologous and homologous DNA probes were tested. Because of higher stringency in hybridization a tendency to use homologous probes can be observed, so that the construction of DNA libraries is promoted for farm animals.

The identification of RFLPs and VNTRs effectively demonstrates genetic variability in farm animals. Sets of probes may be combined for several applications. For the interpretation of the DNA polymorphisms, the segregation and the mutation rates must be studied for single gene loci, and mathematical aspects need attention. The broad application of molecular genetic technology are discussed by Geldermann [52, this volume].

Methodical progress will support both, scientific investigations and practical application. Possible methodical improvements by the use of polymerase chain reaction (PCR) are already available. For this technique very small amounts of DNA are sufficient to produce a high number of copy fragments of a defined DNA region. Another important step will be the use of specific oligonucleotides to screen for specific basepair exchanges on the basis of known alterations between alleles.

Relative to the present knowledge on RFLPs in several species considerable effort will be necessary in farm animals. However, cooperation of scientists in this field takes place and sustain progress.

Table 10-6 Numbers of gene loci marked with RFLPs in various farm animals.

number of gene loci	species	DNA probes used heterologous	homologous
28	cattle	8 human cDNA 1 murine cDNA 1 rat cDNA	8 cDNA 4 genomic DNA
9	pig	6 human cDNA 1 rabbit cDNA	1 cDNA
7	horse	3 human cDNA 2 murine cDNA 1 rabbit cDNA	1 cDNA
5	sheep	1 human genomic DNA 2 caprine cDNA	1 cDNA
5	goat	1 human genomic DNA 1 ovine cDNA	2 cDNA

10.9 References

[1] Cooper, D.N., and Schmidtke, J., *Human Genet.* **1984**, *66*, 1-16.

[2] Jeffreys, A.J., *Proc. Biochem. Soc. Transaction* **1987**, *15*, 309-317.

[3] Zischler, H., Nanda, I., Steeg, C., Schmid, M., and Epplen, J.T.: How to demonstrate genetic individuality in any human and animal subject, in: Geldermann, H., Ellendorff, F. (Eds.): *Genome Analysis in Domestic Animals*. Verlag VCH, Weinheim, **1990**.

[4] Southern, E.M., *J. Mol. Biol.* **1975**, *98*, 503-517.

[5] Andersson, L. (1989): Major histocompatibility genes in cattle and their significance for immune response and disease susceptibility, in: Geldermann, H., Ellendorff, F. (Eds.): *Genome Analysis in Domestic Animals*. Verlag VCH, Weinheim, **1990**.

[6] Andersson, L., Böhme, J., Rask, L., and Peterson, P.A., *Anim. Genet.* **1986**, *17*, 95-112.

[7] Andersson, L., Lunden, A., Sigurdardottir, S., Davies, C.J., and Rask, L., *Immunogenetics* **1988**, *27*, 273-280.

[8] Andersson, L., and Rask, L., *Immunogenetics* **1988**, *27*, 110-120.

[9] Sigurdardottir, S., Lunden, A., and Andersson, L., *Anim. Genet.* **1988**, *19*, 133-150.

[10] Muggli-Cockett, N.E., and Stone, R.T., *Genet.* **1988**, *19*, 213-225.

[11] Lindberg, P.G., and Andersson, L., *Anim. Genet.* **1988**, *19*, 245-255.

[12] Beckmann, J.S., Kashi, Y., Hallermann, E.M., Nave, A., and Soller, M., *Anim. Genet.* *1986, 17,* 25-38.

[13] Hallerman, E.M., Nave, A., Kashi, Y., Holzer, Z., Soller, M., and Beckmann, J.S., *Anim. Genet. 1987, 18,* 213-222.

[14] Hallerman, E.M., Nave, A., Soller, M., and Beckmann, J.S., *J. Dairy Sci. 1988, 71,* 3378-3389.

[15] Georges, M., Lequarré, A.S., Hanset, R., and Vassart, G., *Anim. Genet. 1987, 18,* 41-50.

[16] Ricketts, M.H., Pohl, V., De Martinoff, G., Boyd, C.D., Bester, A.J., Van Jaarsveld, P.P., and Vassart, G., *EMBO J. 1985, 4,* 731-738.

[17] Prokop, C.-M., Plenz, G., Kratzberg, T., and Geldermann, H., *J. Anim. Genet. 1988, 105,* 70-80.

[18] Geldermann, H., Prokop, C.-M., Plenz, G., and Kratzberg, T.: *Methods and application of recombinant DNA techniques in cattle breeding,* 38th Annual Meeting European Association Animal Production, *1987,* Lisbon (Portugal).

[19] Rando, A., DiGregorio, P., and Masina, P., *Anim. Genet. 1988, 19,* 51-54.

[20] Damiani, G., Ferretti, L., Rognoni, G., and Sgaramella, V.: *RFLPs of bovine k-casein gene.* XXIst International Conference on Animal Blood Groups and Biochemical Polymorphisms, *1988,* Turin, July 4-8.

[21] Kühn, C.: *Darstellung von Restriktions-Fragmentlängen-Polymorphismen (RFLPs) des bovinen k-Kasein-Gens als Marker für Milchleistung und -verarbeitungseigenschaften.* Inaug.-Diss., Tierärztliche Hochschule Hannover, *1989.*

[22] Levéziel, H., Méténier, L., Mahé, M.-F., Choplain, J., Furet, J.-P., Pabaeuf, G., Mercier, J.-C., and Grosclaude, F., *Génét. Sél. Evol. 1988, 20,* 247-254.

[23] Adkinson, L.R., Leung, D.W., and Womack, J.E., *Cytogenet. Cell Genet. 1988, 47,* 62-65.

[24] Adkinson, L.R., Skow, L.C., Thomas, T.L., Petrash, M., and Womack, J.E., *Cytogenet. Cell Genet. 1988, 47,* 155-159.

[25] Krawetz, S.A., Bricker, R.A., Connor, W., Church, R.B., and Dixon, G.H., *Theor. Appl. Genet. 1988, 75,* 402-409.

[26] Chardon, P., Renard, Ch., Kirszenbaum, M., Geffrotin, C., Cohen, D., and Vaiman, M., *J. Immunogenetics 1985, 12,* 139-149

[27] Stolt, H., Edfors-Lilja, I., and Andersson, L.: *Analysis of the swine MHC (SLA) by use of human DNA class II probes and class I antisera.* XXIst International Conference on Animal Blood Groups and Biochemical Polymorphisms, *1988,* Turin, July 4-8.

[28] Sachs, D.H., S. Germana, M. El Gamil, K. Gustafsson, F. Hirsch, and K. Pratt, *Immunogenetics 1988, 28,* 22-29.

[29] Lie, W.R., Rothschild, M.F., and Warner, C.M., *J. Immunol. 1987, 139,* 3388-3395.

[30] Rando, A., and Masina, P., *Anim. Blood Groups Biochem. Genet. 1985, 16,* 35-40.

[31] Davies, W., Harbitz, I., and Hauge, J.G., *Anim. Genet. 1987, 18,* 233-240.

[32] Davies, W., Harbitz, I., Fries, R., Stranzinger, G., and Hauge, J.G., *Anim. Genet. 1988, 19,* 203-212.

[33] Bahou, W.F., Bowie, E.J., Fass, D.N., and Ginsburg, D., *Blood 1988, 72,* 308-313.

[34] Kirszenbaum, M., Renard, C., Geffrotin, C., Chardon, P., and Vaiman, M., *Anim. Blood Groups Biochem. Genet. 1985, 16,* 65-68.

[35] Alexander, A.J., Bailey, E., and Woodward, J.G., *Immunogenetics 1987, 25,* 47-54.

[36] Guerin, G., Bertaud, M., Chardon, P., Geffrotin, C., Vaiman, M., and Cohen, D., *Anim. Genet. 1987, 18,* 323-336.

[37] Kay, P.H., Dawkins, R.L., Bowling, A.T., and Bernoco, D., *J. Immunogenet. 1987, 14,* 247-253.

[38] Kay, P.H., Dawkins, R.L., Bowling, A.T., and Bernoco, D., *Vet. Rec. 1987, 120,* 363-365.

[39] Rando, A., DiGregorio, P., and Masina, P., *Anim. Genet. 1986, 17,* 245-253.

[40] Flint, J., Taylor, A.M., and Clegg, J.B., *J. Mol. Biol. 1988, 199,* 427-437.

[41] Rando, A., Ramunno, L., and Masina, P., *Mol. Biol. Evol. 1986, 3,* 168-176.

[42] Rando, A., Di Gregorio, P., and Masina, P.: *Restriction fragment length polymorphisms (RFLPs) in the DNA region of sheep embryonic globin genes.* XXIst International Conference on Animal Blood Groups and Biochemical Polymorphisms, *1988,* Turin, July 4-8.

[43] Di Gregorio, P., Rando, A., and Masina, *P., Anim. Genet. 1987, 18,* 241-247.

[44] Montgomery, G.W., and Hill, D.F.: *DNA variation in New Zealand Romney and Merino sheep.* XXIst International Conference on Animal Blood Groups and Biochemical Polymorphisms, *1988,* Turin, July 4-8.

[45] Andersson, L., Lundberg, C., Rask, L., Gissel-Nielsen, B., and Simonsen, M., *Immunogenetics 1987, 26,* 79-84.

[46] Kodama, H., Saitoh, H., Tone, M., Kuhara, S., Sakaki, Y., and Mizuno, S., *Chromosoma 1987, 96,* 18-25.

[47] Vassart, G., Georges, M., Monsieur, R., Brocas, H., Leuqarré, A.S., and Christophe, D., *Science 1987, 235,* 683-684.

[48] Broad, T.E., Forrest, J.W., Macdonald, P.E., and Pugh, P.A.: *Prospects for a horse genetic fingerprinting system.* XXIst International Conference on Animal Blood Groups and Biochemical Polymorphisms, *1988,* Turin, July 4-8.

[49] Georges, M., Lequarré, A.S., Castelli, M., Hanset, R., and Vassart, G., Cytogenet. *Cell Genet. 1988, 47,* 127-131.

[50] Ryskov, A.P., Jincharadze, A.G., Prosnyak, M.I., Ianov, P.L., and Limborska, S.A., *FEBS Letters 1988, 233,* 388-329.

[51] Buitkamp, J.: *Darstellung von DNA-Fingerprints beim Rind.* Inaug.-Diss., Tierärztliche Hochschule Hannover, *1989.*

[52] Geldermann, H. : Application of genome research in animal breeding, in: Geldermann, H., Ellendorff, F. (Eds.): *Genome Analysis in Domestic Animals.* Verlag VCH, Weinheim, *1990.*

11. Biosynthesis of Seminal Plasma Proteins

K.H. Scheit

11.1 Introduction

The bulk of seminal fluid, the seminal plasma, is provided by the male accessory organs: epididymis, vas deferens, ampulla, prostate, seminal vesicle, Cowper's gland, coagulating glands as well as Littre's gland. Morphology, histology and the nature of secretions of the accessory sex glands have been studied in many mammals. Due to the presence of the blood-testes barrier, the proteins of the seminal plasma are mainly products of the accessory sex glands. In the following, a brief account will be given on the present state of knowledge related to the biosynthesis of secretory proteins of the accessory sex glands.

11.2 Proteins of the Rat

11.2.1 Proteins Secreted by the Epididymis

The epididymis, an extratesticular organ is made up of highly convoluted ducts, is of immense physiological importance since it acts as a sperm reservoir, provides a fluid environment for spermatozoa, promotes maturation of spermatozoa and supports the disposal of ageing spermatozoa. In 1976, Cameo and Blaquier [1] showed the presence of at least four epididymis-specific proteins (ESPs) in rat and designated them as proteins B, C, D, and E, according to their increasing electrophoretic mobility under non-denaturing conditions.

In a recent study using cDNAs corresponding to the mRNAs of proteins B, C, D, and E as probes, the levels of mRNA for these proteins were monitored during development. The mRNAs for all the proteins were detectable in the epididymis at 20 days of age and increased to adult levels by 30 or 40 days of age. The appearance of these proteins closely matches with epididymal differentiation which also takes place at day 20 [2].

Brooks [3] has recently shown that the levels of mRNAs of proteins B-E decline rapidly during the first week after castration. The synthesis of any of the proteins B, C, D, and E varies in their sensitivity to androgen therapy [3].

The cDNAs for two of the closely related ESPs, namely B and C, were cloned [4]. The identity of the cDNA clones was established by matching the amino acid sequence of the N-terminus of the purified protein with that of the cDNA-derived amino acid sequence. The cDNA coded for a protein of 166 amino acids with a MW of 18,500. The initiation codon ATG at position 9 provided for the synthesis of a signal sequence of 18 amino acids, followed by an amino acid sequence which matched that of ESP B [4]. Protein C is about 500 daltons larger in MW than protein B, and has a blocked N terminus; it was therefore not possible to establish, by amino acid sequencing, its identity with the cDNA-derived amino acid sequence.

The primary amino acid sequence of proteins D and E was also derived from the cDNA clones specific for these proteins. These clones were selected from an epididymal cDNA library which was screened with affinity-purified antibodies to proteins D and E [5]. It appeared that the amino acid sequences of both these glycoproteins are identical. Northern blot analysis indicated that the mRNA was ≈ 1250 bp long. The first 57 nucleotides in the cDNA sequence represented the signal sequence which was followed by the sequence for the mature protein of 227 amino acid residues representing a MW of 26,000. The protein contained four potential glycosylation sites (N-X-S/T) located at positions 13, 66, 128 and 194, respectively. It also possessed a high amount of cystine which accounted for 14 out of the 93 residues in the C-terminal half of the molecule. These cysteins may be involved in intrachain disulfide bonds. The sequence showed a high degree of homology with yeast carboxypeptidase Y [6] in the cysteine-rich domain.

11.2.2 Proteins Secreted by the Seminal Vesicles

Seminal vesicles are organs of the male genital tract that secrete large amounts of a fluid which constitutes a substantial portion of the total seminal plasma ejaculated.

The protein content of the secretion is high. Most of the proteins of seminal plasma are specific to this gland and are probably synthesized and secreted only by it in the animal.

In a more detailed study, SDS-PAGE in the presence of ß-mercaptoethanol and DTT resolved rat seminal vesicle secretory proteins (SVS) into 5-7 distinct bands which are designated SVS-I to SVS-VI in the order of increasing electrophoretic mobility [7, 8, 9]. The MW of these proteins was as follows: SVS-I, 80,000-100,000; SVS-II, 42,000-50,000; SVS-IIIA, 37,000; SVS-IIIB, 35,000; SVS-IV, 18,500; SVS-V, 15,000-17,000; and SVS-VI, 15,000 [7, 8, 9]. The amino acid sequence of SVS-IV closely matched with the amino acid sequence deduced from the nucleotide sequence of the coding region of SVS-IV cDNA [10, 11]. McDonald et al. [11] derived the amino acid sequence of both proteins, SVS-IV and SVS-V, from the cDNAs corresponding to the respective mRNAs, and also established the sequence of the signal peptide of the two proteins. Comparison of the amino acid sequences of the two proteins indicated that of the 85 amino acids present in SVS-IV, only 29 amino acids matched with the SVS-V protein which consisted of 101 amino acids [11].

During the period of seminal vesicle maturation, the synthesis of SVS-IV protein increases from 0.8% to 20% of the total tissue protein but the mRNA specific for this protein increases by less than 4-fold. This lack of correspondence between the synthesis of SVS-IV and its mRNA, clearly indicates that factor(s) other than the quantity of SVS-IV mRNA regulate the synthesis of SVS-IV protein [12]. However, mRNA for SVS-II was undetectable in the seminal vesicle of three week-old rats but increased subsequently with increase in the circulating levels of androgens [13].

The synthesis of SVS-IV and SVS-V is under hormonal control [8, 14, 15, 16]. This finding was further substantiated by Mansson et al. [17], and Williams et al. [18], who used cDNA of SVS-IV and SVS-V as a hybridization probe for the SVS-mRNA and observed that, in castrated rats, the SVS-mRNA was significantly reduced when compared to normal rats. The levels of mRNA for the SVS-IV rise from a few molecules to some 100,000 molecules/cell, 96 h after administration of testosterone to castrated rats, indicating that androgens regulate the expression of the gene for this protein [10].

SVS-IV mRNA could be detected in rat seminal vesicles, the target organ for testosterone; it was absent in the non-target tissues, namely liver, brain, skeletal muscle, ventral prostate, kidney, and reticulocytes of the rat [19]. Though under methylation of the SVS-IV gene was anticipated for seminal vesicles, the observation that it is undermethylated in ventral prostate and coagulating gland was unexpected since this protein is not secreted by the latter gland [19, 20, 21, 22]. It is therefore possible that the SVS-IV gene in the ventral prostate is transcribed but the transcript is not functional. If this is the case, then total RNA of the ventral prostate should hybridize with the SVS-IV gene probe. However, hybridization revealed the absence of SVS-IV mRNA.

Based on these results it could be concluded that the SVS-IV gene is demethylated not only in the seminal vesicles in which it is expressed but also in the ventral prostate and coagulating glands where the protein could not be detected [23, 24, 25].

The mRNAs of SVS-IV and SVS-V, like all eukaryotic mRNAs that code for secretory proteins, consist of a short 5'-untranslated region (UTR) ahead of the protein-coding region (PCR) which begins with the N-terminal hydrophobic signal sequence followed by the sequence for the secreted protein. At the 3'-end, there is a long UTR containing the polyadenylation signal. The mRNAs of SVS-IV and SVS-V show an overall homology of 57% indicating that the two genes may have arisen by duplication; this could also be the reason for a similar non-intron arrangement [26, 27]. Calculation of the evolutionary divergence for PCRs of the two mRNAs indicated that SVS-IV and SVS-V must be amongst the fastest evolving proteins known [11].

The SVS-IV gene has been isolated, characterized and its sequence established [20, 22, 26]. The gene coding for SVS-IV was present on a 3.3-3.5 kb Eco RI fragment [20, 22]. McDonald et al. [26] elucidated that the SVS-IV gene occupied only about 2 kb of the rat genome. The first or the leading exon (120 bp) comprises the 5'-UTR, the nucleotides coding for the 21 amino acids of the signal peptide [26] and four amino acids of the pre-SVS-IV protein [20]. The second exon contains the remainder of the coding sequence of the SVS-IV protein plus the first part of 3'-noncoding region. The third exon contains most of the 3'-noncoding region and begins after the end of the second intron [20, 26]. Harris et al. [20] observed that approx. 30% of the second intron was made up of a series of 20 bp direct tandem repeats.

Attempts have also been made to detect the SVS-IV gene in the female. In each case, two distinct bands were observed but, in the males, the upper band (5.2 kb) was more intense than the lower band (4.5 kb) whereas, in the female, the lower band was more intense than the upper. This difference in intensity of the two bands may reflect a difference in the number of copies of these DNA sites in the two sexes which in turn could result from a difference in gene organization [19].

The SVS-V gene of 1.7 kbp consists of two introns and three exons [18, 28]. The 5'intron is 420 bp and the 3'intron 600 bp long [18]. The 5'exon is 99 bp long and corresponds to the 5'-UTR of the SVS-V mRNA; it codes for the signal peptide sequence and the first four amino acids of the protein. The central exon is 335 bp in length and codes for the rest of the protein (98 amino acids); it also contains 41 bp of the 3'-UTR. The remaining part of the 3'-UTR is coded by the 3'exon (186 bp) [28].

Even though both the proteins SVS-IV and SVS-V are highly basic, rich in serine, specific to the same tissue, under the control of the same hormones and participate in the same function (formation of the copulatory plug) a comparison of the cDNA sequences did not show any homology [29].

With a view to identify androgen-regulatory sequence elements, the 5'flanking regions of the SVS genes were compared with the corresponding regions in the rat prostate genes C1, C2 and C3 [30], and two mouse renin genes (Ren 1 and Ren 2) [31, 32]. Kandala et al. [33] carried out a detailed comparison of the upstream sequences of the SVS-IV and the C3 genes. At -300 to -330 on the SVS-IV gene and at -190 to -220 on the C3 gene, a 30 nucleotide-long homologous sequence was present. The 30 bp homology observed in the two genes is unlikely to be circumstantial; further, it has some similarity to the site reported for the estrogen receptor on the vitellogenin gene [24]. It, however, does not show any sequence homology to the hitherto identified steroid-binding sites of genes [34, 35]. To predict, whether this sequence, or the homologies repeated in the upstream region of the SVS-IV gene and other androgen responsive genes, are the main androgen-regulatory sequence elements, is difficult [28, 33]. The SVS-IV gene is probably widely spread - at least in rodents - as Kandala et al. (1983) noticed that a 3.5 Eco RI fragment from the DNA of guinea pig kidney and mouse liver also hybridized with the SVS-IV cDNA probe. Using cDNA to SVS-V mRNA as a hybridization probe, the SVS-V gene was identified in the DNA of rat liver and seminal vesicles [18]. However, when the cellular DNA was digested with Hind III, an additional fragment which hybridized strongly with the cDNA probe was identified; this fragment may be an allele of the SVS-V gene. As yet, this allelic gene of SVS-V has not been isolated.

It is generally known that one or more proteins of the seminal vesicles of rodents are involved in the formation of the seminal plug or with coagulation immediately after ejaculation [27, 36]. The clotting reaction involves crosslinking of proteins through the formation of e-(g-glutamyl)-lysine crosslinks catalyzed by transglutaminase present in the coagulating gland [37]. It is therefore possible that in rat, the SVS proteins discussed above participate in the formation of the seminal coagulum.

11.2.3 Proteins Secreted by the Prostate

Heyns et al. [38] and Heyns and DeMoor [39] isolated a steroid-binding protein from the prostatic cytosol of rat, the prostatic binding protein (PBP). PBP is specific to prostate and constitutes 50% of the prostatic cytosol proteins [40]. PBP is probably secreted by the prostate since the prostatic fluid was found to bind as much steroid as the prostatic cytosol. When PBP was reduced with b-mercaptoethanol and then electrophoresed on SDS-PAGE, three bands, C1, C2, and C3, with MW 7,500, 11,500, and 13,000, respectively appeared; only C3 was a glycoprotein [41].

Using Poly (A)$^+$ RNA from rat ventral prostate, Peeters et al. [42] showed that, on castration, the PBP-mRNA decreased from 54% to 16% of the total mRNA 6 days after castration, and that the level increased to 42% on androgen treatment for 3 days. This confirms that the synthesis and/or activity of mRNA specific to PBP is under androgen homology. Analysis of the specific mRNAs for the three polypeptides (20, 10 and 9 kD) indicated that they contained 930 and 550 nucleotides, respectively [43]. The C1 mRNA contained a 5'non-coding region of 36 nucleotides, a coding region of 383 nucleotides of which 264 coded for the 88-amino acid secretory polypeptide, and a 3'non-coding region of 60 nucleotides including the termination codon UGA. The C2 mRNA, likewise, contained 34 nucleotides in the 5'non-coding region of 294 nucleotides of which 60 coded for the signal peptide and 234 for the secreted polypeptide, and a non-coding region of 105 nucleotides with UAA as the stop codon. Both the mRNAs showed extensive homology with each other [44].

Proteins with similar steroid-binding properties were isolated from rat prostate and designated prostatein [45] and prostate α-protein [46, 47]. The proteins are apparently identical to PBP. The protein is a tetramer consisting of two subunits, S and F [39, 48], which are made up of polypeptides C1 (or I) and C3 (or III), and C2 (or II) and C3 (or III), respectively [48]. A compact protein consisting of a core unit of I and II which is linked to III by disulfide linkage [46], has been suggested for the tetrameric native protein, PBP. The amino acid sequence of the components, C1, C2 and C3, of PBP has been established [49, 50, 51].

The genes for the polypeptides C1 and C2 were identified by screening a rat-liver genomic DNA library using cDNA of C1 and C2 as the probe [52]. Both genes are approx. 3 kb in size and contain three exons separated by two introns about 1.8 kb and 0.9 kb long. The CI genomic clone DNA is identical to the gene in the rat, and the gene does not exhibit polymorphism [52].

C3 genomic clones were used to study the structure and organization of the C3 gene [53]. The C3 gene was also found to be 3.2 kb long, and to consist of three exons, 0.13, 0.21, and 0.21 kb long, separated by two introns 1.72 and 0.77 kb long, respectively. A comparison of the DNA sequences of C1, C2, and C3 showed extensive homology. Very likely, only the latter gene is polymorphic [53]. Of the two genes for C3, the gene coding for the protein was designated as C3 (1) and the other C3 (2) [30]. Both these genes showed identical exon/intron arrangements [30, 53]. The extensive homology observed between the C1, C2, C3 (1) and C3 (2) genes, suggests that they arose by duplication of an ancestral gene. The organization and expression of the C3 gene was studied by ligating the two non-allelic genes, C3 (1) and C3 (2) to the SV2-gpt vector, and then transfecting androgen-responsive S115 cells from mouse mammary glands with the fused gene [54, 55]. In the transfected cells both C3 (1) and C3 (2) were transcribed to the same extent and the transcription was androgen dependent [54, 55]. It appears that under *in vivo* conditions, C3 (2) is hypermethylated [55]. This study also indicated that the C3 (2) gene is not a pseudogene and that the difference between the C3 (1) and C3 (2) genes is probably due to the signals that regulate DNA methylation [55] which in turn regulate gene expression. Attempts were made to identify the androgen receptor-binding sites on the C3 gene [56, 57]. Two fragments with high affinity for the androgen-receptor complex have been identified. One fragment of 300 bp extended from position -220 to 80 of the gene. The other fragment of 500 bp consisted entirely of the first intron [57].

Apart from PBP, another protein called the 20 kD protein has been isolated from the ventral prostate cytosol of rat and characterized [58]. Though, Chamberlin et al. [58] were the first to purify the 20 kD protein, it was already reported by Parker et al. [59] who termed it α-protein. It is a major protein in rat ventral prostate, constituting 5-8% of the total tissue protein [59]. Unlike PBP, it binds both single and double-stranded DNAs from rat liver, calf-thymus, chicken erythrocytes, as well as polyI-polyC and poly(dA)-poly(dT). The expression of the gene for the 20 kD protein is regulated by androgens [43, 58, 59].

Spermine-binding protein is a glycoprotein containing fucose, galactose, mannose and N-acetylglucosamine [60]. The MW of the protein as estimated by SDS-PAGE and gel filtration is 30,000. Castration reduced the level of mRNA by 80% within two days. However, injection of dihydrotestosterone two days after castration restored the level of the mRNA [43, 61]. As yet, the physiological function of this protein is not known. By antibody screening and hybrid-selected translation, two positive clones were identified from a cDNA expression library which hybridized to an mRNA of 1260 bp and synthesized a 34 kD polypeptide identical to the size of spermine-binding protein [60]. The sequence had an open reading frame extending from -81 to 921 which coded for the spermine-binding protein with its signal sequence. The mature protein was 307 amino acids long and had 27 additional N-terminal amino acids representing the signal sequence. Acidic amino acid residues such as aspartic and glutamic acids were preferentially found at the C-terminal region. The protein has two N-glycosylation sites, Asn-Trp-Ser and Asn-Glu-Ser. The "acidic core" exists as a continuous α-helical structure.

Fang and Liao [62] isolated a major glycoprotein accounting for 35% of the total cytosol protein and observed that the protein was capable of binding androgens and other steroids. This protein was named prostate α-protein and it was purified to homogeneity [46]. In the presence of SDS and ß-mercaptoethanol, it resolved into three distinct polypeptide components with MW's of 10,000 (I), 14,000 (II) and 15,000 (III) [46]. Liao et al. [47] sequenced component I of prostate α-protein and verified that the sequence of component I was identical with the sequence of C1, a component of PBP.

Prostate α-protein which consists of socalled subunits A and B is therefore identical to PBP [38, 39]. Though the cDNA-clone PM-40 specific for the dorsal protein doublet has a single insert complementary to the dorsal prostate mRNA, it blocked the synthesis of both the proteins in the cell-free translation studies, implying that both the proteins are coded by a single gene or by two genes having extensive sequence homology. mRNA specific for the doublet protein could be detected even in the ventral prostate using ^{32}P-labelled PM-40 DNA as a probe. The levels of mRNA in the ventral prostate remained unchanged following castration, whereas in the dorsal prostate the levels decreased significantly and increased 16-fold on androgen administration following castration [63].

11.3   The Proteins of the Guinea Pig

Four seminal vesicle secretory proteins have been purified from guinea pig seminal vesicles. These proteins were designated as SVP 1, 2, 3 and 4. Proteins 1, 2, and 4 were homogenous on SDS-PAGE whereas protein 3 contained a band identical to SVP-4 and another fast moving low MW protein [64]. The apparent MW's of these proteins was 25,000 for SVP-I, 14,000 for SVP-2, and 23,000 for SVP-3 and SVP-4 [67]. Moore et al. [65] identified cDNA clones from a seminal vesicle cDNA library of guinea pig, coding for the high MW precursor of SVP-1. The 609 bp coding sequence for SVP-1 contained eight and a half tandem repeats of 72 nucleotides. All the repeats exhibited a high degree of homology. As anticipated, the amino acid sequence of SVP-1 derived from the above sequence showed eight repeats of 24 amino acids in each domain, with the consensus sequence   VTGQDSVKGRLQMKGQDSLAERFS.   In   each   repeat,   the   sequence Gly-Gln-Asp-Ser appears twice [65]. Epithelial cells of seminal vesicles of guinea pig are androgen-dependent with respect to their morphology as well as the synthesis of secretory proteins SVP-1 to 4 [66, 67, 68]. Moore et al. [69] studied the transcriptional activity of the SVP-1, SVP-3, and SVP-4 genes following castration and androgen therapy. Total RNA was isolated from the nuclei of seminal vesicle epithelial cells following *in vitro* translation, and then hybridized with the cDNA probes, M13-76 (representing genes for SVP-3 and SVP-4) and M13 56B (representing the gene for SVP-1). It appears that there is no decrease in the mRNA levels of the SVP proteins even after castration [69].

11.4   The Proteins of the Mouse

Very few studies have been devoted to the characterization of the seminal vesicle secretory proteins of mice [70, 71]. The proteins, on electrophoresis, resolve into more than 10 bands, three of which (bands IV to VI) seem to be major and specific for seminal vesicles [71]. A recent study indicated that the proteins of the mouse seminal vesicle fluid, when analyzed by SDS-PAGE, resolved into four distinct bands M1 (80,000), M2 (37,000), M3

(17,000), and M4 (16,000) [7]. Proteins M3 and M4 were the most abundant. Using cDNA probes for rat SVS-IV and SVS-V, it was concluded that M1 is homologous to SVS-1, M2 to SVS-II, M3 to SVS-IV, and M4 to SVS-5 [7].

11.5 Proteins of the Bull

The major protein with an apparent MW of M_r 15 kD was isolated from seminal plasma as well as from the seminal vesicle secretion of the bull and it was proved by amino acid analysis and tryptic peptide mapping that the two proteins were identical [72]. The protein chemical analysis indicated that the major protein was identical to the protein PDC109 purified from bull seminal plasma and which was sequenced by Esch et al. [73]. Immunochemical studies by Aumüller et al. [74] revealed the specific binding of the major protein to the fossa and the mid-piece region of the tails of bull spermatozoa. Binding of gold-labeled major protein to subplasmalemmal sites at the mid-piece region of spermatozoa was demonstrated by electron microscopy and two membrane proteins which selectively bind the major protein were isolated from bull epididymal spermatozoa [74]. Cell-free translation of poly(A)$^+$ RNA from seminal vesicle tissue resulted in formation of a polypeptide with an apparent MW of M_r 18 kD, immunoreactive with anti-major protein antiserum [72]. Following a suggestion of Mbikay et al. [75] this bull seminal protein be referred to as Seminal Vesicle Secretory Protein of 109 amino acids (SVSP109). To screen poly(A)$^+$ RNA from seminal vesicle tissue for the presence of major protein-specific mRNA, a synthetic DNA probe was designed on the basis of the known amino acid sequence of the protein [73]. The amino acid sequence extending from residue 78 to 94 was chosen and a 51 nucleotide long DNA, complementary to the corresponding mRNA, was synthesized. The selection of codons was based on the codon usage as found in the cDNA sequence for seminal ribonuclease of bovine seminal vesicle tissue [76]. Northern analysis of poly(A)$^+$ RNA from seminal vesicle was carried out with the synthetic probe using standard hybridization conditions at $42°C$ in the presence of formamide. The result indicated the presence of a SVSP109-specific mRNA species of 750 bp. Recombinant clones (10^3) of a cDNA library of bull seminal vesicle poly(A)$^+$ RNA were screened by colony hybridization using the radioactively labeled synthetic probe, yielding 18 positive clones. A positive clone containing the longest cDNA insert (pMP17) was sequenced [77].

The obtained sequence of the cDNA insert of pMP17 contains an open reading frame extending over 405 nucleotides and starting with a leucine residue. The derived polypeptide sequence includes in the C-terminal part the 109 amino acid residue long sequence of the protein PDC109 as published by Esch et al. [73]. The N-terminus of this polypeptide sequence is extended by 23 amino acid residues of a signal sequence which was regarded as incomplete because of a missing methionine start codon. Hence, the PstI-insert of the pMP17 DNA did not represent the complete mRNA of the precursor for SVSP109. Sequence analysis by direct mRNA sequencing employing reversed transcriptase and a sequencing primer complementary to the region 99-118 of the cDNA-insert extended the known sequence of pMP17 by 20 bp. The extension comprised the missing start and a short untranslated 5'-region of the mRNA. The putative initiation methionine occurs within a sequence, CTACCATGG, highly homologous to the consensus sequence CCA(G)CCATGG which controls translational efficiency of mammalian mRNAs [78, 79, 80]. The cDNA-insert of pMP17 possesses a poly(A) tail and the putative polyadenylation signal AATAAA, 13 nucleotides upstream of the poly(A) tail [81]. The M_r of the 134 amino acid residue precursor polypeptide of SVSP109 is 15,480. The precursor sequence of 25 amino acid residues has a hydrophobic character and very likely constitutes a signal peptide, directing the protein towards the secretory pathway [82]. The deduced amino acid sequence contained no consensus sequence indicative of N-glycosylation (N-X-S or N-X-T).

Northern analysis of seminal vesicle poly(A)$^+$RNA with the PstI-insert of clone pMP17 furnished a mRNA species of 750 bp.

Genomic DNA from an individual animal was analyzed by Southern blot after digestion with several restriction enzymes. The enzymes EcoRI, KpnI, PstI and SmaI yielded single radiolabeled bands respectively. In HindIII- and BamHI-digests, three and two fragments respectively were detected which reacted with the PstI-insert of pMP17, indicating that these enzymes very likely cut within the gene of the major protein. From these results it is tentatively inferred that there is one gene for SVSP109 per haploid bovine genome.

A high RNAase activity was first detected in the seminal plasma of bull by D'Allessio and Leone in 1963 [83]; the enzyme hydrolyzed both native RNAs and synthetic polyribonucleotides [84] and was designated RNAase BS-1. RNAase BS-1 is a basic protein with a pI of 10.3; it lacks tryptophan and cysteine and contains galactosamine. The amino acid composition of RNAase BS-1 and pancreatic RNAase A are very much similar.

Native homogeneous RNAase BS-1, on treatment with denaturing agents such as urea and guanidine hydrochloride, does not show any change in electrophoretic mobility or its S value. Addition of thiol-reagents in the presence of urea or guanidine-HCl completely dissociates the protein into two monomers [85, 86, 87]. When the dimeric protein was fractionated on a CM-cellulose column it resolved into three components. This heterogeneity was due to the two subunits α and ß which existed in the subforms as αα, ßß and αß. The subunits varied in their amide content and hence by deamidation it was possible to convert the most cationic ßß into αß which in turn could be isolated directly from the seminal plasma. It would appear that deamidation occurs *in vivo,* before secretion into seminal plasma [88].

Poly (A)$^+$RNA from bull seminal vesicles when translated in a reticulocyte lysate system synthesized a number of proteins. One of these proteins of MW 18,000 could be immunoprecipitated using antibodies to RNAase BS-1. This protein is a precursor of RNAase BS-1 subunit since it could be processed to a polypeptide of MW 14,500 in the presence of a microsomal fraction from dog pancreas. Further, the polypeptide of 14,500 could also be immunoprecipitated using specific antibodies, and forms dimers under *in vitro* conditions. This indicates that RNAase BS-1 is synthesized as a precursor in bull seminal vesicles [89]. Isolated lobules of the seminal vesicles of bull, when cultured, synthesized and secreted a number of proteins [90]. RNAase BS-1 could be purified both from the lobules and the medium in which the lobules were cultured, by acid extraction followed by ammonium sulphate precipitation, RNAase affinity chromatography and cation exchange chromatography [90]. It was concluded that the lobules of the seminal vesicles synthesize the enzyme and secrete it mainly as the amidated form (ßß) which, on secretion, becomes deamidated to generate the isoenzymes αß and αα [90, 91]. RNAase BS-1 is made up of two identical subunits held together by disulfide bridges and noncovalent forces [87]. Each subunit of Mw 14,500 consists of 124 amino acids with lysine as the N terminal residue and valine as the C terminal residue. The subunit shows a great degree of homology with pancreatic RNAase A [92]. A comparison of the sequence of RNAase BS-1 with that of RNAase A indicated substitutions in 23 positions [92]. The position of the intrachain disulfide bridges in RNAase BS-1 involving the eight homologous half cystines was identical to that observed in RNAase A. The remaining two half cystines at positions 31 and 32 were shown to be involved in interchain disulfide bridges between the two subunits of RNAase BS-1. A comparison of angiogenin, a protein which stimulates the formation of blood vessels in humans with RNAase BS-1 indicated that 38 residues occupied identical positions in both the proteins [92]. Among these identical residues were His^{12}, His^{119} and Lys^{41} known to be essential for ribonuclease activity. However, angiogenin does not exhibit any RNAase A-like activity though it has an endonucleolytic activity towards 28$\underline{S}$ and 18$\underline{S}$ RNA [93].

Poly(A)$^+$RNA purified from bull seminal vesicles was transcribed into cDNA, inserted into the PstI site of pAT153 and cloned in *E. coli* C-600. Using a ^{32}P-labelled cDNA coding for rat pancreatic RNAase A as a hybridization probe, ten clones were identified and one of them (p17G3) sequenced [76, 94]. It was observed that the cloned cDNA started at the 47th amino acid residue of the known amino acid sequence and terminated twelve nucleotides beyond the consensus sequence AATAAA, in the 3'- noncoding region of the mRNA. Northern blot analysis indicated that mRNA of RNAase BS-1 was 950 nucleotides long; this value was similar to that observed for the mRNA of rat pancreatic RNAase A [95].

11.6 Proteins of the Human

Human seminal plasma contains many proteins secreted by the prostate [96] but the role of most of these proteins still remains unknown. Recently, a protein with inhibin-like activity was isolated from human seminal plasma [97]; it was found in greater amounts in the prostate than in the testis [98, 99]. The amino acid sequence indicated that it consists of 94 amino acids with 10 of them being cysteine [100, 101, 102].

A cDNA library constructed from human prostatic poly(A)$^+$RNA, was screened using a 63 bp synthetic probe representing amino acids 67 to 87 of the human seminal protein. The cDNA consisted of 483 nucleotides with a poly A tail of 17 bases [75]. The open reading frame was 342 nucleotides long with the initiation codon ATG at position 12 and TAA as the stop codon at position 353. At the 5'end it was flanked by a 11-bp stretch, and at the 3'end by a 130-bp noncoding region. The polyadenylation signal AATAAA common to most eukaryotic mRNA was not found within 30 nucleotides upstream of the polyA tail, but was present at position 444-448. The mRNA of PSP94, as estimated by Northern blot, was 615 bases long. The cDNA-derived amino acid sequence comprised the amino acid sequence of PSP94 preceded by the signal sequence rich in hydrophobic residues. The cDNA-derived amino acid sequence of PSP94 is highly homologous to that of the 94 amino acid-long ß-inhibin [101, 102]. The 93 amino acid long ß-microseminoprotein [100] differed from the former in two positions and from the latter in four positions. Northern blot analysis of the mRNA of PSP94 indicated that it is abundantly present in the human prostate and absent in the testis and placenta. It was also detectable in the testis of rat [75]. The histological distribution of PSP94 was examined by *in situ* hybridization employing PSP94 cDNA [103]. The results demonstrate exclusive localization of PSP94-specific mRNA in the epithelial cells of the human prostate gland.

Mbikay et al. [104] searched for the presence of PSP94 in cyanomolgus monkey (*Macaca fascicularis*) by the use of a PSP94-specific radioimmunoassay as well as Northern hybridization with a PSP94-cDNA probe. The radioimmunoassay was unable to detect any immuno-reactive PSP94 in a variety of tissue extracts as well as secretions of cyanomolgus monkey. However, Nothern analysis demonstrated the presence of a 580 bp mRNA-species which reacted under stringent conditions with the human PSP94-cDNA probe. The presence of a PSP94-homologous gene in monkey was established by Southern analysis of genomic DNA. Expression of this gene appears to be restricted to the prostate tissue. The immunological results, however, indicate that the homology between gene products of the human and the monkey PSP94 gene may be limited. The epitopes recognized by a polyclonal antiserum against human PSP94 are obviously not present in the putative PSP94 analog in monkey.

Prostate specific antigen (PSA) was purified and recognized as a secretory product of prostate first by Wang et al. [105]; it was detected only in the epithelial cells of the prostate ductal element (by immunocytochemstry) and not in other tissues or organs. PSA is a 33 kD glycoprotein, containing approximately 7% carbohydrate; it is secreted into the prostatic fluid in concentrations that may be as high as 3.6 mg/ml. PSA has been demonstrated to be clinically important as a marker for monitoring prostatic cancer. Watt et al. [106] determined the amino acid sequence of PSA from human seminal plasma. The single polypeptide chain contains 240 amino acid residues and possesses a MW of 26,496. A N-linked carbohydrate side chain is predicted at Asn-45, and O-linked carbohydrate residues at Ser-69, Thr-70 and Ser-71. A human prostate-specific cDNA library in the expression vector lgt11 was screened with an antibody against PSA [107], and a positive clone lHPS-1 was sequenced. The deduced amino acid sequence comprised 257 amino acid residues with an incomplete signal peptide and a short propiece at the N-terminus of the mature polypeptide chain; this sequence was in complete agreement with the amino acid sequence of g-SM [108]. Hints for the biological function of PSA came from experiments which demonstrated kallikrein-like serine protease-activities of PSA [109]. PSA cleaves the structural protein of human coagulum, which is the predominant protein in seminal vesicle secretion. The natural substrate for PSA appeared to be the predominant seminal vesicle protein. Lilja et al. [110] showed that PSA cleaves two of the major components of ejaculated fibronectin and seminogenin, thus contributing to the degradation of the seminal gel.

Riegman et al. [111] constructed a cDNA library in lgt10 from poly(A)$^+$RNA obtained from a transplantable human prostate tumor PC82. Screening this library with a synthetic probe specific for PSA furnished three new cDNA clones PA75, PA525 and PA424 which were sequenced. PA75 contains an open reading frame of 771 bp encoding 13 amino acid residues of a signal peptide followed by the complete amino acid sequence for PSA. From the earlier reported cDNA clone for PSA [107] PA75 differs at two positions in the 3'-untranslated region. It is interesting to note that clones lHPSA-1 and PA75 possess identical 5'-sequences and obviously do not represent the complete signal sequence of PSA. Clone PA525 contained a fragment of 442 bp which is not present in PA 75. This fragment is highly homologous to a part of the human glandular kallikrein gene, the last 446 bp of the hGK-1 intron. Translation of PA525 mRNA will lead to a mature protein of 214 amino acids, the terminal 28 residues of which will be different from PSA. The clone PA424 comprises a complete 3'-untranslated region including a polyadenylation signal and a poly A tail. PA424 contains a stretch of 145 bp, absent from the sequence of PA75, with the consensus GT.....AG motif at the boundaries. Thus the internal fragment of 145 bp seems to be a retained intron. The human kallikrein gene hGK-1, which shows a homology to PSA of approximately 80%, has at the same position a 113 bp intron. The putative major protein of PA424 will have 156 amino acids of which the first 140 will be identical to PSA.

Acid phosphatases are a group of iso-enzymes that hydrolyze phosphate monoesters under acidic conditions with liberation of inorganic phosphate. The acid phosphatase secreted by the human prostate (human acid phosphatase, PAP) is an urological tumour marker with primary importance in staging and monitoring patients with prostatic adenocarcinoma [112]. Epithelial cells of the prostate synthesize PAP and secrete it into seminal fluid to approximately 1 mg/ml [113]. PAP is a glycoprotein composed of two identical subunits of identical size of 48 kDA [114, 115]. Vihko et al. [116] have screened a human prostate cDNA library, constructed in lgt11, by means of PAP-specific antibodies. A number of positive overlapping cDNA-clones were sequenced. The size of the PAP cDNA was 3088 bp and covered the entire coding region for PAP-mRNA. The 1065 bp coding region comprised the sequence for a polypeptide of 354 amino acids; the molecular mass of the non-glycosylated protein is 41 kDa. The 5'-end of the cDNA codes for a putative signal peptide of 32 amino acid residues. The starting codon ATG was not part of the consensus sequence CCA(G)CCATGG [78, 79, 80]. The mature form of PAP contains three putative glycosylation signals. The 3'-untranslated region of PAP-cDNA with 1905 bp contained a polyadenylation signal AATAAA 44 bp upstream of the poly A tail. A PAP-cDNA clone was characterized which possessed a 96 bp deletion. This could indicate polymorphism of the PAP-gene locus.

Northern analysis of poly (A)$^+$RNA derived from normal, carcinoma and benign hyperplastic prostatic tissue revealed a discrete mRNA species of 3.3 kb; no hybridization of a PAP-cDNA probe to human placental poly (A)$^+$RNA could be detected.

11.7 References

[1] Cameo, M. S., and Blaquier, J. A., *J. Endocrinol. 1976, 69,* 47-55.

[2] Brooks, D. E., *Biochem. Int. 1987, 14,* 235-240.

[3] Brooks, D. E., *Mol. Cellul. Endocr. 1987, 53,* 59-66.

[4] Brooks, D. E., Means, A. R., Wright, E. J., Singh, S. P., and Tiver, K. K., *J. Biol. Chem. 1986, 261,* 4956-4961.

[5] Brooks, D. E., Means, A. R., Wright, E. J., Singh, J. P., and Tiver, K. K. *Eur. J. Biochem. 1986, 161,* 13-18.

[6] Svendsen, I., Martin, B. M., Viswanatha, T., and Johansen, J. T., *Carsberg Res. Commun. 1982, 47,* 15-27.

[7] Fawell, S. E., McDonald, C. J., and Higgins, S. J., *Mol. Cellul. Endocrinol. 1987, 50,* 107-114.

[8] Ostrowski, M. C., Kistler, M. K., and Kistler, W. S., *J. Biol. Chem. 1979, 254,* 383-390.

[9] Wagner, C. L., and Kistler, W. S., *Biol. Reprod. 1987, 36,* 501-510.

[10] Mansson, P. E., Sugino, A., and Harris, S. E., *Nucl. Acids Res. 1981, 9,* 935-946.

[11] McDonald, C., Elipoulos, E., and Higgins, S. J., *EMBO J. 1984, 3,* 2517-2521.

[12] Kistler, M. K., Ostrowski, M. C., and Kistler, W. S., *Proc. Natl. Acad. Sci. USA 1981, 78,* 737-741.

[13] Dodd, G. J., Kreis, C., Sheppard, P. C., Hamel, A., and Matusik, R. J., *Mol. Cellul. Endocrinol. 1986, 47,* 191-200.

[14] Higgins, S. J., and Burchell, J. M., *Biochem. J. 1978, 174,* 543-551.

[15] Ostrowski, M. C., Kistler, M. K., and Kistler, W. S., *Proc. Int. Congr. Biochem. 1979, 11A,* 207.

[16] Ostrowski, M. C., Kistler, M. K., and Kistler, W. S., *Biochemistry 1982, 21,* 3525-3529.

[17] Mansson, P. E., Carter, D., and Harris, S., *J. Androl. 1982, 3,* 16.

[18] Williams, L., McDonald, C., Jackson, S., McIntosh, E., and Higgins, S., *Nucl. Acids Res. 1983, 11,* 5021-5036.

[19] Abrescia, P., Guardiola, J., Felsani, A., and Metafora, S., *Nucl. Acids. Res. 1982, 10,* 1159-1174.

[20] Harris, S. E., Mansson, P. E., Tully, D. B., and Burkhart, B., *Proc. Natl. Acad. Sci. USA 1983, 80,* 6460-6464.

[21] Higgins, S. J., Burchell, J. M., and Mainwaring, W. I. P., *Biochem. J. 1976, 158,* 271-282.

[22] Kandala, J. C., Kistler, M. K., Lawther, R. P., and Kistler, W. S., *Nucl.Acids Res. 1983, 11,* 3169-3186.

[23] Geiser, M., Mattaj, I. W., Wilks, A. F., Seldran, M., and Jost, J. P., *J. Biol. Chem. 1983, 258,* 9024-9030.

[24] Jost, J. P., Seldran, M., and Geiser, M., *Proc. Natl. Acad. Sci. USA 1984, 81,* 429-433.

[25] Wilks, A. F., Cozens, P. J., Mattaj, I. W., and Jost, J. P., *Proc. Natl. Acad. Sci. USA* **1982**, *79*, 4252-4255.

[26] McDonald, C., Williams, L., McTurk, P., Fuller, F., McIntosh, E., and Higgins, S., *Nucl. Acids. Res.* **1983**, *11*, 917-930.

[27] Williams-Ashman, H. G., *Curr. Top. Cellul. Reg.* **1983**, *22*, 201-275.

[28] Williams, L., McDonald, C., and Higgins, S., *Nucl. Acids Res.* **1985**, *13*, 659-672.

[29] Kistler, M. K., Taylor, R. E., Kandala, J. C., and Kistler, W. S., *Biochem. Biophys. Res. Commun.* **1981**, *99*, 1161-1166.

[30] Hurst, H., and Parker, M. G., *EMBO J.* **1983**, *2*, 769-774.

[31] Field, L. J., Phibrick, W. M., Howles, P. N., Dickinson, D. P., McGowan, R. A., and Gross, K. W., *Mol. Cellul. Biol.* **1984**, *4*, 2321-2331.

[32] Panthier, J. J., Dreyfus, M., Tronik-Le Roux, D., and Rougeon, F., *Proc. Natl. Acad. Sci. USA* **1984**, *81*, 5489-5493.

[33] Kandala, J. C., Kistler, W. S., and Kistler, M. K., *J. Biol. Chem.* **1985**, *260*, 15959-15964.

[34] Dean, D. C., Knoll, M. J., Riger, M. E., and O'Malley, B. W., *Nature* **1983**, *305*, 551-554.

[35] Mulvihill, E. R., Le Pennec, J. P., and Chambon, P., *Cell* **1982**, *28*, 621-632.

[36] Williams-Ashman, H. G., *Mol. Cellul. Biochem.* **1984**, *58*, 51-61.

[37] Williams-Ashman, H. G., Notides, A. C., Pabalan, S. S., and Lorand, L., *Proc. Natl. Acad. Sci. USA* **1972**, *69*, 2322-2325.

[38] Heyns, W., Verhoeven G., and DeMoor, P., *J. Steroid Biochem.* **1976**, *7*, 987-991.

[39] Heyns, W., and DeMoor, P., *Eur. J. Biochem.* **1977**, *78*, 221-230.

[40] Heyns, W., *FEBS Lett.* **1977**, *81*, 43-47.

[41] Mous, J., Peeters, B., and Rombauts, W., *Biochem. Biophys. Res.Commun.* **1977**, *79*, 1111-1116.

[42] Peeters, B. L., Mous, J. M., Rombauts, W. A., and Heyns, W., *J. Biol. Chem.* **1980**, *255*, 7017-7023.

[43] Parker, M. G., White, R., and Williams, J. G., *J. Biol. Chem.* **1980**, *255*, 6996-7001.

[44] Parker, M., Needham, M., and White, R., *Nature* **1982**, *298*, 92-94.

[45] Lea, O. A., Petrusz, P., and French, F. S., *J. Biol. Chem.* **1979**, *254*, 6196-6202.

[46] Chen, C., Schilling, K., Hiipakka, R. A., Huang, I. Y., and Liao, S., *J. Biol. Chem.* **1982**, *257*, 116-121.

[47] Liao, S., Chen, C., and Huang, I. Y., *J. Biol. Chem.* **1982**, *257*, 122-125.

[48] Heyns, W., Peeters, B., Mous, J., Rombauts, W., and DeMoor, P., *Eur. J. Biochem.* **1978**, *89*, 181-186.

[49] Peeters, B., Rombauts, W., Mous, J., and Heyns, W., *Eur. J. Biochem.* **1981**, *115*, 115-121.

[50] Peeters, B., Heyns, W., Mous, J., and Rombauts, W., *Eur. J. Biochem.* **1982**, *132*, 669-679.

[51] Peeters, B., Heyns, W., Mous, J., and Rombauts, W., *Eur. J. Biochem.* **1982**, *123*, 55-62.

[52] Parker, M., Needham, M., White, R., Hurst, H., and Page, M., *Nucl. Acids. Res.* **1982**, *10*, 5121-5132.

[53] Parker, M. G., White, R., Hurst, H., Needham, M., and Tilly, R., *J. Biol. Chem.* **1983**, *258*, 12-15.

[54] Page, M. J., and Parker, M. G., *Cell* **1983**, *32*, 495-502.

[55] Parker, M., Hurst, H., and Page, M. J., *Steroid Biochem.* **1984**, *20*, 67-71.

[56] Mulder, E., Vrij, A. A., Brinkmann, A. O., van der Molen, H. J., and Parker, M. G., *Biochem. Biophys. Acta* **1984**, *781*, 121-129.

[57] Rushmere, N. K., Parker, M. G., and Davies, P., *Mol. Cellul. Endocrinol.* **1987**, *51*, 259-265.

[58] Chamberlin, L. L., Mpanias, U. D., and Wang, T. Y., *Biochemistry* **1983**, *22*, 3072-3077.

[59] Parker, M. G., Scrace, G. T., and Mainwaring, W. I. P., *Biochem. J.* **1978**, *170*, 115-121.

[60] Chang, C., Saltzman, A. G., Hiipakka, R. A., Huang, I. Y., and Liao, S., *J. Biol. Chem.* **1987**, *262*, 2826-2831.

[61] Lesser, B., and Burchovsky, N., *Biochem. Biophys. Acta* **1979**, *308*, 426-437.

[62] Fang, S., and Liao, S., *J. Biol. Chem.* **1971**, *246*, 16-24.

[63] Dodd, G. J., Sheppard, P. C., and Matusik, R. J., *J. Biol. Chem.* **1983**, *258*, 10731-10737.

[64] Venziale, C. M., and Deering, L. C., *Andrologia* **1976**, *8*, 73-82.

[65] Moore, J. T., Hagstrom, J., McCormick, D. J., Harvey, S., Madden, B., Holicky, E., Stanford, D. R., and Wieben, E. D., *Proc. Natl. Acad. Sci. USA* **1987**, *84*, 6712-6714.

[66] Burns, J. M., Winberger, M. J., and Venziale, C. M., *J. Biol. Chem.* **1979**, *254*, 2258-2264.

[67] Moore, J. T., Norvitch, M. E., Wieben, E. D., and Venziale, C. M., *J. Biol. Chem.* **1984**, *259*, 14750-14756.

[68] Venziale, C. M., Burns, J. M., Lewis, J. C., and Buchi, K. A., *Biochem. J.* **1977**, *166*, 167-173.

[69] Moore, J. T., Venziale, C. M., and Wieben, E. D., *Mol. Cellul. Endocrinol.* **1986**, *46*, 205-214.

[70] Mintz, B., Domon, M., Hungerford, D. A., and Morrow, J., *Science* **1972**, *175*, 657-659.

[71] Platz, R. D., and Wolfe, H. G., *J. Hered.* **1969**, *60*, 187.

[72] Kemme, M., Madiraju, M. V. V. S., Krauhs, E., Zimmer, M., and Scheit, K. H., *Biochem. Biophys. Acta* **1986**, *884*, 282-290.

[73] Esch, F. S., Ling, N. C., Bohlen, P., Ying, S. Y., and Guillemin, R., *Biochem. Biophys. Res. Comm.* **1983**, *113*, 861-867.

[74] Aumüller, G., Vesper, M. Seitz, J., Kemme, M., and Scheit. K. H., *Cell Tissue Res.* **1988**, *252*, 377-384.

[75] Mbikay, M., Nolet, S., Fournier, S., Benjannet, S., Chapdelaine, P., Paradis, G., Dubé, J. Y., Tremblay, R., Lazure, C., Seidah, N. G., and Chretien, M., *DNA* **1987**, *6*, 23-29.

[76] Palmieri, M., Carsana, A., Furia, A., and Libonati, M., *Eur. J. Biochem.* **1985**, *152*, 275-277.

[77] Kemme, M., and Scheit, K. H., *DNA* **1988**, *7*, 595-599.

[78] Kozak, M., *Nature* **1984**, *308*, 241-246.

[79] Kozak, M., *Nuc. Acids Res.* **1984**, *12*, 857-872.

[80] Lütcke, H. A., Chow, K. C., Mickel, F. S, Moss, K. A., Kern, H. F., and Scheele, G. A., *EMBO J.* **1987**, *6*, 43-48.

[81] Proudfoot, N. J., and Brownlee, G. G., *Nature* **1976**, *263*, 211-214.

[82] Austen, B. M., *FEBS Lett.* **1979**, *103*, 308-313.

[83] D'Allessio, G., and Leone, E., *J. Biochem.* **1963**, *89*, 7.

[84] D'Allessio, G., Floridi, A., DePrisco, R., Pignero, A., and Leone, E., *Eur. J. Biochem.* **1972**, *26*, 153-161.

[85] D'Allessio, G., Parente, A., Guida, C., and Leone, E., *FEBS Lett.* **1972**, *27*, 285-288.

[86] D'Allessio, G., Parente, A., Farina, B., La Montagna, R., Deprisco, R., Demma, G. B., and Leone, E., *Biochem. Biophys. Res. Commun.* **1972**, *47*, 293-299.

[87] D'Allessio, G., Malorni, M. C., and Parente, A., *Biochemistry* **1975**, *14*, 1116-1122.

[88] Di Donato, A., and D'Allessio, G., *Biochemistry* **1981**, *20*, 7232-7237.

[89] Furia, A., Palmieri, M., and Libonati, M., *Biochem. Biophys. Acta* **1983**, *741*, 303-307.

[90] Quarto, N., Tajana, G. F., and D'Allessio, G., *J. Reprod. Fertil.* **1987**, *80*, 81-89.

[91] Tamburrini, M., Piccoli, R., DePrisco, R., di Donato, A., and D'Allessio, G., *Ital. J. Biochem.* **1986**, *35*, 22-32.

[92] Suzuki, H., Parente, A., Farina, B., Greco, L., La Montagna, R., and Leone, E., *Biol. Chem. Hoppe-Seyler* **1987**, *368*, 1305-1312.

[93] Shapiro, T., Riordan, J. F., and Vallee, B. L., *Biochemistry* **1986**, *25*, 3527-3532.

[94] Furia, A., Confalore, E., Carsana, A., Palmieri, M., and Libonati, M., *Ital. J. Biochem.* **1986**, *35*, A193-A194.

[95] MacDonald, R. J., Stacy, R. J., and Swift, G. H., *J. Biol. Chem.* **1982**, *257*, 14582-14585.

[96] Lizana, J. D., and Eneroth, P. In: *Proteins in Body Fluids, Amino Acids and Tumor Markers:* Ritzmann, S. E., and Killingworth, L. M. (eds.) New York: Alan R. Liss Inc., **1983**.

[97] Sheth, A. R., Moodbiri, S. B., Bandivdekar, A. H., Vanage, G. R., Hirkadli, K. S., and Arbatti, N. J. In: *Gonadal Proteins and Peptides and their Biological Significance:* Sairam, M. R., and Atkinson, L. E. (eds.) Singapore: World Scientific Publishing Co., **1985**; pp. 39-45.

[98] Beksac, M. S., Khan, S. A., Eliasson, R., Shakkeback, N. E., Sheth, A. R., and Diczfalusy, E., *Int. J. Androl.* **1984**, *7*, 389-397.

[99] Vaze, A. Y., Thakur, A. N., and Sheth, A. R., *J. Reprod. Fertil. (Suppl.)* **1979**, *26*, 135-146.

[100] Akiyama, K., Yoshioka, Y., Schmid, K., Offner, G. D., Troxler, R. F., Tsuda, R., and Hara, M., *Biochim. Biophys. Acta* **1985**, *829*, 288-294.

[101] Johansson, J., Sheth, A. R., Cederlund, E., and Jornvall, H., *FEBS Lett.* **1984**, *176*, 21-26.

[102] Seidah, N. G., Arbatti, N. J., Rochemont, J., Sheth, A. R., and Chretien, M., *FEBS Lett.* **1984**, *175*, 349-355.

[103] Brar, A., Mbikay, M., Sirois, F., Furnier, S., Seidah, N. G. and Chretien, M., *J. Androl.* **1988**, *9*, 253-260.

[104] Mbikay, M., Linard, C. G., Sirois, F., Lazure, C., Seidah, N. G. and Chretien, M., *Cell. Mol. Biol.* **1988**, *34*, 387-398.

[105] Wang, M. C., Valenzuela, L. A., Murphy, G. P., and Chu, T. M., *Invest. Urol.* **1979**, *17*, 159-163.

[106] Watt, K. W. K., Lee, P. J., Timkulu, T. M., Chan, W. P., and Loor, R., *Proc. Natl. Acad. Sci. USA* **1986**, *83*, 3166-3170.

[107] Lundwall, A. and Lilja, H., *FEBS Letters* **1987**, *214*, 317-322.

[108] Schaller, J., Akiyama, K., Tsuda, K., Hara, M., Marti, T., and Rickli, E. E., *Eur. J. Biochem.* **1987**, *170*, 111-120.

[109] Lilja, H., *J. Clin. Invest.* **1985**, *76*, 1899-1903.

[110] Lilja, H., Oldbring, J., Ramnevik, G., and Laurell, C. B., *J. Clin. Invest.* **1987**, *80*, 281-285.

[111] Riegman, P. H. J., Klaassen, P., van der Korput, J. A. G. M., Romijn, J. C., and Trapman, J., *Biochem. Biophys. Res. Commun.* **1988**, *155*, 181-188.

[112] Vihko, P., Konturri, M., Lukkarinen, O., and Vihko, R. *J., Urol.* **1985**, *133*, 979-982.

[113] Rönneberg, L., Vihko, P., Sajanti, E., and Vihko, R., *Int. J. Androl.* **1981**, *4*, 372-378.

[114] Vihko, P., *Clin. Chem.* **1978**, *24*, 1783-1787.

[115] Derechin, M. Ostrowksi, W., Galka, M., and Barnard, E. A., *Biochim. Biophys. Acta* **1971**, *250*, 143-154.

[116] Vihko, P., Virkkunen, P., Henttu, P., Roiko, K., Solin, T., and Huhtala, M.-L., *FEBS Letters* **1988**, *236*, 275-281.

12. Genes for Egg Formation in Poultry

D. Baum, G. Graser, M. Heib, S. Schüler and G. Krampitz

12.1 Introduction

Formation and development of eggs belong to the most complicated phenomena in biology with their origins and roots in the evolution of the animal kingdom. There is a diversity in morphology of eggs. However, all of them have essential features in common. In modern research the interest is focused on genes encoding functional mechanisms for the formation of eggs. Genes participating in the construction of egg can be classified according to the function of their expression products. One group of genes contributes to the synthesis of components of the egg, the other group of genes controls processes, e.g., of egg formation. The knowledge of these mechanisms and processes is of considerable theoretical interest and of practical importance. For studies of this kind chicken eggs are preferable objects because they are large enough and available in sufficient numbers. Furthermore chicken eggs can be investigated in a much easier and less expensive way than eggs from other species including mammals. The results of studies on chicken eggs can in essence be compared with the situation in other species. Regarding the formation of chicken egg, spatial and temporal phases must be distinguished: (1) yolk, (2) egg white, (3) egg shell. Moreover, it should be born in mind that (1) proteins of yolk are synthesized in the liver, (2) egg white proteins are produced in the magnum portion of the oviduct and (3) proteins of the egg shell are also synthesized in the avian liver [1]. Yolk proteins have intensively been studied in the past. The family of lipovitellin polypeptides and the family of the heavily phosphorylated phosphovitins have been far better investigated than the water soluble livetins. The reason for this particular interest in lipovitellin polypeptides and phosphovitins is based on their ability to transport lipids, phosphate and metal ions. Furthermore these polypeptides can serve as a source of nutrients for the growing embryo [2]. Both yolk protein families originate from a common precursor, the vitellogenin.

12.2   Vitellogenin Gene

Vitellogenin is synthesized in the liver of oviparous vertebrates [3] in response to estrogenic stimulation. In the hen, vitellogenin is transported to the ovary and deposited in the yolk fluid after proteolytic cleavage to α-lipovitellin, ß-lipovitellin, and phosvitins. Both lipovitellins contain two or more polypeptide chains which arise from regions of vitellogenin containing little or no phosphorus [4]. Phosvettes are thought to derive from an additional cleavage site in the serine-rich portion of one of the forms of vitellogenins [5]. According to the state of the art chicken vitellogenin includes at least three components. Vitellogenins (VTG) I, II and III can be distinguished by their immuno-chemical properties, amino acid compositions and peptide (fingerprint) patterns. The vitellogenin genes from *Xenopus laevis* have been studied extensively by Wahli and coworkers. Analysis of cDNA clones indicate that at least four different vitellogenin genes are expressed in this frog [6,7]. In contrast to *Xenopus* the situation in chicken is less clear in this respect. However, there is no convincing evidence for the existence of more than one chromosomal gene [8]. The chicken gene [VTG-II] consists of at least 28 exons and 27 introns and has a total length of 23,6 kbp [9].

12.3   Regulation of Vitellogenin Gene Activity

In recent years the working group of Jost and Saluz in Basel has detected that estrogen-dependent regulatory mechanisms of the VTG-II-gene involve methylation and demethylation of some cytosine molecules. Demethylation triggers a turning on of the activity of VTG-II-gene while methylation of the same cytosine residues results in switching off the active gene. There are many examples for the regulation of gene expression by methylation of C and/or G molecules in the regulatory range of eukaryotic genes. DNA of *Herpesvirus saimiri* is methylated in nonproducer cells where virus proliferation has been terminated. In contrast, the same DNA is undermethylated in cells producing virus [17]. DNA of adenovirus 2 in transformed hamster cells is undermethylated in expressed regions, whereas silent regions are highly methylated [18]. Such differences in methylation between active and inactive genes have also been observed in cellular genes, e.g., ovalbumin, conalbumin, ovomucoid [19], globin [20, 21], metallothioneine [22] and the J-chain of IgM [23]. Prerequisite for the understanding of the hormonal regulation of chicken vitellogenin synthesis at the molecular level is the determination of the fine structure of the 5'end region of the gene as well as the possible regulatory signals in the 5'end-flanking region.

The presence of methylated cytosine residues within the regulatory regions of viral and eukaryotic genes has been shown to result in reduced transcription rates of these genes [10]. In mammalian DNA 2-7 % of cytosine residues are converted to 5'-methylcytosine shortly after DNA replication [11, 12]. More than 90% of these 5-methylcytosine residues occur in the dinucleotide sequence CpG. It has been postulated [11, 13], that changes in

DNA methylation could provide a means of controlling gene expression in a hereditable manner. A major breakthrough for distinguishing between symmetrically methylated and hemimethylated sites came from direct genomic sequencing, first described by Church and Gilbert [14] and further developed by Saluz and Jost [15, 16].

In the livers of immature chicken and *Xenopus* exposed to estradiol for the first time (primary stimulation) there is a lag period preceding the onset of vitellogenin synthesis [3, 24-26]. About 10 days after primary stimulation, the amount of VTG-mRNA in the chicken liver cells returns to background level. If a second dose of estradiol is administered at this time (secondary stimulation), the VTG-mRNA is synthesized at the maximum rate without any time lag [3, 25, 27]. During the primary stimulation of the chicken liver by estradiol, the VTG-II gene is marked at the chromatin level by two nuclease-hypersensitive sites located at the 5'end, which persists even after estradiol withdrawal [28, 29].

Furthermore, during the primary stimulation a hypomethylation site exists where the estradiol-receptor complex binds [30, 31]. Further estradiol-dependent hypomethylation sites have recently been detected on the upstream region of the gene.

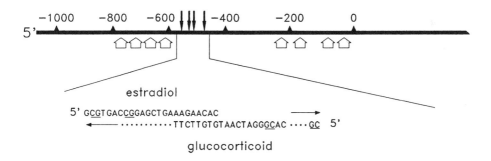

Fig. 12-1 Organization of the 5'end of chicken vitellogenin II gene. Four CpGs are located in between the nucleotide coordinates -531 to -511 and within the overlapping glucocorticoid and estradiol-receptor binding sites, respectively. Open arrows represent DNaseI-hypersensitive sites. All CpGs on the sequence are marked by vertical arrows pointing downwards (after Saluz, Jiricny and Jost [10]).

The results of chicken VTG-II gene analyses revealed four CpG-positions within the regulatory section (Fig. 12-1). All four CpG-sequences are located 500-600 base pairs upstream of the cap-sequence of the avian VTG-II gene. In this gene segment both estradiol and glucocorticoid binding sites overlap. Two methylation sequences are positioned exactly within the sequence of the estradiol binding site. Another CpG-sequence is arranged close to them, and a fourth methylation sequence can be found in the overlapping sequences of the binding sites of both steroid hormones [10]. Three CpG-sequences are completely methylated in the VTG-II gene of immature hens and juvenile as well as mature roosters. The fourth CpG-sequence (in the overlapping region) is hemimethylated. Demethylation of these sites occurs initially in one DNA strand only. This demethylation correlates well with the induction of VTG-mRNA synthesis. The demethylation of the complementary DNA strand lags approximately 24 h behind [10]. The hemimethylated state remains until the full transcription rate is reached.

The presence of an active estradiol-receptor complex is required for demethylation processes. However, the detailed mechanism is not very well understood as yet. Mutations will have to be introduced experimentally in the 5'-end-flanking region of the gene and the mutated gene will have to be introduced by conventional techniques in chicken liver hepatocytes, where its expression could be studied [9].

12.4   Egg White Proteins and Their Genes

Typical proteins of the egg white are ovalbumin, conalbumin, ovomucoid and lysozyme which are synthesized in the tubular gland cells of the magnum portion of the oviduct. Ovalbumin is the predominant protein of these cells (50-60% of the total protein synthesis rate; ca. 40% of the total mRNA population) [33].

Genomic chicken ovalbumin DNA is a set up of about 15.7 kbp including 7 intervening sequences [33, 34]. The occurrence of polymorphisms has been reported [35]. The ovalbumin gene is linked with two pseudogenes (X, Y) in the sequence 5'-X-Y-ovalbumin-3' [36].

For about 20 years the chicken oviduct has served as a model to studies on the regulation of the eukaryotic gene expression, in part because a few gene regulatory molecules other than steroidal receptors have been identified in higher eukaryots. Therefore the biology of the steroidal regulatory system is well defined [38]. In juvenile birds, estrogen initiates differentiation of the tubular gland cells of the magnum portion of the oviduct and induces the transcription process of genes encoding ovalbumin, transferrin, conalbumin, lysozyme, ovomucoid and minor constituents of egg white. Estrogen requires the permissive effects of insulin [39] and corticosteroids [38] to act on transcription processes. After primary exposure to estrogen three other classes of steroidal hormones (androgens, glucocorticoides and progestins) may activate genes coding for egg white proteins. These observations emphasize the particular significance of studies how the specificity of induction of target genes by different steroid hormones can be achieved [37].

There are relative few informations available on regulatory genetic elements controlling the ovalbumin gene. More informations have been accumulated on the classical TATA and CAAT-boxes which are typical for very active promoters [37]. However, not very much is known on sequences necessary for the regulation of gene activity by estrogens and other steroid hormones. Recently positive and negative regulatory elements controlling the steroid-sensitive ovalbumin promoter have been detected [37]. A steroid-responsive element is located between the nucleotide coordinates -800 and -585 of the ovalbumin gene. A negative regulatory element is arranged between nucleotides -350 and -248 of that gene. A simplified model of the regulation of the ovalbumin promoter is demonstrated in Fig. 12-2.

The DNA coding for chicken ovomucoid comprises ca. 15 kbp and at least six introns [34]. The chicken gene for lysozyme consists of about 4 kbp including three introns [40].

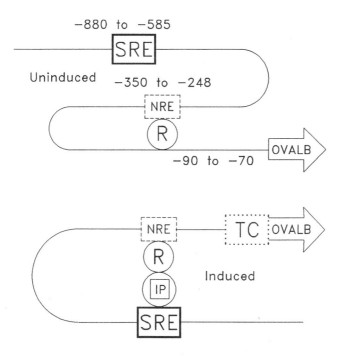

Fig. 12-2 Proposed model for the regulation of the ovalbumin promoter. Two hypothetical states of the ovalbumin gene are shown, the basal and the steroid-induced states. It is suggested that the repressor protein (R) binds to a negative regulatory element (NRE), thereby suppressing transcription by preventing the interaction of the transcription complex (TC) with the promoter. Upon treatment with steroids, a protein with a short half-life (IP) is induced that binds to the steroid response element (SRE). The interaction of the IP with the SRE initiates contact with R such that the TC can now initiate transcription of ovalbumin gene [37].

12.5   Egg Shell Proteins

In the past the existence of egg shell proteins has been denied. This prejudice delayed research in this field for some time. However, recently not only the existence of egg shell proteins, but also their significance and mode of action in shell formation has been demonstrated. Details have already been reported [41, 42]. It should be mentioned that genes encoding egg shell proteins are activated by estrogens in the liver [1] and that high concentrations of estrogens cause liver damages thus reducing the rate of egg shell protein synthesis drastically [42, 43].

So far no informations are available on the isolation and characterization of genes encoding egg shell proteins.

12.6   Chicken Growth Hormone

Apart from steroidal hormones the formation of egg depends on the presence of avian growth hormones [44]. For this and other reasons the knowledge of the structure of growth hormone (GH) gene of chicken and other poultry is of particular interest. The successful syntheses and cloning of bovine and chicken GH-cDNAs have been reported [45, 46]. Meanwhile at least two nucleotide sequences of the chicken GH-cDNA have been established [46, 47]. Fig. 12-3 shows the results of base sequence analysis of GH-cDNA synthesized from corresponding pituitary gland mRNA of White Leghorn hybrids compared with base sequence analysis data of chicken GH-cDNA reported by a group in Zagorsk, USSR [47]. Differences have been observed in particular in the noncoding sequence near the 3'-end of the structure. These differences cannot be made visible and distinguished by DNA-restriction fragment analysis because the fragments are too small, and they vary only in size by a few nucleotides.

The reasons for these differences in the 3'-noncoding range are not known. Probably they are due to another chicken race used by the Zagorsk group. If this assumption is correct base sequence analysis of GH-cDNA permits insights into variations of base sequences caused by breeding.

Comparison of chicken GH-cDNA with duck GH-cDNA [48] reveals minor changes in the nucleotide sequence of the coding region which result in the exchange of amino acids. Major changes again can be observed in the 3'-noncoding section.

The biological significance of base changes in the 3'-noncoding area of cDNA is not known as yet. Much more information is needed for the understanding of this phenomenon and judging its consequences. The structures of avian GH-genes are not known at all. However, the publication of the chicken GH-gene can be expected soon. The chicken GH-gene can be regarded as an important marker gene in the chicken genome. The same is true for other poultry.

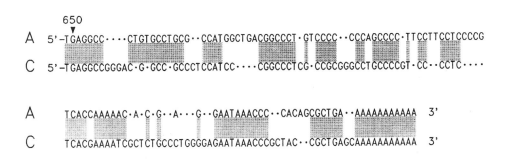

Fig.12-3 Comparison of nucleotide sequences of the 3'-non-coding end of poultry GH-cDNAs. Nucleotides are numbered in the 5'to 3'direction beginning with the first noncoding base pair. Gaps (represented by dots) are introduced to increase the homology. Stippled boxes indicate homologous nucleotide sequences. The nucleotide sequences of duck GH-cDNA (C) is cited from Chen et al. [48] and that of the "Zagorsk" chicken GH-cDNA (B) is from Zhvirbilis et al. [47]. The chicken GH-cDNA sequence (A) has been determined by Baum [46].

12.7 Conclusions and Prospects

So far some, but not all genes for egg formation have been characterized. Most of these genes code for constituents of the egg, only very few genes which control egg formation have been studied. However, some genes for egg constituents contain regulatory sequences controlling the activity of the corresponding gene by binding steroid hormones with consecutive reactions such as, e.g., demethylation or methylation of predominant cytosine residues.

Some of the genes already known for egg formation could serve as marker genes in order to study polymorphisms. The knowledge of polymorphisms or of major changes in nucleotide sequences at important strategic sections of the gene could help to identify animals carrying undesirable polymorphisms. This does not require waiting for the development of certain criteria of production, e.g., growth, formation of eggs, egg quality, meat quality etc. Genomic analysis simply could be carried out by taking avian blood samples followed by extraction of DNA from erythrocytes and analysis of polymorphismus by known probes.

Other considerations, e.g., transfer of selected genes into fertilized avian eggs cannot be realized at the time being. Reliable techniques as they can be applied with mammalian eggs are not available as yet. May be in the future techniques will be available which can be used this way. Indications of successful germline transmission of exogenous genes in the chicken are just appearing [49].

12.8 References

[1] Eckert, J., Schade, R., Glock, H., Krampitz, G., Enbergs, H., and Petersen, J., *J. Anim. Physiol. Anim. Nutri.* **1986**, *56*, 258-265.

[2] Wang, S.Y., Smith, D.E., and Williams, D.L., *Biochemistry* **1983**, *22*, 6206-6212.

[3] Bergink, E.W., Wallace, R.A., Van den Berg, J.A., Bos, E.S., Gruber, M., and Ab, G., *Am. Zool.* **1974**, *14*, 1177-1193.

[4] Wang, S.Y., and Williams, D.L., *Biochemistry,* **1980**, *19*, 1557-1563.

[5] Wahli, W., Dawid, I.B., Ryffel, G.V., and Weber, R., *Science,* **1981**, *212*, 298-304.

[6] Wahli, W., Dawid, I.B., Wyler, T., Jaggi, R.B., Weber, R., and Ryffel, G.V., *Cell,* **1979**, *16*, 535-549.

[7] Wahli, w., Dawid, I.B., Wyler, R:, and Ryffel, G.V., *Cell,* **1980**, *20*, 107-117.

[8] Arnberg, A.C., Meijlink, F.C.P.W., Mulder, J., Van Bruggen, Gruber, M., and Ab, G., *Nucleic Acid Res.,* **1981**, *9*, 3271-3285.

[9] Geiser, M., Mattaj, J.W., Wilks, A.F., Seldran, M., and Jost, J.P., *J. Biol. Chem.,* **1983**, *258*, 9024-9030.

[10] Saluz, H.P., Jiricny, J., and Jost, J.P., *Proc. Natl. Acad. Sci. USA,* **1986**, *83*, 7167-7171.

[11] Razin, A., and Riggs, A.D., *Science,* **1980**, *210*, 604-610.

[12] Ehrlich, M., and Wang, R.Y.-H., *Science, 1981, 212,* 1350-1357.

[13] Holliday, R., and Pugh, J.E., *Science, 1975, 187,* 226-232.

[14] Church, G.M., and Gilbert, W., *Proc. Natl. Acad. Sci. USA, 1984, 81,* 1991-1995.

[15] Saluz, H.P., and Jost, J.P., *Gene, 1986, 42,* 151-157.

[16] Saluz, H.P., and Jost, J.P., *A Laboratory Guide to Genomic Sequencing,* Basel, Boston: Birkhäuser, *1987.*

[17] Desrosiers, R.C., Mulder, C., and Fleckenstein, B., *Proc. Natl. Acad. Sci. USA, 1979, 76,* 3839-3843.

[18] Vadirman, L., Neumann, R., Kohlmann, I., Sutter, D., and Doerffler, W., *Nucleic Acids Res., 1980, 8,* 2461-2473.

[19] Mandel, J.L., and Chambon, P., *Nucleic Acids Res., 1979, 7,* 2081-2103.

[20] Shen, C.K.J., and Maniatis, T., *Proc. Natl. Acad. Sci. USA, 1980, 77,* 6634-6638.

[21] van den Ploeg, L.H.T., and Flavell, R.A., *Cell, 1980, 19,* 947-958.

[22] Compere, S.P., and Palmiter, R.D., *Cell, 1981, 25,* 233-240.

[23] Yagi, M., and Koshland, M.E., *Proc. Natl. Acad. Sci. U.S.A., 1981, 78,* 44907-4911.

[24] Deeley, R.G., Gordon, J.I., Burns, A.T.H., Mullinix, K.P., Bina-Stein, M., and Goldberger, R.F., *J. Biol. Chem., 1977, 252,* 8310-8319.

[25] Ryffel, G.v., Wahli, W., and Weber, R., *Cell, 1977, 11,* 213-221.

[26] Wang, S.Y., and Williams, D.L., *Biochem. Biophys. Res. Commun., 1983, 112,* 1049-1055.

[27] Jost, J.P., Ohno, T., Pangim, S., and Schörch, A.R., *Eur. J. Biochem., 1978, 84,* 355-361.

[28] Burch, J.B.E., and Weintraub, H., *Cell, 1983, 33,* 65-76.

[29] Burch, J.B.E., *Nucleic Acids Res., 1984, 12,* 1117-1135.

[30] Jost, J.P., Seldran, M., and Geiser, M., *Proc. Natl. Acad. Sci. USA, 1984, 81,* 429-433.

[31] Jost, J.P., Moncharmont, B., Jiricny, J., Saluz, H.P., and Heitner, T., *Proc. Natl. Acad. Sci. USA, 1986, 83,* 43-47.

[32] Cohrs, R.J., Goswami, B.B., and Sharma, O.K., *Biochemistry, 1988, 27,* 3246-3252.

[33] Breathnach, R., Mandel, J.L., and Chambon, P., *Nature, 1977, 270,* 314-319.

[34] Catteral, J.F., Stein, J.P., Lai, E.C., Woo, S.L.C., Dugaiczyk, A., Mace, M.L., Means, A.R., and O'Malley, B.W., *Nature, 1979, 278,* 323-327.

[35] Weinstock, R., Sweet, R., Weiss, M., Cedar, H., and Axel, R., *Proc. Natl. Acad. Sci. USA, 1978, 75,* 1299-1303.

[36] Colbert, D.A., Knoll, B.J., Woo, S.L.C., Mace, M.L., Tsai, M.J., and O'Malley, B.W., *Biochemistry, 1980, 19,* 5586-5592.

[37] Sanders, M.M., and McKnight, G.S., *Biochemistry, 1988, 27,* 6550-6557.

[38] Sanders, M.M., and McKnight, G.S., in: *Molecular Genetics of Mammalian Cells:* Malacinski, G.M. (ed.) New York: MacMillan *1986;* pp. 183-216.

[39] Evans, M.E., and McKnight, G.S., *Endocrinology, 1984, 115,* 368-377.

[40] Jung, A., Sippel, A.E., Grez, M., and Schütz, G., *Proc. Natl. Acad. Sci. USA, 1980, 77,* 5759-5763.

[41] Schade, R., *Arch. Geflügelkd., 1987, 51,* 81-83.

[42] Krampitz, G., and Graser, G., *Angew. Chem., 1988, 100,* 1181-1183.

[43] Wieser, D., and Krampitz, G., *Zuchthygiene, 1987, 22,* 267-271.

[44] Scanes, C.G., Lauterio, T.J., and Buonomo, F.C., in: *Avian Endocrinology: Environmental and Ecological Perspectives:* Mikami, S. et al. (eds.), Tokyo: Japan. Sci. Soc. Press, *1983,* Springer Verlag, pp. 307-326.

[45] Heib, M., Dissertation, Universität Bonn, 1988.

[46] Baum, D., Dissertation, Universität Bonn (in preparation).

[47] Zhvirbilis, G.S., Gorbulev, V.G., Rubtsov, P.M., Karapetyan, R.V., Zhuravlev, I.V., Fisnin, V.I.,Skryabin, K.G., and Baev, A.A., *Molekulyarnaya Biologiya, 1987, 21,* 1620-1624.

[48] Chen, H.-T., Pan, F.-M., and Chang, W.-C., *Biochim. Biophys. Acta, 1988, 949,* 247-251.

[49] Bosselman, R.A., Hsu, R.-Y., Boggs, T., Hu, S., Broszewski, J., OU, S., Kozar, L., Martin, F., Green, C., Jacobsen, F. Nicolson, M., Schultz, J.A., Semon, K.M., Rishel, W., and Steward, R.G., *Science, 1989, 243,* 533-535.

13. The Structure and Function of Peptide Hormone Genes

R. Ivell

13.1 Introduction

Although natural selection primarily has to act at the phenotypic level of polypeptides rather than on the DNA itself, nevertheless, indirectly, every nucleotide involved in the genotype is subject to Darwinian scrutiny related to optimising the functioning of a given gene. Polypeptide hormone genes are particularly useful models in which to analyze this relationship between DNA structure in the genome and the physiological function.

Firstly, the majority of protein hormones are very ancient occurring in various homologous forms throughout evolution, and therefore the changes in the genome linked with the adapting of a species to a new environment can be followed over a long time scale, with many intermediate steps.

Because they are rapidly degraded by circulating proteases, protein hormones are generally used in the body for conveying short-term changes in the internal milieu. Information is therefore transmitted in the form of a rapid on-off pulse. This is why peptides are used to report acute changes in nutritive uptake or blood glucose levels (insulin, glucagon), or to convey a milk ejection signal (oxytocin), or to react quickly to stress (ACTH, opiates). Information which needs to be conveyed in a chronic fashion makes use of less degradable hormones, classically the steroids. This distinction is evidently primitive, and selection has succeeded in qualifying these characteristics. Secondly, therefore, protein hormones are often *secondarily modified* to increase their serum lifespan - amidationat the C-terminus (e.g. vasopressin, substance P, etc.), a pyroglutamate residue at the N-terminus (e.g. LHRH) or perhaps glycosylation (e.g. the glycoproteohormones of the anterior pituitary). They may also become associated with carrier proteins (e.g. somatostatin [1]). Another strategy that may be speculated upon is for some of the normally short peptide hormones to be secreted with N- or C-terminal extensions. So, for example, somatostatin can be released as a 28-residue precursor form, which has a longer serum half-life than the normal 14-residue hormone, but still binds equally well to the specific receptors [2]. All these and other possibilities for modification are encoded in the genome.

Thirdly, hormone genes have to possess *controlled expression* in time (chronic differentiation, as in puberty, or acute as in the vasopressin thirst response) and in space (tissue specific expression). These control systems also have to be encoded in the genome.

Finally, the protein or peptide hormones can be divided into four *different structural categories*. There are (a) those which are hormonal as the single gene product without subsequent cleavage (e.g. prolactin); (b) those which have to be cleaved into a smaller physiologically functional unit (e.g. LHRH or insulin); (c) the so-called polyprotein hormones where more than one functional unit exists on the same polypeptide precursor (e.g. proopiomelanocortin); and (d) there is a small group of hormones where physiological activity results only when there is coordinated expression of two independent gene products (subunits) (e.g. inhibin, the gonadotropins).

The importance of studying protein hormone genes lies therefore not only in their immediate endocrinological relevance, but in their reflecting a variety of genotypic mechanisms and the ways in which these have evolved.

Vasopressin and oxytocin are two closely related neuropeptides which illustrate well many of the points summarized, and a study of their genetic origin has introduced a number of novel concepts in our understanding of how the genome works.

13.2   Evolution of the Vasopressin and Oxytocin Hormone Systems

Vasopressin/oxytocin-like molecules have been detected in a broad variety of phyla including the *Cnidaria (Hydra)* [3] and *Mollusca (Helix* [4], *Octopus* [5]) as well as in all orders of vertebrates. Vasopressin and oxytocin are the two eutherian representatives of a gene pair which arose by duplication at least 400 million years ago, during the evolution of the fish. The modern nonapeptides differ from one another in only two of their nine amino acids, at residues 2 and 8, indicating a very high selection pressure not only conserving the hormones but also their receptors.

Most protein hormone systems appear to have evolved along the pattern illustrated in Fig. 13-1. In unicellular organisms peptides are among the first molecules, besides mutually required substrates, which are used as carriers of information; an example is the alpha-mating factor in yeast [6]. This transfer of information between different cells was a prerequisite for the evolution of multicellularity. As the primitive hormonal system diversified so more complex tissues and organs could differentiate. The primitive system requires local synthesis of a peptide and the production of a cognitive receptor on a neighbouring cell as well as the post-receptor signalling mechanisms. This situation is what we still have in so-called paracrine or autocrine control systems, or, within nervous tissues, the neuromodulatory function.

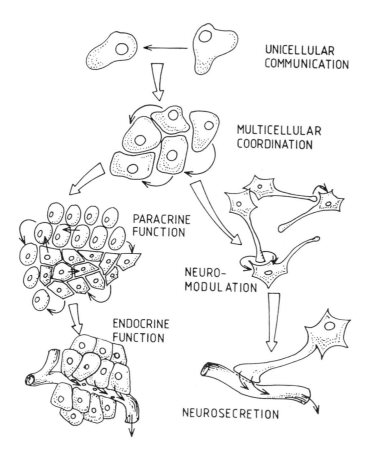

UNICELLULAR
COMMUNICATION

MULTICELLULAR
COORDINATION

PARACRINE
FUNCTION

NEURO-
MODULATION

ENDOCRINE
FUNCTION

NEUROSECRETION

Fig. 13-1 Probable evolution of peptidergic hormonal systems.

A quantum leap was achieved by the evolution of a secretory hormonal system (neurosecretory, endocrine). Here the peptide hormone is produced in much larger quantities and is released directly into the bloodstream to search for target receptors a considerable distance away. Vasopressin and oxytocin exhibit both paracrine (for example in the rat testis [7, 8]) and endocrine functions. The latter are represented for vasopressin by the classic antidiuretic response to thirst stress, or by vasoconstriction, and for oxytocin by the milk ejection reflex and uterine contraction at birth. Both hormones are produced in the hypothalamo-neurohypophyseal complex and exert their effects at distant target organs.

There are several consequences or prerequisites of this evolutionary development. For paracrine systems, only low levels of peptide need to be produced so that local concentrations in the nanomolar range can be reached. Dilution effects and peptidase activity outside that local range are such as to eliminate physiological cross-talk with other regulating systems. For endocrine functions, firstly much greater production is needed. For example, the vasopressinergic and oxytocinergic cells of the rat hypothalamus, though numbering less than 10000, produce together more than 1000 times the amount of oxytocin- and vasopressin-mRNA produced in the rat testis [9], which contains many more peptidergic cells. Secondly, greater specificity is required at the receptor to prevent incorrect signalling in tissues which might otherwise be activated by the peptide. This could be achieved by evolution of high affinity and selectivity or by regulating receptors to appear on the cell surface only when needed: endometrial receptors for oxytocin appear only at the end of the oestrous cycle in sheep where luteal oxytocin is involved with endometrial prostaglandin to regulate cycle length [10], or in the human uterine wall only immediately before parturition [11]. Alternatively, the human placenta produces high levels of "vasopressinase" and "oxytocinase" [12], presumably to prevent preemptive induction of labour by sub-threshold levels of oxytocin.

13.3   Vasopressin and Oxytocin Gene Structures

Essentially, vertebrate species possess one gene for vasopressin and one for oxytocin [13]. Exceptions are provided by marsupials where evidently further gene duplication has occurred [14]. In the species where the gene structure has been analyzed in detail (bovine [15], rat [13], human [16], mouse [17], toad [18]) a remarkable conservation of organization is apparent. Typically, each gene comprises three exons separated by two intervening sequences. All DNA and protein data, also from other species, indicate for both hormones the same structure for the encoded precursor. Namely, the nonapeptide hormone shares a precursor polypeptide with a ca. 95 amino acid cyteine-rich protein called neurophysin, and for vasopressin-like genes there is also at the C-terminus of this precursor another ca. 39 amino acid polypeptide, which is absent in the oxytocin-like precursors.

The two bovine genes are shown schematically in Fig. 13-2 and are typical of those now known from other species. Although the genes arose by a duplication early in vertebrate evolution, they clearly have not evolved and mutated at random. Considering the first exon, the homology of up to 80% might suggest a strong selection pressure for a highly conserved sequence. Yet it is significant (a) that in the rat hypothalamus oxytocin- and vasopressin-mRNAs are up-regulated together during both thirst [19] and lactation [20] although the functions for the two hormones are apparently different; (b) the hormones still cross-react at their receptors at the 1% level or better [21]. Both observations suggest that each hormone can substitute for the other in extreme physiological situations, implying an additional selection pressure not just to conserve but also to maintain slight but significant similarity to one another.

The 5' non-coding region which precedes this first exon contains as one might expect the control elements for transcription. These appear as blocks differing between the genes but highly conserved within any one gene between different species [13]. Proteins binding to these conserved sequences are responsible for the highly tissue- and cell-specific expression of each gene as well as for its quantitative regulation upon stimulation of appropriate intracellular signalling mechanisms. Although consensus sequences have now been obtained for regulatory elements linked with either cAMP [22], phorbol ester [23] or steroid regulation [24], no clear-cut homologies with any of these sequences have been observed in the vasopressin or oxytocin genes. The latter does contain elements which are similar to estrogen receptor binding sequences, though experiments in culture imply that estradiol is not a major regulatory effector (McArdle, Holtorf, Furuya & Ivell, unpublished). Similarly, stimulators of cAMP (LH) are without effect on oxytocin gene expression. The low homology in this 5' control region (<35%) is similar to that in most of the intron and 3' non-coding sequences and probably reflects random similarity of two GC rich sequences (Fig. 13-2).

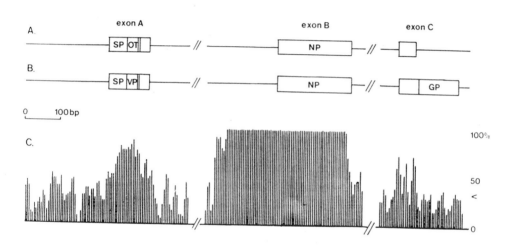

Fig. 13-2 Structural organization of the bovine oxytocin (A) and vasopressin (B) genes (SP, signal peptide; OT, oxytocin; VP, vasopressin; NP, neurophysin; GP, glycopeptide). (C) is the corresponding smoothed homology profile comparing bovine gene sequences using a homology window of 10 nucleotides. Introns have been abbreviated to account for their differing lengths between genes. The arrow (<) indicates the level of background homology (ca. 35%) due to random coincidence of nucleotides for two sequences of approximately 70% GC content.

Extremely surprising is the 100% homology of the second exon and much of the first intron (Fig. 13-2); if it was related to strict conservation of the protein phenotype one would expect there to be silent changes at the third position of some codons. And it should be noted that this homology is not bounded by exon-intron junctions and cannot therefore be related to an event at the post-DNA level. Hence reverse transcription from an mRNA or its use as template to correct the gene is also ruled out since this would be limited to exon sequences only. Therefore, this high homology must be attributed to what is called a gene conversion event [25]. This occurs where two closely related but non-allelic genes become more similar to one another, and probably involves one gene acting as template to "correct" the other gene. Whether this correction is analogous to the DNA excision/repair mechanisms seen to operate after UV irradiation, or whether it involves an actual DNA strand exchange (unequal crossing-over) is not known. It has, however, been described for other genes, e.g. the immunoglobulins [26]. Since the evidence for a gene conversion generally comes from sequencing genomic DNA it has often been assumed that it is an event in the germ line probably associated with meiosis. This need not be true. Recent curious findings from the Brattleboro strain of rats, which have inherited diabetes insipidus due to a single nucleotide deletion in their vasopressin gene, imply that a gene conversion might also be occurring in somatic cells after they have stopped dividing [27]. The result is the appearance, increasing with age, of neurones with a heterozygous phenotype within the hypothalamus of these homozygous mutants [27].

Why is this happening? The exon 2 sequence encodes for some 67 of the ca. 95 amino acids of the neurophysins, which are thus extremely homologous, differing only at their N- and C-termini. The function of these molecules is still unknown though they may play a role in maintaining a conformation of the hormone precursors within the secretory granules which is optimal for post-translational processing events [28]. Other functions outside the neurones after the neurophysins have been released together with the nonapeptide hormones are not known. What may be significant is that in the three species (rat, mouse, human) where this has been analyzed, the oxytocin and vasopressin genes are less than 10 kb apart on the same chromosome and in an inverted repeat arrangement [16, 29]. From work in Fungi this configuration would appear to demand less free energy to undergo gene conversion than one where the genes were separated by a greater distance or on separate chromosomes [25]. Nevertheless, gene conversion is not a very frequent event: both the rat and human sequences indicate subsequent mutations within an obvious region of gene conversion, and the Brattleboro rat whose deletion in the vasopressin gene is within the converted sequence does not appear to revert with any regularity, at least not in the germ line.

The third exon exhibits very little homology (Fig. 13-2) most of which is associated with the single basic amino acid which serves as post-translational cleavage signal or with proline and cysteine residues in the neurophysin moieties (see later). The 39 amino acid C-terminal glycopeptide (Fig. 13-2, GP) which is found in the vasopressin precursor has no counterpart in the oxytocin precursor. Recently, evidence has been presented which implicates this polypeptide in the control of prolactin release from the anterior pituitary [30]. This result is strongly supported by the observation that in Brattleboro rats where this moiety is also missing there is no prolactin release associated with suckling [31].

The vasopressin and oxytocin genes belong to what are called the polyprotein type of hormone genes. Different physiological moieties are present in the same precursor. In fact each exon approximately corresponds to one of these moieties and would provide an example for the exon shuffling hypothesis proposed by Gilbert [32]. The lack of precise matching between exon-intron junctions and physiological peptide may be due to a selective disadvantage for the post-translational cleavage signals to be near to such junctions, assuming that crossing-over events are more likely to occur in intron sequences, as proposed by the hypothesis.

13.4   Allelic Expression of Hormone Genes

Eutherian mammals appear to possess one copy each of the vasopressin and oxytocin genes, based on restriction analysis of genomic DNA [13]. Both alleles of the vasopressin gene, at least, are expressed. This has been demonstrated in several ways. Firstly, in the Brattleboro rat, the defective vasopressin gene is inherited as a simple recessive Mendelian trait: only in the homozygous mutants is diabetes insipidus extant [33]. In the cow the vasopressin-associated neurophysin (NpII) exhibits a simple polymorphism with an exchange of one amino acid at residue 89 (isoleucine/valine). Hope and colleagues were able to separate the two forms of neurophysin on HPLC and showed that individual pituitaries had either all Val_{89}-NpII, or all Ile_{89}-NpII or 50% of each type [34]. Finally, in the pig family, whose modern domestic descendants only have Lys_8-vasopressin in their pituitaries, warthogs have been shown to have either all Arg_8- vasopressin, or all Lys_8-vasopressin or again 50% of each in single pituitaries [35]. Probably, ancestral pigs underwent first a mutation in one allele to become polymorphic, and then during domestication lost the original allele completely.

Recently, it has been suggested that the alleles may be differentially regulated. Using a nuclease protection assay with an oligonucleotide probe spanning the single base deletion in the Brattleboro vasopressin gene, it was shown that on thirst stress in heterozygotes, proportionally more of the functional allele was being expressed [36].

For another hormone gene, that for proopiomelanocortin, we have also identified an allelic variant [37]. mRNA was isolated and sequenced from a cDNA library made from a single bovine corpus luteum. Of only three independent clones isolated two were identical and conformed with the pituitary sequence. The third clone included one silent change and a deletion of four small neutral amino acids in a region of the exon 3 between putative hormonal sequences (Fig. 13-3). At this stage it is difficult to judge whether such allelic variants are purely neutral mutations or whether they serve the selective advantage of a true genetic polymorphism.

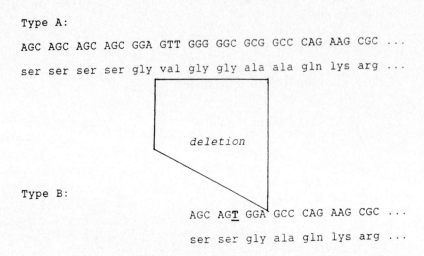

Type A:

AGC AGC AGC AGC GGA GTT GGG GGC GCG GCC CAG AAG CGC ...

ser ser ser ser gly val gly gly ala ala gln lys arg ...

deletion

Type B:

AGC AG**T** GGA GCC CAG AAG CGC ...

ser ser gly ala gln lys arg ...

Fig. 13-3 Nucleotide and amino acid sequences of part of the normal and allelic variants of the bovine proopiomelanocortin precursor obtained by sequence analysis of luteal cDNA clones.

13.5 Tissue Specific Expression

The principal sites of synthesis of vasopressin and oxytocin are the magnocellular nuclei of the hypothalamus. In the rat brain, where it has been most intensively studied, no cells appear to transcribe both oxytocin and vasopressin genes [38]. This implies that the two superficially similar cell types (oxytocinergic and vasopressinergic neurones) must maintain mutually exclusive transcriptional control systems. Whether this refers to exclusive populations of DNA-binding proteins or to more physical discrimination, possibly linked to the inverse, rather than tandem alignment of the genes on the same chromosome, is not known. Despite the cell specificity, expression of the two genes does appear to be linked, at least in the rat, with levels of both mRNAs increasing together on thirst stress or during lactation.

More recently, sites of gene expression outside the brain have also been implicated for these hormones (Table 13-1). These assertions have been based mostly on Northern blot analysis, and as shown later may be erroneous.

Quantitatively, the most significant of these extra- hypothalamic sites of synthesis is the ruminant *corpus luteum* where the oxytocin gene is expressed on an organ basis at levels up to 100-fold higher than the hypothalamus. Since the latter is contributed from only a few thousand cells, whereas the corpus luteum comprises many millions of oxytocin-producing cells, the maximum output per cell may be comparable. Unlike the steady state in the hypothalamus, however, oxytocin gene expression in the bovine corpus luteum is ephemeral. On the day of ovulation, the granulosa cells of the ovulatory follicle, and no other, switch up specific oxytocin gene transcription about 100-fold from a low background level to a maximum at about day 3 of the estrous cycle [40, 41].

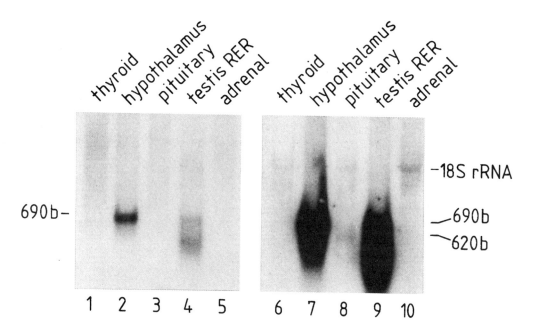

Fig.13-4 Northern blot hybridization of mRNA from various bovine tissues as indicated, using the oxytocin gene specific probe (OT-3')[43]. Except for the testis rough endoplasmic reticulum (RER) RNA, which was obtained by sucrose gradient density centrifugation of polysomes [44], mRNA was obtained by the guanidinium thiocyanate procedure, electrophoresed, transferred to nylon membranes and hybridized as described previously [9]. The left panel represents a short (2 h) autoradiographic exposure and the right panel a longer (48 h) exposure of two similar but independent experiments. The testicular polysomal RNA is probably degraded giving the smeared appearance of the signal. Both this and hypothalamic OT- mRNA are about 690 b long. A shorter (620 b) OT-mRNA is just detectable on long exposure (right panel) in the pituitary. This shorter size is similar to that from the bovine corpus luteum [43]. Thyroid and adrenal glands appear not to have any OT-mRNA.

Between days 3 and 5 the gene is then apparently switched off again, levels of oxytocin-mRNA declining exponentially through the remainder of the cycle. Essentially the same situation has been observed for oxytocin-mRNA in the sheep [42] (Ivell & McCracken, unpublished), but not in the rat, pig or human. In the latter species oxytocin-mRNA remains very low or undetectable throughout the cycle and pregnancy. Although oxytocin-mRNA drops to low levels in the later luteotrophic part of the cycle, it can still be detected *in corpora albicantia* in the sheep (Ivell & McCracken, unpublished) and sometimes in the corpus luteum of pregnancy (Ivell, unpublished). Although oxytocin- and vasopressin-mRNAs have been detected in several peripheral tissues in the rat (Table 13-1), the only other bovine tissue where oxytocin-mRNA can be detected are the testis and possibly the anterior pituitary (Fig. 13-4).

Table 13-1 Occurrence of oxytocin and vasopressin synthesis in different mammalian tissues.

Tissue	Oxytocin			Vasopressin		
	Bovine	Human	Rat	Bovine	Human	Rat
Hypothalamus	+++	+++	+++	+++	+++	+++
EH Brain #		+				+
Spinal Ganglia			(+)			(+)
Thymus		(+)	(+)		(+)	(+)
Ant. Pituitary	(+)					+
Ovary/CL	+++	?	+	-*	?	+
Testis	+	-	+		-	+
Adrenal	-		+	-		+
Placenta	-	?		-		
Thyroid	-					
Liver			-			-
Kidney	-		-	-		-
Endometrium	-			-		
Oviduct	-			-		

+ Evidence of synthesis is considered to be a cDNA sequence or an unambiguous hybridization signal on a Northern blot.
(+) Synthesis is strongly indicated but not absolutely confirmed.
 ? Some evidence for synthesis is available but not substantiated. - In spite of intensive searching via Northern hybridization or in cDNA libraries no positive signals can be obtained. Blanks indicate that for this tissue no search has been made. -
* As discussed in the text an mRNA is produced which is non-functional.
# EH Brain: extrahypothalamic brain areas. Data are abstracted from various sources reviewed in Ivell [39] or from unpublished data.

13.6 Regulation of Oxytocin Production in the Bovine Corpus Luteum

Unlike the hypothalamus with its very few hormone-producing cells, the bovine corpus luteum offers an excellent model system in which to examine the expression of a protein hormone gene and its regulation. *In situ* hybridization showed that the oxytocin mRNA was uniquely present in the large luteal cells of granulosa origin. However, although granulosa cells are relatively easy to culture and can be shown to secrete some oxytocin [45], until now it has not been possible to induce an up-regulation of the oxytocin gene. Instead in these cultures the low background levels of oxytocin-mRNA decline exponentially in culture with a half-life of about 24 hours (Ivell, unpublished). Thus, although these cells appear to luteinize as judged by progesterone production, their oxytocin secretion does not reflect the in vivo events accompanying ovulation. Evidently the culture conditions used are still not optimal and further work is necessary to induce proper luteinization, as judged by oxytocin gene up-regulation *in vitro*. As an alternative, luteal cells have been cultured from early corpora lutea (days 2-5), where not only large, granulosa-derived luteal cells are present but also those derived from the theca layers, as well as endothelial cells and fibroblasts. In the large luteal cells the oxytocin gene is already maximally stimulated *in vivo* [40]. Using serum-free cultures it was shown that insulin and IGF-I (but not IGF-II) were able to stimulate and maintain oxytocin release and that these effects were inhibited by sustained exposure to prostaglandin [46]. The effect of insulin and IGF-I on peptide release were mirrored by similar effects on oxytocin mRNA in these cells (Furuya, McArdle, Holtorf & Ivell, unpublished). Experiments are currently in progress to characterize precisely those factors responsible for the up-regulation, maintenance and down-regulation of oxytocin gene expression in these bovine ovarian cells.

13.7 Vasopressin Gene Expression in the Bovine Corpus Luteum

Dot blot analysis of bovine luteal mRNA gives a weak but positive signal for vasopressin gene expression [43]. It has not been possible to obtain an equivalent signal using Northern hybridization. Recently, nine cDNA clones were identified, isolated and sequenced from a complex ($>2 \times 10^6$ independent clones) cDNA library from early phase bovine *corpus luteum*. None of the clones revealed a normal functional vasopressin gene transcript [47]. Two clones were copies of RNA evidently started from an alternative polymerase initiation site within the first intron; the second intron had been correctly spliced out. The remaining seven clones all indicated RNA transcripts initiating within the second intron. None of the transcripts contained exon 1 (vasopressin-encoding) sequences; none could have yielded a functioning polypeptide.

The results can be interpreted as follows. It seems likely that in the cow as in other mammals the oxytocin and vasopressin genes are physically very close on the same chromosome, whereby the former gene is highly active in this tissue and therefore the DNA neighbourhood is likely to be in an ideally transcribable configuration (ie, in regard to DNA methylation and other chronic modifications). To block unwanted transcription of the vasopressin gene, an inhibitor molecule is possibly occupying the normal initiation site. There might be sequences, however, within the first and second introns which can bind (weakly) other (non-oxytocin) promoter proteins and allow low level polymerase activity to commence from these sites.

A low level expression of vasopressin peptide in this tissue would anyway be without physiological relevance, since any local vasopressin receptors would be swamped by the much higher levels of cross-reacting oxytocin.

A similar occurrence has also been shown for proopiomelanocortin transcripts in testis (rat [48], human [49]) and *corpus luteum* (bovine [37]), where in all three tissues transcription evidently did not initiate at the normal position but within the coding region of exon 3, also to give rise to non-functional RNA molecules.

13.8   Translation and Post-Translational Processing

Transcription of the vasopressin and oxytocin genes is qualitatively similar in all tissues so far examined, except that a differential polyadenylation occurs, with fewer adenosine residues in the poly(A) tails of the non-neural mRNAs [9, 43]. Thus, unlike genes for some protein hormones such as that for substance P or for calcitonin where the same gene may produce different mRNA molecules [50], vasopressin or oxytocin are always synthesized via the same protein precursors (Fig. 13-5).

There appears to be a minimum length for any secreted polypeptide synthesized on ribosomes. This length is determined by the number of residues which need to be synthesized before the nascent chain extends beyond the ribosome plus that amount spanning the membrane of the endoplasmic reticulum, thus guaranteeing a co-translational transfer of the growing polypeptide into the lumen of the rough endoplasmic reticulum. Altogether this dictates a minimum size of at least 60 amino acids. For vasopressin and oxytocin the primary translation product, which is also that observed on translating mRNA in a cell-free in vitro system, comprises 19-21 amino acids at the N- terminus forming the signal sequence. This leader sequence is responsible for attaching to the signal recognition nucleoprotein particle, which in turn links the ribosome to the endoplasmic reticulum and facilitates the traverse of these N-terminal and hydrophobic residues across into the reticular lumen. On the inner lumen side a signal peptidase cleaves off the signal sequence at a serine or alanine residue, to expose vasopressin or oxytocin now at the extreme N-termini of their respective precursors. These are now completely drawn into the lumen.

Once within the endoplasmic reticulum the precursors are exposed to the enzyme systems responsible for transferring the mannose-rich glycosylation core to asparagine residues within protein sequences containing the triplet -Asn-X-Thr/Ser-. This n-glycosylation is subsequently modified in the Golgi apparatus with pruning of some mannose groups and their replacement and extension with other sugars. The vasopressin precursor contains such a triplet sequence within the C-terminal GP moiety (Fig. 13-5). Supplementing a cell-free translation system with microsomal membranes allows the core-glycosylation to be performed in vitro, with the result that the bovine 21000 Mr prepro-vasopressin is first shortened to the 19000 Mr unglycosylated pro-hormone by cleavage of the signal sequence, and then n-glycosylated to give a larger molecule of 23000 Mr [51].

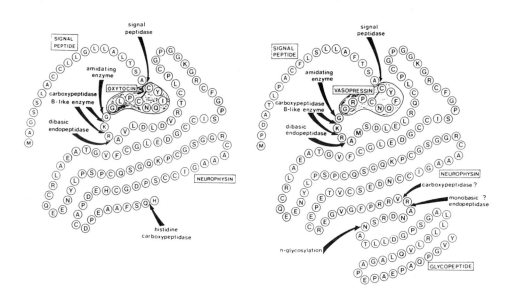

Fig. 13-5 Schematic structures of the oxytocin (left) and vasopressin (right) precursor polyproteins indicating the sites of post translational cleavage and modification. To date, the enzymes responsible for the single basic amino acid cleavage at the carboxy-terminal end of the neurophysin are speculative and poorly characterized.

The Golgi apparatus effectively sorts the newly synthesized proteins into those destined for lysosomes (enzymes, defective proteins) and those for secretion (hormones, cell surface antigens). How this sorting works is not known but appears also to be encoded within the amino acid sequence of the precursor itself. The fully glycosylated pro-hormone is conveyed into budding secretory granules within the trans-Golgi. Within these granules the final cleavages and modifications take place. For vasopressin and oxytocin, this means that there is cleavage to separate the nonapeptide from the neurophysin, and then a further cleavage to separate the neurophysin from the GP moiety (for pro-vasopressin). The former cleavage has been analyzed in the bovine *corpus luteum* and shown to involve the amino acid triplet -Gly-Lys-Arg- just C-terminal of the nonapeptide sequence [52]. First an endopeptidase cleaves after the arginine, this is followed by a carboxypeptidase cleavage to remove the basic amino acids, exposing the C-terminal glycine. This in turn is substrate for another, the so-called amidating enzyme, which uses the nitrogen of the glycine to replace this with the amide group characteristic of the C-termini of the nonapeptide hormones. This arrangement of using pairs of basic amino acids is a ubiquitous feature of many hormone precursors throughout the animal kingdom. What is less common is the other cleavage in these polyproteins, the single arginine residue for the vasopressin precursor, separating off the GP glycopeptide, or its equivalent C-terminal histidine in the bovine oxytocin precursor. There is evidence to suggest that the neurophysin moiety with its cysteine-rich globular core is important for presenting the precursor in a correct conformation to the processing enzymes [28]. Finally, the cleaved physiological peptides are stored in the electron-dense granules until the secreting cells receive the appropriate stimulation to allow membrane fusion of the granules with the plasma membrane and hence release of granular contents.

13.9 Summary

Taking the somewhat simple example provided by the hormones vasopressin and oxytocin, we can break down the genome sequence into the following overlapping informational features:

Non-Coding Sequence   1. Distance effects related simply to the closeness or otherwise of two sequences, e.g. the inverse repeat arrangement of the vasopressin and oxytocin genes.
2. Sequence responsible for a particular DNA conformation.
3. DNA- binding protein sequences.
4. Polymerase initiation sites.
5. Intron and splicing signals.
6. Transcription stop sites.
7. 5' Translational control signals.
8. 3' RNA regulatory signals.
9. Polyadenylation signals.

Coding Sequence
10. Signal peptide sequence.
11. Signal peptidase cleavage.
12. Physiological peptide.
13. Precursor cleavage signals.
14. Amidation signal.
15. Glycosylation signal.
16. Protein sequence providing conformational or spacer functions.

At present we have superficial knowledge about some of these features. The availability of regulatable expression in cell culture will now make it possible to analyze in more detail some of the otherwise less accessible aspects, particularly those sequence elements involved in the regulation of RNA transcription itself.

Acknowledgements

My gratitude is due to Drs. Craig McArdle and Norbert Walther for their helpful criticism of the manuscript and to Drs. Holtorf, McArdle, Furuya and McCracken for permission to cite the unpublished results of our collaborative studies. I am also grateful to the Deutsche Forschungsgemeinschaft (Forschergruppe Ho 388/6-1) for their financial support of some aspects of the studies presented.

13.10 References

[1] Ogawa, N., Thompson, T., Friesen, H.G., Martin, J.B., and Brazeau, P., *Biochem. J.* **1977**, *165*, 269-277.

[2] Srikant, C.B., and Patel, Y.C., *Nature* **1981**, *294*, 259-260.

[3] Grimmelikhuijzen, C., Dierickx, K., and Boer, G. *Neuroscience* **1982**, *7*, 3191-3199.

[4] Levitan, I.B., Harmar, A.J., and Adams, W.B., *J. Exp. Biol.* **1979**, *81*, 131-151.

[5] Martin, R., Froesch, D., Kiehling, C., and Voigt, K.H., *Neuropeptides* **1981**, *2*, 141-150.

[6] Liao, H., and Thorner, J., *Proc. Natl. Acad. Sci. USA* **1980**, *77*, 1898-1901.

[7] Adashi, E.Y., and Hsueh, A.J.W., *Endocrinology* **1981**, *109*, 1793-1795.

[8] Worley, R.T.S., Nicholson, H.D., and Pickering, B.T. in: *Recent Progress in Cellular Endocrinology of the Testis:* Saez, J.M., Forest, M.G., Dazord, A., and Bertrand, J. (eds.) Paris: Colloques de l'INSERM, *1985*, *123*, pp. 205-212.

[9] Ivell, R., Schmale, H., Krisch, B., Nahke, P., and Richter, D., *EMBO J.* **1986**, *5*, 971-977.

[10] Roberts, J.S., McCracken, J.A., Gavagan, J.E., and Soloff, M.S., *Endocrinology* **1976**, *99*, 1107-1114.

[11] Fuchs, A.R., and Fuchs, F., *Br. J. Obstet. Gynecol. 1984, 91,* 948-967.

[12] Burd, J.M., Davison, J., Weightman, D.R., and Baylis, P.H., *Acta Endocrinol. 1987, 114,* 458-464.

[13] Ivell, R., and Richter, D., *Proc. Natl. Acad. Sci. USA 1984, 81,* 2006-2010.

[14] Acher, R., and Chauvet, J., *Biochemie 1988, 70,* 1197-1207.

[15] Ruppert, S., Scherer, G., and Schuetz, G., *Nature 1984, 308,* 554-557.

[16] Sausville, E., Carney, D., and Battey, J., *J. Biol. Chem. 1985, 260,* 10236-241.

[17] Gainer, H., Altstein, M., and Hara, Y., *Proceedings of the 10th International Symposium on Neurosecretion,* Bristol *1987,* (in press).

[18] Nojiri, H., Ishida, I., Miyashita, E., Sato, M., Urano, A., and Deguchi, T., *Proc. Natl. Acad. Sci. USA 1987, 84,* 3043-3046.

[19] Sherman, T.G., Day, R., Civelli, O., Douglass, J., Herbert, E., Akil, H., and Watson, S.J., *J. Neurosc. 1988, 8,* 3785-3796.

[20] Zingg, H.H., and Lefebvre, D.L., *Mol. Brain Res. 1988, 4,* 1-6.

[21] Jard, S., *J. Physiol. Paris 1981, 77,* 621-628.

[22] Roesler, W.J., Vandenbark, G.R., and Hanson, R.W.,*J. Biol. Chem. 1988, 263,* 9063-9066.

[23] Chiu, R., Imagawa, M., Imbra, R.J., Bockoven, J.R., and Karin M., *Nature 1987, 329,* 648-651.

[24] Green, S., and Chambon, P., *Trends in Genetics 1988, 4,* 309-314.

[25] Kourilsky, P., *Trends in Genetics 1986, 3,* 60-63.

[26] Ollo, R., and Rougeon, F., *Cell 1983, 32,* 515-523.

[27] Van Leeuwen, F.W., Van Der Beek, M.P., Seger, M., Burbach, J.P.H., and Ivell, R., *Proc. Natl. Acad. Sci. USA 1989, 86,* 6417-6420.

[28] Ando, S., Murthy, A.S.N., Eipper, B.A., and Chaiken, I.M. *J. Biol. Chem. 1988, 263,* 769-775.

[29] Mohr, E., Schmitz, E., and Richter, D., *Biochimie 1988, 70,* 649-654.

[30] Nagy, G., Mulchahey, J.J., Smyth, D.G., and Neill, J.D., *Biochem. Biophys. Res. Commun., 1988, 151,* 524-529.

[31] Nagy, G., Makara, G., Banky, Z., Neill, J.D., and Halasz, B. in: *Progress in Neuropeptide Research:* Döhler, K.D., and Pawlikowski, M. (eds.) Basel: Birkhauser Verlag, *1989* (in press).

[32] Gilbert, W., *Nature 1978, 271,* 501.

[33] Valtin, H., Sokol, H.W., and Sunde, D., *Rec. Progr. Horm. Res. 1976, 31,* 447-486.

[34] Moore, S.E.H., and Hope, D.B., *British Neuroendocrine Group Inaugural Meeting,* Oxford *1985,* Abstract No. 32.

[35] Ferguson, D.R., *Gen. Comp. Endocrinol. 1969, 12,* 609-613.

[36] Sherman, T.G., and Watson, S.J., *J. Neurosc. 1988, 8,* 3797-3811.

[37] Ivell, R., Walther, N., and Morley, S., *Nucl. Acids Res. 1988, 16,* 7747.

[38] Mohr, E., Bahnsen, U., Kiessling, C., and Richter, D. *FEBS Lett. 1988, 242,* 144-148.

[39] Ivell, R., in: *Neuropeptides and their Peptidases:* Turner, A.J. (ed.) Chichester: VCH Horwood, *1987,* pp. 31-64.

[40] Ivell, R., Brackett, K., Fields, M.J., and Richter, D. *FEBS Lett. 1985, 190,* 263-267.

[41] Fehr, S., Ivell, R., Koll, R., Schams, D., Fields, M., and Richter, D., *FEBS Lett. 1987, 210,* 45-50.

[42] Jones, D.S.C., and Flint, A.P.F., *J. Endocr. 1988, 117,* 409-414.

[43] Ivell, R., and Richter, D., *EMBO J. 1984, 3,* 2351-2354.

[44] Richter, D., Schmale, H., Ivell, R., and Schmidt, C. in: *Biosynthesis, Modification and Processing of Cellular and Viral Polyproteins:* Koch, G., and Richter, D. (eds.) London: Academic Press, *1980,* pp. 43-66.

[45] Luck, M.R., and Jungclas, B., *J. Endocr. 1987, 114,* 423-430.

[46] McArdle, C.A., and Holtorf, A.P., *Endocrinology 1989* (in press).

[47] Morley, S.D., and Ivell, R., *Mol. Cell. Endocrinol., 1987, 53,* 255-258.

[48] Jeannotte, L., Burbach, J.P., and Drouin, J., *Mol. Endo. 1987, 1,* 749-757.

[49] Lacaze-Masmonteil, T., de Keyzer., Y., Luton, J.P., Kahn A., and Bertagna, X., *Proc. Natl. Acad. Sci. USA 1987, 84,* 7261-7265.

[50] Leff, S.E., and Rosenfeld, M.G., *Ann. Rev. Biochem. 1986, 55,* 1091-1117.

[51] Ivell, R., Schmale, H., and Richter, D., *Biochem. Biophys. Res. Commun. 1981, 102,* 1230-1236.

[52] Clamagirand, C., Camier, M., Fahy, C., Clavreul, C., Creminon, C., and Cohen, P., *Biochem. Biophys. Res. Comm. 1987, 143,* 789-796.

14. Bovine Somatotropin Gene Expression in Mammary Cells Directed by Various Eucaryotic Promoters

R. C. Gorewit

14.1 Introduction

Bovine somatotropin (bSTH) is a potent galactopoietic factor when injected into lactating cattle [1, 12]. Sejrson et al. [16] have recently shown that bSTH potentiates mammary development in virgin cattle. The mechanism by which bSTH effects mammary gland function has not been clearly defined. The hormone appears to repartition nutrients to the mammary gland when it is administered to dairy cows exogenously [1]. To date, bSTH receptors have not been characterized in bovine mammary cells. It is believed that somatotropin produces its effects on mammary tissue secondarily by stimulating the production of somatomedins (IGF1 and IGF2) [12]. Recombinant derived bSTH is as effective in stimulating milk secretion as the natural pituitary derived peptide [12].

Use of recombinant DNA technology has resulted in the cloning of a variety of somatotropin genes and their cDNAs including those derived from human, rat, bovine, and porcine tissue [5, 10, 14, 15]. Somatotropin genes have been expressed in both procaryotic and eucaryotic cells. DNAs complementary to human, bovine and porcine somatotropin mRNA have been expressed in *Escherichia coli* [5, 15]. Rat, human, and bovine somatotropin have been expressed in cultured mammalian cells [2].

Recombinant DNA molecules containing an avian retroviral promoter attached to the bSTH gene have been shown to direct expression of biologically active bSTH by cultured mouse fibroblasts [8, 9, 13]. Moreover, Pasleau et al. [11] were able to compare the relative levels of bSTH expression in mouse L cells and rat GH3 cells directed by the Rous sarcoma virus long terminal repeat or cytomegalovirus immediate early promoter. Also, others have used human somatotropin gene expression as a transcriptional reporter in eucaryotic cells to compare levels of gene expression directed by various eucaryotic promoters [17]. No attempts have been made to direct recombinant bSTH DNA expression in mammary cell types.

The objectives of our study were to [1] determine if transcriptional regulatory sequences which direct somatotropin synthesis in non mammary cell types are effective in bovine mammary cells, [2] compare somatotropin gene expression in primary cells versus subcultures, and [3] determine if cells grown on collagen matrices express higher levels of somatotropin than cells grown on plastic.

14.2 Materials and Methods

Secretory epithelial and myoepithelial cell types were isolated from mammary tissue of two lactating Holstein cows by an adaption of the methods described by Soloff et al. [18] for rat mammary tissue. Primary cultures of these cells were established directly after their isolation. Stable cultures were considered established after 25 consecutive trypsinizations and reculture of the primary cells. Each culture and reculture was about 7 days. Culture medium (maintenance medium) consisted of Delbecco's modified Eagles medium containing 5% bovine serum collected from the cows at slaughter, 2.38 mg/ml Hepes, 0.35 mg/ml L-glutamine, 0.2 U/ml penicillin.

A transient eucaryotic expression assay system was employed. The technique used has been described previously [7, 8]. Briefly, approximately 5.0×10^5 cells were obtained after trypsinization of stock cultures and plated onto 35 mm tissue culture plates in the absence or presence of bovine epidermal collagen (Vitrogen 100, Collagen Corporation, Palo Alto, CA) in maintenance medium. Following an overnight incubation, either cells established directly as primary cultures or as subcultures were washed 3 times with plain Delbecco's medium, minus serum. A 1.0 ml solution containing DNA (1 µg) and DEAE-dextran (200 µg) in phosphate buffered saline, pH 7.4, was added to the cells. Cells were transfected with bovine somatotropin (bSTH) rDNA, containing the following eucaryotic promoters: human cytomegalovirus immediate-early promoter (pCMVIE-bGHb). Simian virus 40 early promoter (pBGH-20), the mouse metallothionein I promoter (pBGH-10) and the Rous sarcoma virus long terminal repeat (pBGH-5). Following incubation, cells were washed with Delbecco's medium without serum. Cells treated with pCMVIE-bGHb, pBGH-5, and pBGH-20 were then incubated in maintenance medium. Cells transfected with the mouse matallothionein I promoter were divided into 2 subtreatment groups. One group of cultures was incubated with maintenance medium containing 5% bovine serum and 100 µM $ZnCl_2$ while the other was incubated with maintenance medium containing 5% bovine serum without $ZnCl_2$. The purpose of the $ZnCl_2$ was to activate the promoter. Medium was collected every 24 hrs for 3 days and then every third day for 20 to 28 days. Bovine somatotropin in medium was assayed by the method of Leung et al. [9]. The concentration of bSTH was determined in fresh maintenance medium. The bSTH concentrations measured from cells post transfection were all adjusted. This was done by subtracting the bSTH concentration contained in the maintenance medium.

Constructs of the plasmids are shown in Fig. 14-1. Each construct contains a BamHI-coRI restriction fragment (approx. 1.9 kb) encoding the five exons (closed boxes), four introns, and a 3'-untranslated flanking sequence of the bSTH gene cloned in PBR322 (striped lines). Translational initiation (ATG) and termination (TAG) codons are indicated

(Fig. 14-1). The transcriptional regulatory sequence, TATA box, is shown. The SV40 72 base pair and 21 base pair repeats are indicated (Fig. 1B). pBGH-5 contained an NCOI-BSTEII restriction fragment (about 1.0 kb) derived from a plasmid clone of a Schmidt Ruppin B strain of RSV pL397. This DNA fragment encodes a portion of the viral envelope gene. pCMVIE-bGHb contained a 1.1 kb DNA fragment isolated from a plasmid clone derived from the Esienhardt strain of human cytomegalovirus and encodes a viral immediate early promoter [11].

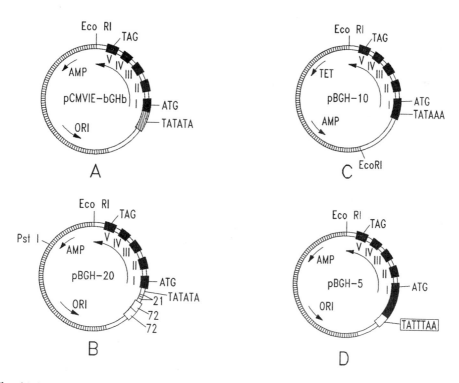

Fig. 14-1 Constructs of plasmid DNA's: (A) pCMVIE-bGHb, (B) pBGH-20, (C) pBGH-10 and (D) pBGH-5. All plasmids contain a BamHI-EcoRI fragment from the bSTH gene.

The SV40 early promoter was ligated to the bSTH gene in the following manner. pSVO (from Peter Lamedro, Hoffman La Roche) was cleaved with HindIII and EcoRI and a 415 base pair fragment containing the SV40 early promoter was isolated. pBGH-2 DNA was cleaved with SalI and BamHI as described previously [7, 15]. All restriction endonucleases used in these cleavages generate 5'protruding termini. The ends were made flush by addition of deoxynucleotides using DNA polymerase I. The resulting blunt-end molecules were ligated and transfected into *E. coli* I cells. A plasmid containing the SV40 early sequence attached to the bGH gene was isolated (pBGH-20).

The mouse metallothionein I promoter was ligated to the bSTH gene as follows: pMK DNA (from R. Ellis, Merck & Co.) was cleaved with BglII and EcoRI and a 750 bp fragment containing the promoter isolated. Similarly pBGH-2 was cleaved with EcoRI and BamI and a 2.2 linear fragment isolated. pBR322 was linearized with EcoRI. Following a three fragment ligation, *E. coli* RRI cells were transfected using standard procedures. A plasmid containing the metallothionein I promoter ligated to the bGH gene and was isolated (pBGH-10).

Statistical Analysis: Significance of means were tested using a Student's test [4].

14.3   Results and Discussion

14.3.1 Gene Expression in Primary Cultures

Myoepithelial cell types: Mean concentrations of bGH in medium from myoepithelial cell types, in primary culture, treated with the four transcriptional regulatory sequences are shown in Table 14-1. The pCMVIE-bGHb was effective in directing bSTH gene expression in myoepithelial cell types grown on plastic and collagen. Bovine somatotropin reached significant (p < .001) concentrations by 2 days post transfection from cells grown on both matrices when compared with day 1 sample means. Somatotropin concentrations were higher from those cells grown on plastic, for the first 8 days after transfection, when compared with those grown on collagen. After this, bSTH concentrations were higher in medium from cells grown on collagen (days 11 and 17). Somatotropin concentrations ranged from 5 to 50 ng/ml. The Simian virus 40 early promoter (pBGH-20) directed bovine somatotropin gene expression in primary cultures of myoepithelial cells grown on both plastic and collagen. Somatotropin concentrations were significantly (p < .001) elevated by 2 days post transfection (Table 14-1). Somatotropin in medium from cells grown on plastic was higher than bSTH from cells grown on collagen at days 2 to 8. At days 11 to 21, bSTH concentrations were higher for cells grown on collagen when compared to those grown on plastic.

The mouse metallothionein I promoter (pBGH-10) only directed bSTH gene expression in those cells grown on plastic, but is was weakly effective compared to pCMVIE-bGHb and pBGH-20. Peak concentrations of bSTH were 3.4 ng/ml on day 8 post transfection (Table 14-1).

The *Rous sarcoma virus* long terminal repeat (pBGH-5) was ineffective in directing detectable levels of bSTH synthesis in myoepithelial cells grown on plastic and collagen (Table 14-1).

The four eucaryotic transcriptional regulatory sequences could be ranked in terms of their effectiveness to direct bovine somatotropin gene expression in primary cultures of myoepithelial cell types as follows: pCMVIE-bGHb > pBGH-20 > pBGH-10 > pBGH-5.

Table 14-1 Mean concentrations of bSTH (ng/ml) in medium of myoepithelial cell types (primary cultures)[1].

Promoter	Substrate	Days Post Transfection									
		1	2	3	4	5	8	11	17	21
pCMVIE-pGHb	Plastic	5.0	48.3	36.0	30.0	50.0	38.0	19.3	9.4	3.5
	Collagen	3.5	23.3	25.0	19.3	19.0	33.0	20.6	21.8	-3.5
pBGH-20	Plastic	0.2	5.0	7.7	6.0	7.3	16.8	8.4	1.6	-3.9
	Collagen	0.2	3.9	3.2	2.7	3.1	6.6	11.5	11.1	7.8
pBGH-10[2]	Plastic	-0.2	-1.1	1.2	1.6	1.6	3.4	1.7	-1.9	-2.9
	Collagen	-0.9	0.6	1.1	-0.9	-1.0	-1.3	-1.9	-2.8	-2.5
pBGH-5	Plastic	0.4	0	-0.2	-0.3	0	-0.1	-1.6	-2.9	-2.4
	Collagen	-1.4	-1.1	-1.0	-1.3	-0.3	-0.8	-1.8	-2.4	-3.4

[1] Data are corrected for concentrations of bSTH in maintenance medium as described in the Methods section. Means were calculated from quadruplicates. Standard errors of the means ranged from 3 to 7 percent.

[2] Data are shown from cell cultures incubated in the presence of 100 µM $ZnCl_2$. No difference was seen between cells cultured in the presence or absence of $ZnCl_2$.

Secretory epithelial cell types: Table 14-2 lists mean concentrations of somatotropin in medium of mammary secretory epithelial cell types, in primary cultures, treated with the four promoters. pCMVIE-bGHb was effective in directing bSTH gene expression in primary cultures of mammary secretory epithelial cell types grown on plastic and collagen. Peak concentrations of bSTH from cells grown on plastic were seen on day 8 post transfection. Somatotropin concentrations adjusted for the content of hormone in the maintenance medium were 28.0 ng/ml at this time. Somatotropin concentrations remained elevated for 14 days post transfection. Maximum bSTH concentrations for cells grown on collagen were reached on day 11. Concentrations of bSTH were maintained up to 14 days post transfection (Table 14-2).

pBGH-20 directed bSTH gene expression in primary cultures of mammary gland secretory epithelial cell types. Concentrations of bSTH were similar for cells grown on either matrix up to 3 days post transfection. Concentrations of bSTH were higher in medium of cells on collagen, on days 5 and 8, when compared to those cells grown on plastic. Peak expression occurred at day 8 post transfection for cells grown on either substrate.

The mouse metallothionein I promoter (pBGH-10) was effective in directing gene expression in primary cultures of epithelial cell types grown on plastic. However, expression was weak compared to pCMVIE-bGHb and pBGH-20.

The *Rous sarcoma virus* promoter (pBGH-5) did not direct bSTH expression in primary cultured epithelial cell types on either substrate (Table 14-2). The promoters could be ranked as far as their effectiveness in directing bSTH gene expression bSTH gene expression as follows: pCMVIE-bGHb > pBGH-20 > pBGH-10 > pBGH-5.

Table 14-2 Mean concentrations of bSTH (ng/ml) in medium of secretory cell types (primary cultures)[1].

Promoter	Substrate	Days Post Transfection									
		1	2	3	4	5	8	11	14	17	20
pCMVIE-											
pGHb	Plastic	1.1	9.3	13.0	15.0	15.0	28.0	18.0	5.0	1.1	-0.1
	Collagen	1.6	-0.3	0.5	8.0	4.0	20.0	29.4	29.0	-0.7	-0.1
pBGH-20	Plastic	0.9	2.3	3.2	3.1	3.2	7.7	5.0	0.9	0.3	-1.2
	Collagen	-0.4	2.1	2.5	8.0	19.5	21.1	5.7	1.8	1.8	-0.7
pBGH-10[2]	Plastic	-0.3	0.4	0.7	1.0	1.8	4.6	2.8	0.3	-0.8	-1.6
	Collagen	-0.2	-0.3	-0.2	0	-0.2	-0.3	0	-0.1	0	-0.3
pBGH-5	Plastic	-2.2	-1.0	-2.0	-2.3	-2.2	-2.6	-0.2	-0.1	-0.1	0
	Collagen	-1.0	0.1	0	-0.1	-0.1	-0.1	0	-0.1	-0.2	-0.1

[1] Data are corrected for concentrations of bSTH in maintenance medium as described in the Methods section. Means were calculated from quadruplicates. Standard errors of the means ranged from 4 to 10 percent.
[2] Data are shown from cell cultures incubated in the presence of 100 μM $ZnCl_2$. No difference was seen between cells cultured in the presence or absence of $ZnCl_2$.

14.3.2 Gene Expression in Mammary Subcultures

Myoepithelial cell types: The cytomegalovirus immediate early promoter (pBGH-20) acted as a potent regulatory sequence in directing bSTH gene expression in myoepithelial cell cultures (Table 14-3). Cells grown on plastic showed maximum gene expression on day 7 post transfection. At this time, the peak concentration of bSTH was 59 ng/ml. Cells grown on collagen responded to pCMVIE-bGHb in a similar fashion. However, peak concentrations of bSTH in medium were nearly twice (115 ng/ml) that when compared to these cells grown on plastic. Somatotropin concentrations remained elevated in cells grown on both substrates up to 25 days post transfection (Table 14-3).

The SV40 early promoter also was effective in directing bSTH gene expression in myoepithelial cell types grown on plastic and collagen. For cells grown on plastic, bSTH concentrations fluctuated somewhat from days 7 to 16, but a peak in bSTH was reached during this interval on day 13 post transfection. Somatotropin remained elevated up to 25 days post transfection when compared to those cells grown on collagen. Peak bSTH concentrations (46 ng/ml) were seen by day 7 post transfection for cells grown on collagen (Table 14-3). Concentrations of bSTH remained elevated for 22 days post transfection.

Table 14-3 Mean concentrations of bSTH (ng/ml) in medium of myoepithelial cell types (subcultures)[1].

Promoter	Substrate	Days Post Transfection											
		1	2	3	4	7	10	13	16	19	22	25	28
pCMVIE-pGHb	Plastic	-1.6	1.2	16.8	22.9	59.2	18.6	11.5	10.1	12.2	5.4	3.0	-8.6
	Collagen	2.7	35.0	82.0	93.0	115.0	42.0	31.0	20.0	21.0	14.1	14.3	-5.8
pBGH-20	Plastic	-0.1	-0.9	2.0	1.8	26.0	15.0	122.0	28.0	16.0	1.3	15.0	-5.6
	Collagen	2.8	9.8	23.9	31.0	46.0	26.0	20.0	17.0	11.0	9.4	-5.1	-5.1
pBGH-10[2]	Plastic	0	-0.9	-0.6	-0.6	-1.9	0	0.1	-0.8	-0.6	-0.5	0	2.0
	Collagen	1.3	1.0	-0.8	-1.4	-0.1	-0.9	1.0	0	-0.8	-0.1	-0.9	-1.1
pBGH-5	Plastic	-0.4	0	-0.9	-0.3	0	0	0	-0.1	-0.1	-0.1	0	0
	Collagen	0.9	0.4	0.3	-0.1	0	-0.1	0	-0.1	0	-0.1	-0.1	0

[1] Data are corrected for concentrations of bSTH in maintenance medium as described in the Methods section. Means were calculated from quadruplicates. Standard errors of the means ranged from 7 to 10 percent.

[2] Data are shown from cell cultures incubated in the presence of 100 µM $ZnCl_2$. No difference was seen between cells cultured in the presence or absence of $ZnCl_2$.

The mouse metallothionein I promoter did not direct detectable levels of somatotropin expression in subcultured mammary myoepithelial cells on either substrate.

Myoepithelial cell types also did not respond to the *Rous sarcoma* virus LTR on either plastic or collagen.

Secretory Epithelial Cell Types: The cytomegalovirus immediate early promoter was effective in directing gene expression in subcultured secretory epithelial cells (Table 14-4). pCMVIE-bGHb appeared to be most effective in directing gene expression in cells grown on collagen, when compared to those cells grown on plastic. Peak concentrations of bSTH were seen at day 11 in those cells on plastic and on day 8 in those cells on collagen, respectively. Concentrations of bSTH were roughly twice those achieved by cells growing on plastic.

The SV40 early promoter was about as effective as pCMVIE-bGHb in directing bSTH gene expression in subcultured mammary secretory epithelial cells (Table 14-4). Higher levels of expression were seen in cells grown on collagen.

The mouse metallothionein I and pBGH-5 promoters were ineffective in directing bovine somatotropin gene expression in subcultured secretory epithelial cells.

Pasleau et al. [11] have shown that the cytomegalovirus immediate early gene regulatory region was three to fourfold more efficient than the *Rous sarcoma* retroviral long terminal repeat in promoting expression of the bovine growth hormone gene in rat GH3 cells.

    In the present study, the cytomegalovirus and SV 40 early regulatory sequences were equally effective in directing gene expression in both primary cultures and subcultured secretory epithelial and myoepithelial cell types. The mouse metallothionein I promoter was able to direct gene expression in primary cultures of both cell types. It was not as strong a promoter as the cytomegalovirus and SV-40 sequences. The *Rous sarcoma* virus long terminal repeat was not effective in directing synthesis and secretion of detectable levels of bSTH in either cell type. The matrix that the cells were grown on made no difference.

Table 14-4   Mean concentrations of bSTH (ng/ml) in medium of secretory cell types (subcultures)[1]

Promoter	Substrate	Days Post Transfection								
		1	2	3	4	5	8	11	17	20
pCMVIE-										
pGHb	Plastic	1.1	-0.7	0.7	2.6	3.9	4.2	4.8	2.0	1.0
	Collagen	2.1	3.0	1.5	5.0	10.0	11.2	9.0	4.0	1.0
pBGH-20	Plastic	1.0	3.6	2.7	3.3	4.4	3.9	4.3	4.7	2.0
	Collagen	0.9	5.0	5.6	6.1	8.0	8.1	7.9	2.0	2.1
pBGH-10[2]	Plastic	-0.8	-1.0	-1.0	-0.9	-0.9	-1.0	-1.9	-1.1	0
	Collagen	-0.9	-1.0	-0.9	-1.0	-1.3	-1.9	-1.3	2.8	-0.1
pBGH-5	Plastic	0.3	0.2	0.3	-3.0	-3.1	-0.4	-3.2	-1.1	-0.9
	Collagen	0.5	-0.9	-0.1	-0.2	-0.1	0	0.1	0	0

[1] Data are corrected for concentrations of bSTH in maintenance medium as described in the Methods section. Means were calculated from quadruplicates. Standard errors of the means ranged from 5 to 10 percent.
[2] Data are shown from cell cultures incubated in the presence of 100 μM ZnCl$_2$. No difference was seen between cells cultured in the presence or absence of ZnCl$_2$.

    The extent of bSTH expression appeared to be dependent upon the type of promoter used and the substrate which the cells were grown on. This may be related to the degree of cellular differentiation at the time of transfection. Primary cultures of bovine mammary cells secrete lactose and beta-lactoglobin up to 72 hours of culture on glass and are thought to dedifferentiate biochemically over time [3]. We have shown that the fine structure of the bovine mammary myoepithelial cell can change depending on the substrate on which they are cultured. Cultures of myoepithelial cells grown on collagen appear to be highly differentiated when compared to those on plastic [6]. The matrix on which the cells are grown and the degree of their subcellular differentiation, at the time of plasmid insertion, appear to be important factors to be considered when introducting recombinant genetic material into eucaryotic cell types. There may also be differences in translation of promoter sequences and/or post translational packaging of peptides for secretion, which are related to the type of matrix on which the cells are grown.

To our knowledge, no reports have been published describing the insertion of genetic material into cultures of bovine mammary cells. Moreover, this represents the first study where eucaryotic regulatory sequences have been evaluated in primary cultures and compared with established cell lines.

The techniques described in this paper will be useful in studying the physiological and biochemical mechanisms whereby mammotropic agents, such as somatotropin, influence the development and function of mammary cell types.

14.4 Abstract

Mammary gland cells were isolated from lactating cows. Primary cultures and subcultures were established. Cells were plated onto dishes with or without collagen gels. Cells were transfected with bovine somatotropin recombinant deoxyribonucleic acid containing the following eucaryotic promoters: human cytomegalovirus immediate-early, simian virus 40 early, mouse metallothionein I and the *Rous sarcoma* virus long terminal repeat. Medium was collected every 24 hrs for 3 days and every third day for 20 to 28 days. Transcriptional regulatory sequences were evaluated for their ability to direct recombinant bovine somatotropin deoxyribonucleic acid expression in the cultured cells.

All regulatory sequences except the *Rous sarcoma* virus directed bovine somatotropin gene expression in primary cultures of secretory epithelial cells and myoepithelial cells. The cytomegalovirus and simianvirus 40 early transcriptional sequences directed somatotropin gene expression in subcultured mammary cells. The mouse metallothionein I promoter and the *Rous sarcoma* long terminal repeat did not direct detectable levels of somatotropin. Somatotropin expression was greatest in subcultured myoepithelial cells. The substrate on which the cells were grown appeared to influence expression.

Acknowledgments

This work was supported, in part, by the New York State Agricultural Experiment Station. The author wishes to express his appreciation to Drs. Howard Y. Chen and John J. Kopchick for their technical help in carrying out this work.

14.5 References

[1] Bauman, D.E., and McCutcheon, S.N., The effects of growth hormone and prolactin on metabolism. In: *Control of Digestion and Metabolism of the Rumen*. L.P. Milligan, W.L. Groven and A. Dobson, editors. Prentice-Hall, Englewood Cliffs 436, *1986*.

[2] Cathala, G., Erhardt, N.C., Lan, N.C., Gardner, D.G., Gutierrez-Hartmann, A., Mellon, S.H., Karin, M., and Baxter, J.D., Structure and expression of growth hormone-related genes. In: *Perspectives on Genes and the Molecular Biology of Cancer*. Roberson, D.L., and Sanders, G.F (eds.) Raven Press, New York 169, *1983*.

[3] Ebner, K.E., Hoover, C.R., Hageman, E.C., and Larson, B.L., *Experimental Cell Research 1961, 23,* 373.

[4] Federer, N.T., *Experimental Design, Theory and Application*. The Macmillan Co., New York, *1955*.

[5] Goeddel, D.V., Heyneker, H.L., Ozumi, T., Arentzen, R., Itakura, K., Yansura, D.G., Ross, J.J., Miozari, G., Crea, R., and Seeburg, P.H., *Nature 1979, 281,* 544.

[6] Gorewit, R.C., and Rizzo, N.W., *J. Dairy Sci. 1986, 69,* 204.

[7] Kopchick, J.J. and Stacy, D.W., *Mol. Cell Biol. 1984, 4,* 240.

[8] Kopchick, J.J., Malavarca, R.H., Livelli, T.J., and Leung, F.C., *DNA 1985, 4,* 23.

[9] Leung, F.C., Jones, B., Steelman, S.C., Rosenblum, C.I., and Kopchick, J.J., *Endocrinology 1986, 119,* 1489.

[10] Miller, W.L., Martial, J.A., and Baxter, J.D., *J. Biol. Chem. 1980, 255,* 7521.

[11] Pasleau, F., Tocci, M.J., Leung, F., and Kopchick, J.J., *Gene 1985, 38,* 227.

[12] Peel, C.J., and Bauman, D.E., *J. Dairy Sci. 1987, 70,* 474.

[13] Robins, D.M., Paek, I., Seeburg, P.H., and Axel, R., *Cell 1982, 29,* 623.

[14] Seeburg, P.H., Shine, J., Martial, J.A., Baxter, J.D., and Goodman, H.M., *Nature 1977, 270,* 486.

[15] Seeburg, P.H., Sias, S., Adelman, J., DeBoer, H.A., Hayflick, J., Jhurani, J., Goeddel, V., and Heyneker, A.L., *DNA 1983, 2,* 37-45.

[16] Sejrsen, K., Foldager, J., Sorensen, M.T., Akers, R.M., and Bauman, D.E., *J. Dairy Sci. 1986, 69,* 1528.

[17] Seldin, R.F., Howie, K.B., Rowe, M.E., Goodman, H.M., and More, D.D., *Mol. Cell Biol. 1986, 6,* 3173.

[18] Soloff, M.S., Chakraborty, J., Sadhukan, P., Sanitzer, D., Wieder, M., Fernstrom, M.A., and Sweet, P. *Endocrinology 1980, 106,* 887.

15. Major Histocompatibility Genes in Cattle and Their Significance for Immune Response and Disease Susceptibility

L. Andersson

15.1 Introduction

The major histocompatibility complex (MHC) was discovered about 50 years ago as a major locus controlling the rejection of foreign tissue grafts (transplantation) in the mouse. The molecules encoded by the MHC therefore have been referred to as transplantation antigens. The MHC in man, designated HLA, was discovered about thirty years ago as a blood-group-like system detected on lymphocytes [1]. A major histocompatibility complex has subsequently been described in a number of mammalian, avian, and amphibian species including all domestic animal species [1]. The MHC systems in domestic animals have been reviewed by Newman and Antczak [2] and more recently at a symposium on the molecular biology of the MHC in domestic animals [3].

There are two classes of MHC molecules, denoted class I and class II, which are both integral cellsurface glycoproteins. The class I molecule is found on virtually all nucleated somatic cells as a 45.000 dalton polypeptide chain non-covalently associated with $ß_2$-microglobulin (which is not encoded in the MHC). The class II molecule is a dimer composed of an $α$ and $ß$ chain with molecular sizes of about 34.000 and 29.000 daltons, respectively. The class II molecules are expressed primarily in B lymphocytes, macrophages, and some other antigen presenting cells (see Klein [1] for a comprehensive review of the MHC). The MHC class I, class II, and $ß_2$-microglobulin molecules all belong to the immunoglobulin superfamily which among other members also includes immunoglobulin and T-cell receptor molecules of the immune system; the genes encoding these molecules are assumed to have evolved from a common ancestral gene [4].

The human and murine MHC regions have been characterized extensively through gene mapping [5, 6, 7, 8, 9]; the complete MHC regions in these species span more than $2x10^6$ base pairs of DNA. Multiple MHC class I and class II genes are located in the region as well as a number of non-MHC genes such as the genes for some complement factors, 21-hydroxylase, and tumor necrosis factor. The human and murine class I regions include 20 or more genes but only a minority (about three) encode the classical transplantation antigens. A similar basic organization of the class II region has been found in several mammalian species. The class II region in man is divided into three major subregions designated DP, DQ, and DR. Since a class II molecule is composed of an α and ß chain, each of these subregions comprises one or more copies of α and ß genes. No less than six α and nine ß class II genes are known in man. However, some of the genes are evidently pseudogenes and protein products have so far only been found for three α chain and four ß chain genes.

Several of the class I and class II genes in the MHC region are extremely polymorphic in almost all species studied so far. These loci are in fact the most polymorphic protein loci known. The MHC polymorphism differs in several respects from other protein polymorphisms. Firstly, the number of alleles at a single locus is often extremely high; more than 100 alleles are assumed to exist at some class I loci in the mouse [1]. Secondly, the number of amino acid replacements between alleles is extremely high; alleles at MHC loci may differ by more than 10% in their amino acid sequence [10, 11]. Thirdly, there is often an even allele-frequency distribution without any predominant allele. The allele-frequency distribution at MHC loci in man has been reported to deviate significantly from the one expected for selectively neutral alleles [12]. There is strong support for the concept that MHC polymorphism is maintained by natural selection.

The genetic polymorphism at MHC loci is clearly associated with phenotypic effects. It is well established that the humoral and cellular immune responses to foreign antigens are influenced by the MHC; the effect is particularly pronounced for simple antigens such as short peptides [1, 13]. Furthermore, numerous associations between MHC polymorphism and the susceptibility to various diseases have been reported in different species [1, 14, 15]. The basis for the MHC effects on disease susceptibility and on the immune response was for a long time obscure. However, the important role of the MHC molecules became clear when it was discovered that T cells of the immune system recognize foreign antigens in association with self MHC molecules [16]. Cytotoxic T cells elicit their immune response (i.e. killing) after they have recognized foreign antigens on the cell surface together with self class I molecules. T helper cells recognize foreign antigens together with self class II molecules on antigen presenting cells. T helper cells, in turn, influence the activity of other cells in the immune system e.g. B lymphocytes which produce antibodies.

A possible explanation for the immune response effects associated with MHC polymorphism is that MHC variants differ in their ability to bind foreign antigens before presenting them to T cells [17, 18]. This concept was supported when the threedimensional structure of a class I molecule was recently reported [19, 20]. A presumed antigen binding site, formed by side chains from two α helices and a group of antiparallel ß-strands, was recognized in this molecule. According to the current model of antigen presentation, this antigen binding site binds short peptides of foreign antigens which have been processed intracellularly. Interestingly, the majority of polymorphic amino acid residues in MHC molecules are located in the antigen binding site. Thus, it is likely that the MHC polymorphism has a direct influence on the binding and presentation of foreign antigen. Recent statistical analyses of allelic DNA sequences of human and murine MHC genes showed that there is selection for polymorphism in the codons for the antigen binding site of both class I [21] and class II genes [22].

15.2 The MHC in Cattle

The BoLA system (for Bovine Lymphocyte Antigens) has been studied since the early seventies. It is well established that BoLA is the MHC of cattle. For instance, matching for BoLA polymorphism results in prolonged allograft survial [23]. It appears to be a striking similarity between BoLA and the MHC of other species both as regards its genetic organization as its functional role in the immune system.

15.3 Serological and Cellular Studies of BoLA

Genetic polymorphism of bovine class I molecules has been extensively studied by serological methods [24, 25]. In fact, the majority of the published reports on the BoLA system deals with the serological identification of polymorphic class I molecules. A number of different strategies have been used to obtain class I typing sera [2]. These include lymphocyte immunizations, skin grafting, and the collection of sera from parous animals. These procedures give polyclonal alloantisera that are made up of complex mixtures of antibodies to many different epitopes. It may therefore be necessary to absorb the sera before they are used for typing. Statistical evaluations of typing results are carried out to find clusters of antisera which define antigenic specificities. Several international workshops have been organized to compare typing reagents produced in different laboratories. As a result of the second BoLA workshop, 17 class I specificities were accepted and given workshop (w) designations [26]. Additional class I specificities were established in the third BoLA workshop but the results have not yet been published. A single locus, designated BoLA-A, has been assumed to control class I specificities in cattle.

It has been more difficult to apply the serological method for studying the polymorphism of bovine class II molecules. An obvious explanation for these difficulties is that class I molecules are expressed on all nucleated cells while class II is only expressed on a restricted subset of cells such as B cells and macrophages. No allelic series of class II specificities has yet been established at the BoLA workshop. However, Davies [27] were able to identify five class II specificities after immunizing cattle with class I compatible-class II incompatible lymphocytes.

Cellular studies on the genetic control of mixed lymphocyte reactions (MLR) have been used to show the presence of a bovine class II region [27, 28]. Evidence for close genetic linkage between the class II region, designated BoLA-D, and serological class I specificities, BoLA-A, has been shown [29]. The use of alloreactive bovine T cell clones for detecting polymorphism of class II molecules has also recently been described [30].

15.4   Biochemical Characterization of BoLA Molecules

Immunoprecipitations and SDS-polyacrylamide gel electrophoresis have been used to characterize the subunit composition and molecular sizes of bovine MHC molecules [31, 32, 33]. The bovine class I molecule is a glycoprotein with a molecular weight of about 44.000 which is associated with a smaller subunit, most likely ß$_2$-microglobulin, having a molecular weight of about 12,000. Class II molecules were found to be composed of subunits with molecular weights of approximately 27,000 to 34,000. These biochemical characteristics are thus in good agreement with those found for the corresponding molecules of man and mouse (see 15.1).

One-dimensional isoelectric focusing (IEF) has recently been used to study the polymorphism of bovine class I and class II molecules [34, 35]. IEF is performed on immunoprecipitated MHC molecules after they have been metabolically labeled with ^{35}S-methionine. IEF separates proteins according to charge and is expected to be an efficient method for resolving allelic MHC molecules which in many cases differ by multiple amino acid substitutions (see 15.1). Joosten and co-workers [34, 35] were able to resolve allelic series of both class I and class II molecules by this method.

15.5   Molecular Characterization of BoLA

Many class I and class II genes in man and mouse have been isolated as cDNA and/or genomic clones [1, 5, 6, 7, 8, 9]. This rich source of molecular probes has been utilized for the characterization of BoLA genes by Southern blot analysis and by gene cloning.

15.5.1 Organization of the BoLA Complex

The present status of the genetic map of the BoLA complex is shown in Fig. 15-1 in comparison with the much more detailed map of the HLA complex in humans. BoLA has been assigned to chromosome 23 in cattle by *in situ* hybridization [38]. It should be noted that the HLA region is huge in molecular terms spanning more than 3×10^6 base pairs [9], which implies most likely many additional genes not yet identified in this region. Very little is known so far about the linear order of genes along the chromosome in cattle and the map has been drawn according to the location of homologous genes in the HLA complex. There is no data contradicting such an organization and the order of genes is well conserved in the MHC of man and mouse. As shown in Fig. 15-1 the MHC is divided into three regions harboring class I, class II, and class III genes.

The class I region in humans comprises about 20 genes. However, only three loci (HLA-A, -B, and -C) encode the classical transplantation molecules. The function of remaining class I genes, if any, is still obscure and several of them are evidently pseudogenes. There is about 90% nucleotide sequence similarity among HLA-A, -B, and -C genes [39]. A comparison of class I nucleotide sequences from species representing different mammalian families shows that the interspecies similarity is in the range 80-85% [39]. Thus, class I sequences from different loci within species are in general more similar to each other than they are to any class I sequence from other mammalian species.

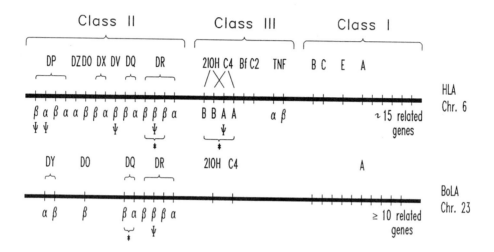

Fig. 15-1 Comparison of the MHC regions in man (HLA) and cattle (BoLA). ψ marks known pseudogenes and * marks regions where the number of genes has been shown to vary between haplotypes. The HLA map is adapted from Trowsdale and Campbell [9]. The BoLA map is adapted from Andersson [36] including information on the assignment of 210H to this chromosome [37].

Bovine class I genes have been investigated by Southern blot analysis using a human cDNA probe [40]. The probe cross-hybridized to a large number of genomic restriction fragments. The result indicated the presence of at least 10 bovine genes which is comparable to the number of human class I genes (Fig. 15-1). A much better knowledge on this region can be obtained when bovine class I genes have been isolated. In fact, two groups have recently isolated bovine cDNA clones [41, 42]. In one of these studies [41], two quite distinct cDNA clones, with only 92% similarity in nucleotide sequence, were found. The result strongly suggests the presence of at least two expressed class I genes in cattle.

The class II region harbors the genes encoding the α and ß subunits of class II molecules (Fig. 15-1). In man, it is divided into three major subregions designated DP, DQ, and DR which includes all class II genes which have been found to be expressed at the protein level. In addition, there are a DX α and ß gene pairs, which are very closely related to the DQ genes, and three individual genes designated DZα, DOß, and DVß; no protein product of any of these genes has yet been described. α and ß genes of different subregions show about 60% similarity in nucleotide sequence [11] and have evolved from common ancestral α and ß genes.

Probes representing all types of human class II genes, except DVß, have been used for Southern blot analysis in cattle [43,44,45,46]. All probes, except DPα and DPß, cross-hybridized strongly to bovine genes and gave specific hybridization patterns; an example where different human class II α probes have been used is shown in Fig. 15-2. By this approach bovine homologous of DQα, DQß, DRα, DRß, DOß, and DZα genes were identified. Two additional genes were also found, and they were designated DYα and DYß since their relationship to human class II genes is not yet clear [46]. It appears that bovine DP homologous have either been lost during evolution or have diverged from the human genes to such an extent that good cross-hybridization is not obtained.

The number of different class II genes in cattle has been estimated by careful Southern blot analyses using very short (about 200 base pairs) exon-specific probes [45]. These studies indicated the presence of single DRα, DOß, DYα, DYß, and DZα genes, and at least three DRß genes. As regards DQ, the number of genes was found to vary between haplotypes. Some BoLA haplotypes carry a single DQα and a single DQß gene while other haplotypes carry two copies of each gene. Similarly, the number of human DRß genes varies between haplotypes [47]. Thus, there is a minimum number of four α and six ß genes in the bovine class II region. However, similar to the situation in man, it is likely that not all of these genes are expressed at the protein level. In fact, the only published DNA sequence of a bovine class II gene so far was found to represent a bovine DRß pseudogene [48]; in accordance with this finding one of the DRß genes in Fig. 15-1 is depicted as a pseudogene.

The DQ, DR, DO and DY genes have all been assigned to the BoLA complex by linkage analysis of RFLPs (RFLP - restriction fragment length polymorphism) [46]; DZα is also most likely located in the BoLA complex but this has not yet been possible to test. In contrast to the situation in man and mouse a high frequency of recombination (about 20%) was found in the class II region. Recombination was found to occur in the interval between DO/DY and DQ/DR genes (Fig. 15-1). The recombination frequency between genes within groups appears however to be very low in particular among DQ-DR genes which are in very strong linkage disequilibrium [49]. The high recombination frequency in the bovine class II region compared with the situation in man and mouse, is either explained by a much larger molecular distance between these loci in cattle or by the presence of a recombinational "hot spot".

Genes for the complement components C4, C2, and Bf, as well as the ones for the enzyme 21-hydroxylase, and for tumor necrosis factor α and ß, are all located in the class III region in man and mouse (Fig. 15-1). It should be noted that these genes are not structurally or functionally related to the "true" MHC genes i.e. class I and class II genes.

The fact that they are located between class I and class II genes may just be a coincidence and have no functional significance. However, it is important to notice that there is a number of "non-MHC" genes located within the MHC region and that some of the traits associated with MHC polymorphism may be explained by variation in these linked genes.

In cattle, C4 and 21OH are the only genes which have been studied in the class III region. C4 was assigned to the BoLA region by linkage analysis [46] while 21OH was assigned on the basis of synteny in somatic cell hybrids [37].

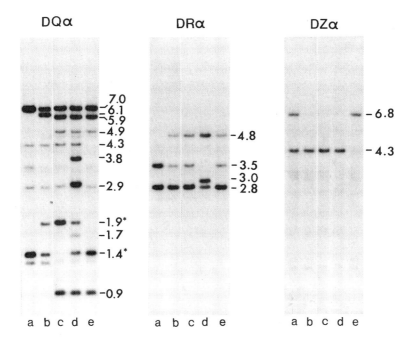

Fig. 15-2 Southern blot analysis of class II α chain genes in cattle. Genomic DNA samples were digested with TaqI and consecutively hybridized with human DQα, DRα, and DZα probes. The animals were RFLP typed as follows: a: DQα 1/1, DRα 1A/1A, DYα 1/1, DZα 1/2; b: DQα 1/2, DRα 1A/2, DYα 1/2, DZα 1/1; c: DQα 1/9, DRα 1A/2, DYα 2/2, DZα 1/1; d: DQα 7/9, DRα 1B/2, DYα 1/2, DZα 1/1; e: DQα 1/9, DRα 1A/2, DYα 1/1, DZα 2/2. The estimated sizes of fragments are given in kilobases and DYα fragments are indicated by an asterisk. From Andersson et al. [46].

15.5.2 BoLA Polymorphism Revealed by DNA Methods

Restriction fragment length polymorphism (RFLP) of bovine class I and class II genes has been investigated by the use of cross-hybridizing human probes. Hybridizations with a human class I cDNA probe resolved very complex and highly polymorphic restriction fragment patterns in cattle [40]. More than 20 different RFLP pattern types were identified in a limited family material and some pattern types were closely associated with serological BoLA-A specificities. The RFLP patterns were so complex that the method did not appear to be suitable for class I typing at the population level. However, the high degree of polymorphism implies that the system could be used for parentage control in cattle.

RFLP's for each of the eight different bovine class II genes have been found [43, 44, 45, 46, 49]; RFLP's of bovine class II α genes are shown in Fig. 15-2. Limited polymorphism, comprising two to five different RFLP types, was identified for DRα, DOß, DZα, DYα, and DYß. In contrast highly polymorphic restriction fragment patterns have been found for DQα, DQß, and DRß. For each of these genes, 20 or more different RFLP types have been identified. Interestingly, DQα, DQß, and DRß are also the most polymorphic class II genes in humans both as revealed by RFLP analysis and by sequence comparisons. There is a very strong linkage disequilibrium among DQ and DR genes [43, 44, 49]; there are only a few rare exceptions to the rule that a given DQα allele is always associated with the same DQß allele. RFLP typing of DQα, DQß, DRß, and DYα genes in a population sample of about 200 Swedish breeding bulls showed that the interpretations of RFLP types, established by family studies, can be used as a key when typing unrelated animals [49]. The genotype of virtually all individuals could be determined with confidence. The very few exceptions were individuals apparently heterozygous for rare haplotypes.

The majority of RFLP studies on MHC genes in domestic animals have been carried out using human or murine probes. However, the trend is now definitely towards the development of species-specific MHC probes. The use of such probes will give stronger hybridization signals due to the perfect homology between probe and target sequence. Thus, there will be a higher signal-to-noise ratio which means the hybridization will be more simple technically and it will be more easy to utilize non-radioactive probe labeling. Furthermore, more specific restriction fragment patterns may be obtained although in many cases the patterns will be identical or very similar to those obtained using heterologous probes. The development of locus-specific probes from non-repetitive intron or flanking sequences may be very useful when the interpretation of RFLP data is complicated due to cross-hybridization between coding sequences from different loci. Such probes may be particularly useful for class I genes which apparently share a high degree of sequence similarity between loci (see 15.5.1). Thus, the cloning of species-specific probes is worthwhile and should be pursued. The cloning work is now facilitated as cloning kits and DNA libraries of domestic animals are commercially available.

An interesting possibility for the future development of more efficient methods for studying MHC polymorphism will be to utilize the polymerase chain reaction (PCR). This is a method for primer-directed enzymatic amplification of specific DNA sequences *in vitro* [50]. The specificity of amplification is based on two oligonucleotide primers which flank the DNA sequence to be amplified. PCR involves repeated cycles of (i.) denaturation of double-stranded DNA, (ii.) annealing of primers to their complementary sequences, and (iii.) extension of the annealed primers by DNA polymerase. By this method microgram quantities of a specific DNA sequence can be amplified in a few hours from small amounts of starting materials such as genomic DNA samples or a few nucleated cells. Sequencing of PCR-amplified DNA is currently used to characterize e.g. sequence variability in human MHC class II genes [51, 52]. This strategy should also be adopted to rapidly generate information on the nucleotide sequence of allelic MHC genes in domestic animals. With this information in hand it will be possible to develop allele specific oligonucleotide (ASO) probes. An ASO is a short oligonucleotide (15-20 base pairs long) which has been constructed so that it will only hybridize to a specific allele under stringent conditions. Hybridizations of ASOs to PCR-amplified genomic DNA would provide a very efficient and rapid typing method of MHC polymorphism. It should be possible to apply this method to the screening of large samples of animals since the hybridization may be performed as simple dot-blots [53].

15.6   Functional Studies of BoLA

There is so far only limited information on the functional importance of BoLA but all data are consistent with the view that BoLA plays the same role in the immune system as the one established for the MHC in other species.

15.6.1 Significance for the Immune Response

The antibody responses to human serum albumin and to the polypeptide (T,G)-A-L in young bulls have been reported to be MHC-linked [54]. Similarly, the immune response to ovalbumin, as measured by antigen-specific T cell proliferation, has also been found to be MHC-linked [55]. However, the most informative studies on the immunological function of BoLA concern the cell-mediated immune response of cattle to *Theileria parva* [56]. *Theileria parva* is a protozoan parasite infecting bovine lymphocytes and causing an acute lymphoproliferative disease known as East Coast Fever. It has been clearly shown that anti-Theileria cytotoxic T cells are restricted by class I MHC molecules. Further, in a given animal there may be a strong bias in the immune response towards one of the two BoLA-A encoded class I molecules.

15.6.2 Significance for Disease Resistance

Compared with the long list of well-established MHC-disease associations in man and mouse [1, 14, 15], the information on this subject in cattle is still rudimentary. The lack of well-established disease associations is at least partly explained by the limited number of serious studies carried out, the difficulties to obtain good clinical material on cattle diseases, and the fact that cattle MHC typing reagents is not as well developed as those in man and mouse.

Lewin and co-workers [57, 58] have reported that the subclinical progression of bovine leukemia virus infection varies between animals carrying different BoLA types. BoLA polymorphism has been weakly associated with the susceptibility to the level of infestation with the tick *Boophilus microplus* [59]. Weak associations between the incidence of mastitis and BoLA-A [60] and BoLA-DQ polymorphism [61] have also been reported. However, further studies are needed to confirm or reject these tentative associations. One of the most important topics for future MHC research in domestic animals will be to develop good diseases models and evaluate the significance of MHC polymorphism for disease susceptibility.

15.7 Possible Practical Applications of MHC Research in Domestic Animals

The present knowledge concerning the functional role of MHC molecules in the vertebrate immune system implies that detailed information on the MHC polymorphism is necessary for a clear understanding of the genetic basis for immune response differences between animals. This fact is a major impetus for the characterization of MHC genes and gene products, in domestic animals. This knowledge may be utilized e.g. in the development of vaccination programs. This application will become more important with the development of new types of vaccines composed of synthetic peptides exhibiting one or a few epitopes.

A central question for MHC research in domestic animals rests with the future possibility to improve disease resistance by selective breeding for certain advantageous MHC alleles/haplotypes. To answer this question we must consider why MHC molecules are so polymorphic in almost all species studied so far. It is noteworthy that extensive MHC polymorphism has been documented in both avian and mammalian species which implies that MHC polymorphism has been present in vertebrate species for more than 250 million years of evolution. However, after more than 50 years of MHC research the evolutionary advantage of MHC polymorphism is still not clearly understood (see Andersson et al. [62] for discussion). The most favored view is that MHC polymorphism is maintained due to significant differences between allelic MHC molecules as regards their influence on the immune response and thereby on the resistance to infectious diseases. Although this hypothesis is consistent with available data there is no compelling evidence for it. Most of the well-documented MHC-associated diseases are rare and non-infectious, having little effect on variability during reproductive years [1, 14, 15]. The most prominent example of an MHC-linked resistance to an infectious disease is in fact the one reported for Marek's disease in chicken carrying the B^{21} haplotype [63].

Although the selection mechanism is still unknown, there is accumulating evidence strongly suggesting that MHC polymorphism is maintained by Darwinian selection (see 15.1). Extensive MHC polymorphism has been documented in all domestic animals of major importance, such as cattle, sheep, pigs, horses, and chicken. There are good reasons to assume that MHC polymorphism also are advantageous in these species. If there was no selection for polymorphism in these species we would expect MHC polymorphism to be reduced by genetic drift since the effective population size (N_e) is often limited in breeds of domestic animals. Furthermore, it has been found that the allele frequency distribution at the BoLA-DQ locus in cattle deviates significantly from the one expected for selectively neutral alleles [49].

An important contribution to our understanding of the evolution of MHC polymorphism was recently given when the sequences of MHC genes in related rodent and primate species were compared [39, 64]. These studies showed that MHC alleles may be maintained over long evolutionary periods (10 million years or more). If the same type of alleles are shared between man and chimpanzee and between mouse and rat, as these studies show, it strongly indicates that the MHC provides a beautiful example of a balanced polymorphism maintained by natural selection.

A reasonable conclusion from available information on MHC polymorphism is that it is unlikely that it will be possible to identify MHC alleles which are associated with superior general disease resistance. In contrast, it is the maintenance of genetic diversity in the MHC system which appears to be of importance.

Acknowledgments

This work was supported by a grant from the Swedish Council for Forestry and Agricultural Research.

15.8 References

[1] Klein, J., *Natural History of the Major Histocompatibility Complex,* New York: John Wiley and Son, *1986.*

[2] Newman, M.J., and Antczak, D.F., *Adv.Vet.Comp.Med. 1983, 27,* 1-76.

[3] Warner, C.M., Rothschild, M.F., and Lamont, S.J. (eds.), *The Molecular Biology of the Major Histocompatibility Complex of Domestic Animal. Species,* Ames: Iowa State University Press, *1988.*

[4] Williams, A.F., *Immunol. Today 1987, 8,* 298-303.

[5] Rask, L., Gustafsson, K., Larhammar, D., Ronne, H., and Peterson, P.A., *Immunol. Rev.* **1985**, *84,* 123-143.

[6] Hardy, D.A., Bell, J.I., Long, E.O., Lindsten, T., and McDevitt, H.O., *Nature* **1986**, *323,* 453-455.

[7] Mellor, A., *Immunol. Today* **1986**, *7,* 19-24.

[8] Steinmetz, M., Stephan, D., and Fischer, K., *Cell* **1986** *44,* 895-904.

[9] Trowsdale, J., and Campbell, R.D., *Immunol. Today* **1988**, *9,* 34-35.

[10] Schenning, L., Larhammar, D., Bill, P., Wiman, K., Jonsson, A.-K., Rask, L., and Peterson, P.A., *Embo J.* **1984**, *3,* 447-452.

[11] Trowsdale, J., Young, J.A.T., Kelly, A.P., Austin, P.J., Carson, S., Meunier, H., So, A., Ehrlich, H.A., Spielman, R.S., Bodmer, J., and Bodmer, W.F., *Immunol. Rev.* **1985**, *85,* 5-43.

[12] Hedrick, P.W., and Thomson, G., *Genetics* **1983**, *104,* 449-456.

[13] Benacerraf, B., and Germain, R.N., *Immunol. Rev.* **1978**, *38,* 70-118.

[14] Ryder, L.P., and Sveijgaard, A., Dausset, J., *Ann. Rev. Genet.* **1981**, *15,* 169-187.

[15] Hedrick, P.W., Thomson, G., and Klitz, W., *Biol. J. Linn. Soc.* **1987**, *31,* 311-331.

[16] Zinkernagel, R.M., and Doherty, P.C., *Adv. Immunol.* **1979**, *27,* 221-292.

[17] Babbit, B.P., Allen, P.M., Matsueda, G., Harber, E., and Unanue, E.R., *Nature* **1985**, *317,* 359-360.

[18] Marrack, P., and Kappler, J., *Sci. Amer.* **1986**, *254,* 28-37.

[19] Bjorkman, P.J., Saper, M.A., Samraoui, B., Bennett, W.S., Strominger, J.L., and Wiley, D.C., *Nature* **1987**, *329,* 506-512.

[20] Bjorkman, P.J., Saper, M.A., Samraoui, B., Bennett, W.S., Strominger, J.L., and Wiley, D.C., *Nature* **1987**, *329,* 512-518.

[21] Hughes, A.L., and Nei, M., *Nature* **1988**, *335,* 167-170

[22] Jonsson, A.-K., Andersson, L., and Rask, L., *Scand. J. Immunol.* **1989**, *30,* in press.

[23] Amorena, B., and Stone, W.H., *Tissue Antigens* **1980**, *16,* 212-225.

[24] Amorena, B., and Stone, W.H., *Science* **1978**, *201,* 159-160.

[25] Spooner, R.L., Leveziel, H., Grosclaude, F., Oliver, R.A., and Vaiman., M., *J. Immunogenetics* **1978**, *5,* 335-346.

[26] Anonymous, *Anim. Blood Grps. Biochem. Genet.* **1982**, *13,* 33- 53.

[27] Davies, C.J., Dissertation, Cornell University, Ithaca, New York, **1988**.

[28] Usinger, W.R., Curie-Cohen, M., and Stone, W.H., *Science* **1977**, *196,* 1017-1018.

[29] Usinger, W.R., Curie-Cohen, M., Benforado, K., Pringnitz, D., Rowe,R., Splitter, G.A., and Stone, W.H., *Immunogenetics* **1981**, *14,* 423-428.

[30] Teale, A.J., and Kemp, S.J., *Animal Genetics* **1987**, *18,* 17-28.

[31] Hoang-Xuan, M., Leveziel, H., Zilber, M.-T., Parodi, A.-L., and Lévy, D., *Immunogenetics* **1982**, *15,* 207-211.

[32] Hoang-Xuan, M., Charron, D., Zilber, M.-T., and Lévy, D., *Immunogenetics* **1982**, *15,* 621-624.

[33] Letesson, J.-J., Loppe, P., Lostrie-Trussart, N., and Depelchin, A., *Anim.Blood Grps. Biochem.Genet.* **1983**, *14,* 239-250.

[34] Joosten, I., Oliver, R.A., Spooner, R.L., Williams, J.L., Hepkema, B.G., Sanders, M.F., and Hensen, E.J., *Anim. Genet.* **1988**, *19,* 103-113.

[35] Joosten, I., Sanders, M.F., van der Poel, A., Williams, J.L., Hepkema, B.G., and Hensen, E.J., *Immunogenetics 1989, 29*, 213-216.

[36] Andersson, L., In: *The Molecular Biology of the Major Histocompatibility Complex of Domestic Animal Species:* Warner, C.M., Rothschild, M.F., and Lamont, S.J. (eds.) Ames: Iowa State University Press, *1988*; pp. 39-52.

[37] Skow, L.C., Womack, J.E., Petresh, J.M., and Miller, W.L., *DNA 1988, 7*, 143-149.

[38] Fries, R., Hediger, R., and Stranzinger, G., *Anim. Genet. 1986, 17*, 287-294.

[39] Lawlor, D.A., Ward, F.E., Ennis, P.D., Jackson, A.P., and Parham, P., *Nature 1988, 335*, 268- 271.

[40] Lindberg, P.-G., and Andersson, L., *Anim. Genet. 1988, 19*, 245-255.

[41] Ennis, P.D., Jackson, A.P., and Parham, P., *J. Immunol. 1988, 141*, 642-651.

[42] Brown, P., Spooner, R.L., and Clark, A.J., *Immunogenetics 1989, 29*, 58-60.

[43] Andersson, L., Böhme, J., Rask, L., and Peterson, P.A., *Anim. Genet. 1986, 17*, 95-112.

[44] Andersson, L., Böhme, J., Peterson, P.A., and Rask, L., *Anim. Genet. 1986, 17*, 295-304.

[45] Andersson, L., and Rask, L., *Immunogenetics 1988, 27*, 110-120.

[46] Andersson, L., Lundén, A., Sigurdardottir, S., Davies, C.J., and Rask, L., *Immunogenetics 1988, 27*, 273-280.

[47] Böhme, J., Andersson, M., Andersson, G., Möller, E., Peterson, P.A., and Rask, L., *J. Immunol. 1985, 135*, 2149-2155.

[48] Muggli-Cockett, N.E., and Stone, R.T., *Anim. Genet. 1988, 19*, 213-225.

[49] Sigurdardottir, S., Lundén, A., and Andersson, L., *Anim. Genet. 1988, 19*, 133-150.

[50] Saiki, R.K., Gelfand, D.H., Stoffel, S., Scharf, S.J., Higuchi, R., Horn, G.T., Mullis, K.B., and Erlich, H.A., *Science 1988, 239*, 487-491.

[51] Horn, G.T., Bugawan, T.L., Long, C.M., and Erlich, H.A., *Proc. Natl. Acad. Sci. USA 1988, 85*, 6012-6016.

[52] Gyllensten, U.B., and Erlich, H.A., *Proc. Natl. Acad. Sci. USA 1988, 85*, 7652-7656.

[53] Saiki, R.K., Burawan, T.L., Horn, G.T. Mullis, K.B., and Erlich, H.A., *Nature 1986, 324*, 163-166.

[54] Lie, O., Solbu, H., Larsen, H.J., and Spooner, R.L., *Anim. Genet. 1988, 19 Suppl.1*, 73-74.

[55] Glass, E.J., and Spooner, R.L., *Anim. Genet. 1988, 19, Suppl. 1*, 74-77.

[56] Morrison, W.I., Goddeeris, B.M., Teale, A.J., Baldwin, C.L., Bensaid, A., and Ellis, J., *Immunol. Today 1986, 7*, 211-216.

[57] Lewin, H.A., and Bernoco, D., *Anim. Genet. 1986, 17*, 197-207.

[58] Lewin, H.A., Wu, M.-C., Stewart, J.A., and Nolan, T.J., *Immunogenetics 1988, 27*, 338-344.

[59] Stear, M.J., Newman, M.J., Nicholas, F.W., Brown, S.C., and Holroyd, R.G., *Austr. J. Exp.Biol. Med. Sci. 1984, 62*, 47- 52.

[60] Solbu, H., Spooner, R.L., and Lie, O., *Proc. 2nd Word Congr. Genet. Appl. Livestock Prod., Madrid, 1984*, pp. 1-6.

[61] Lundén, A., Sigurdardottir, S., and Andersson, L., *Anim. Genet. 1989, 20 Suppl. 1*, 31.

[62] Andersson, L., Pääbo, S., and Rask, L., *Immunol. Today* **1987**, *8,* 206-209.
[63] Briles, W.E., Stone, H.A., and Cole, R.K., *Science* **1977**, 193-195.
[64] Figueroa, F., Günther, E., and Klein, J., *Nature* **1988**, *335,* 265-271.

16. A Reverse Genetic Approach Towards the Bovine "Muscular Hypertrophy" Gene, Based on the Use of DNA Fingerprints

M. Georges

16.1 Introduction

So-called "double muscled" animals are known for many years in a variety of cattle breeds, particularly the Belgian Blue Cattle Breed. These animals are characterized by 1) a generalized muscular hypertrophy of about 20 %, all other organs being reduced in size; 2) a muscular tissue showing a 40 % reduction in fat content as well as a reduction in connective tissue, both traits particularly appreciated by the consumers; 3) a 9 % reduction in feed conversion ratio (kg.feed/kg.gain; 5.85). Despite a high incidence of dystocia mainly due to a 15 % weight increase at birth, and a 1 % increase of mortality during the first year, the economical context in Belgium has been such that the number of double muscled animals has been increasing steadily [1].

The determinism of the "double muscled" entity has been shown to involve an autosomal locus characterized in the concerned populations by a partially recessive "mh" (muscular hypertrophy) allele, having a major effect on muscular development at the homozygous state. This allele is splitting the population into two subgroups: the conventional animals (genotype +/+ or mh/+) and the double muscled animals (genotype mh/mh). Criteria measuring muscular development such as price per kg liveweight, or more objectively, dressing percentage as well as red cell and plasma concentrations of creatine and creatinine, show a clear bimodal distribution in those cattle populations [2, 3].

Red cell and plasma concentrations of creatine and creatinine are particularly useful as quantitative markers of muscular development because they can be determined very easily from a simple venous blood puncture. Remember that creatine is built up in the liver from the three amino acids methionine, glycine and arginine; released into the circulation where it exists in equilibrium between red cells and plasma, from which it is absorbed into the muscles where it is trapped as phospho- creatine, the muscle major energy storage form. Phospho-creatine is slowly degraded into creatinine, which is released into the circulation and excreted in the urine. Compared to conventional animals, double muscled animals have significantly lower creatine and higher creatinine concentrations in red cell and plasma [4].

We are attempting to map the "mh" gene by linkage analysis. Establishing such a linkage relationship between the "mh" locus and a genetic marker would 1) validate the hypothesis itself of the existence of the "mh" locus; 2) provide breeders with an efficient tool to select for or against the gene; 3) be a good starting point towards the isolation of the gene itself, allowing to study the mechanism by which it is exerting its spectacular effect.

16.2   Pedigree Material

At the present time our family material consists of 540 animals composing 6 large paternal half-sib pedigrees segregating for the "mh" locus. The 6 bulls were chosen according to their phenotype as well as the proportion of double-muscled animals on their offspring. A rough idea of the number of individuals required can be obtained assuming an expected lod score (ELOD) of 0.16 for a phase known meiosis informative for the test locus (in this case "mh" locus) as well as a marker locus at 10 cM, a threshold of 3 to demonstrate linkage, and knowing the Polymorphism Information Content (PIC) of the used markers [5].

16.3   A Set of Genetic Markers

Using Ott's formula [6] $P_g \approx 1-(1-0.6\Theta)^g$ (g = number of markers; Θ = recombination fraction) one can estimate the probability to find a linkage at $\leq$ 10 cM as a function of the number of randomly chosen markers included in the analysis. These probabilities are 0.006, 0.058, 0.259 and 0.452 for 1, 10, 50 and 100 randomly chosen markers respectively. These figures clearly indicates that a substantial number of markers are needed to have a reasonable chance to find one associated with the gene of interest. Until recently however, the markers available in cattle were limited to a dozen blood group systems, as well as to a few biochemical polymorphisms. Sixteen of these classical markers, for which at least one of our bulls is heterozygous were included in our analysis. Moreover, one of the bulls is blue and consequently heterozygous at the roan locus. This marker is particularly interesting regarding the role that has been assigned to this gene in the determinism of White Heifer Disease [7].

The renewed interest in linkage analysis in mammals stems of course from the possibility to identify a theoretically unlimited number of RFLP markers at the DNA level. RFLPs at 24 bovine loci were recently reported [8] and 4 of them were included in this analysis: thyroglobulin [9] growth hormone [10], low density lipoprotein receptor [11] and BoLA [12].

However, accumulating several dozens of RFLPs informative in our pedigree material turned out to be a very tedious, time consuming and expensive task. Moreover, theoretical calculations have shown the nucleotide diversity (probability for an individual to be heterozygous at a specific nucleotide site) in the Belgian Blue cattle breed is 0.0006 compared to 0.002 value found previously in man [9]. This reduced genetic variation reflects most probably a higher level of inbreeding resulting from the population structure imposed by breeding strategies. It indicates that search for RFLPs could be less rewarding in domestic animals compared to man. One could face this problem either by working on methodologies identifying a higher proportion of the existing genetic variation, such as the methods described by Myers et al. [13], associated or not with the PCR, and/or to look for hypervariable sequences in the genome.

The recent identification of a new component of the eukaryotic genome, the families of hypervariable minisatellites, is particularly promising in this respect. The human genome, for instance is known to be interspersed with several thousands of hypervariable minisatellite sequences, tandem repetions of short sequence motive. Variation in the number of repeats generates an extensive genetic polymorphism with heterozygosities ≥ 90 % being common. Hypervariable minisatellite sequences can be grouped into families, whose members show cross-hybridization. Different members of such a family can be simultaneously visualized in Southern blots, yielding individual specific DNA fingerprints [14, 15]. Moreover, the different members of a family of hypervariable minisatellites can be fished out of a genomic library and used to generate locus specific hypervariable RFLPs, also known as VNTRs [16]. Several hundreds of such human VNTRs have been isolated in man in a very short time-span, contributing considerably to the coverage of the human genome with DNA markers, despite a certain bias exhibited by the VNTRs towards chromosomal proterminal regions [17].

Several minisatellite sequences known to recognize families of hypervariable minisatellites in man were shown to do the same in a variety of animal species [18]. Different laboratories are now using this information to isolate locus specific VNTRs in a variety of animal species and promising results have been obtained (Marcotte et al., in preparation). In the meantime, animal geneticists can benefit from the advantage that linkage fingerprints can be used much more efficiently for linkage analysis in animals than it is in man. Indeed, and due to the large pedigrees that can be built almost at will, statistically significant results can be obtained on a single pedigree, eliminating the need to cumulate information obtained on different families as is almost always required in the human and which is of course impossible with DNA fingerprints. In cattle, for instance, at least 6 minisatellite probes have been shown to yield DNA fingerprints each characterized by up to 15 resolvable bands per individual, with more than 85 % of these being informative for linkage analysis. The polymorphism being essentially independent of the used enzyme, membranes obtained for instance with Hinf I can be successively hybridized to the different minisatellite probes allowing a significant proportion of the genome to be scanned in a very quick and cheap way. We have chosen for this approach in our search for the bovine "mh" gene (Georges et al., in preparation).

16.4 Computer Programs for Linkage Analysis

Very powerful computer programs for linkage analysis have recently been developed, based on the Elston-Stewart algorithm [19]. Briefly, these programs either determine the likelihood of a set of pedigree data with fixed values for a variety of parameters: phenotypic distribution for the different genotypes, population distribution of the different genotypes, genotypic distribution of offspring given the parents'genotype and, when considering two or more genetic systems simultaneously, from the recombination fractions, θ; or estimates the parameter values maximizing the likelihood of the pedigree data.

These programs were used to search for linkage relationships between the different genetic markers, including the fingerprint bands. Several linkage relationships were found between classical markers, between classical markers and fingerprint bands, and between fingerprint bands (Georges et al., in preparation). Finding linkages between classical markers and fingerprint bands particularly, illustrates the effectiveness of the chosen approach.

We then looked for linkage relationships between the different markers and the "mh" locus. First, animals were either classified as double muscled, conventional or undetermined, when they were hard to classify. Then, creatine and creatinine concentrations in plasma and red cells were included in the analysis. Maximum likelihood means and standard deviations for the 3 genotypes were estimated using the linkage programs. The obtained parameter values were compatible with previous reports (Hanset and Michaux, 1986). Those quantitative measurements were expected to yield supplementary information particularly in 2 situations; 1) to help in the classification of the previously "undetermined" animals; the program considers the relative probabilities to be of either of the 3 genotypes when performing the linkage analysis; 2) to extract information from conventional offspring having a conventional mother. Without quantitative measurements these offspring contain nearly no information, while with the quantitative values, the transmission of either the mh or the + allele by the sire will be assigned a different probability.

Candidate markers for the "mh" locus showing lod scores ≥ 2 have been identified and are under investigation.

16.5 References

[1] Hanset, R., *Exploiting New Technologies in Animal Breeding*. Ed. Smith, C., King, J.W.B., and McKay, J.C., Oxford University Press, *1986*.

[2] Hanset, R., and Michaux, Ch., *Génétique, Sélection, Evolution 1985, 17(3)*, 352-359.

[3] Hanset, R., and Michaux, Ch., *Génétique, Sélection, Evolution, 1985, 17(3)*, 360-369.

[4] Hanset, R., and Michaux, Ch., *Journal of Animal Breeding and Genetics 1986, 103*, 227-240.

[5] Lander, E.S., and Botstein, D., *Cold Spring Harbor Symposia on Quantitative Biology 1986, L I*, 49-62.

[6] Ott, J., *Analysis of Human Genetic Linkage*. Johns Hopkins University Press, *1985*.

[7] Hanset, R., *Génétique, Sélection, Evolution 1985, 17(4)*, 443-458.

[8] Fries, R., Beckmann, J.S., Georges, M., Soller, M., and Womack, J., The Bovine Gene Map, 1988. *Animal Genetics 1988*, in press.

[9] Georges, M., Lequarre, A.S., Hanset, R., and Vassart, G., *Animal Genetics 1987, 18*, 41-50.

[10] Georges, M., *Recherches de liaisons génétiques au moyen de Polymorphismes de Fragments de Restriction dans l'espèce bovine*. Mémoire de licence en Biologie Moléculaire, ULB, *1985*.

[11] Georges, M., Lequarre, A.S., Hanset, R., and Vassart, G., *Animal Genetics 1987, 18 (suppl 1)*, 77-78.

[12] Andersson, L., Lunden, A., Sigurdardottir, S., Davies, C.J., and Rask, L., *Immunogenetics 1988, 27*, 273-280.

[13] Myers, R.M., and Maniatis, T., *Cold Spring Harbor Symposia on Quantitative Biology 1986, LI*, 275-284.

[14] Jeffreys, A.J., Wilson, V., and Thein, S.L., *Nature 1985, 314*, 67-73.

[15] Vassart, G., Georges, M., Monsieur, R., Brocas, H., Lequarre, A.S., and Christophe, D., *Science 1987, 235*, 683-684.

[16] Nakamura, Y., Leppert, M., O'Connel, P., Wolff, R., Holm, T., Culver, M., Martin, C., Fujimoto, E., Hoff, M., Kumlin, E., and White, R., *Science 1987, 235*, 1616-1622.

[17] Royle, N.J., Clarkson, R.E., Wong, Z., and Jeffreys, A.J., *Genomics 1988, 3*, 352-360.

[18] Georges, M., Lequarre, A.S., Castelli, M., Hanset, R., and Vassart, G., *Cytogenetics and Cell Genetics 1988, 47*, 127-131.

[19] Lathrop, G.M., and Lalouel, J.M., *American Journal of Human Genetics 1984, 36*, 460-465.

17. Structure and Function of Milk Protein Genes

J.-C. Mercier, J.-L. Vilotte and C. Provot

17.1 Introduction

Over the last few years, DNA recombinant techniques have been used to study the structure and expression of milk protein mRNAs and genes. In this chapter, we will attempt to summarize the previous knowledge of the biochemistry and biosynthesis of the major milk proteins and to present the most pertinent information provided by the current research on the cognate mRNAs and genes before reviewing briefly the more specific data available on each particular cDNA and gene. The practical applications of these studies will not be developed here and only a brief comment will be made about their obvious interest for the dairy field.

17.2 General Features of the Major Milk Proteins

Milk protein composition can differ quantitatively and/or qualitatively from one species to another (Fig. 17-1). Milk proteins fall into two classes, caseins which coagulate at pH 4.6 and whey proteins which remain in the supernatant. Caseins, which amount to nearly 80 % of the total protein output in cow's milk, undergo O-phosphorylation in the Golgi organelle of the mammary gland where they are packaged through calcium bridges as large and stable copolymers called micelles (Fig. 17-2). k-casein plays a major role in preventing, through micelle formation, the precipitation of other caseins by calcium. The function of caseins is to provide the progeny with a source of amino acids, phosphate and calcium, and possibly biologically active peptides (reviewed in [1]). Furthermore, they are the raw material for the cheese-making industry in the classical manufacturing processes. The main whey proteins are ß-lactoglobulin, a possible carrier of small hydrophobic molecules such as retinol, α-lactalbumin which is involved in lactose synthesis and the Whey-Acidic-Protein (WAP) recently discovered in rodent's milk and whose function is unknown.

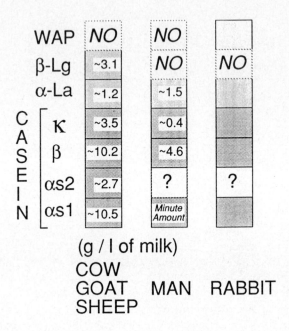

Fig. 17-1 Quantitative and qualitative differences between milks of five species. The protein content of cow's, goat's, sheep's, human and rabbit's milk averages 34, 29, 55, 10 and 136 g/l, respectively. Numbers refer to the specific protein composition of cow's and human milks.

Ruminant's milk proteins were extensively studied and their amino acid sequences as well as the structural differences of a score of known genetic variants, were first elucidated in the bovine species ([2] and citations therein). The knowledge of the primary structures provided a wealth of information about the relationship and evolution of the different caseins, and the mechanism of phosphorylation. It was clearly established that the twenty five or so caseins detected by electrophoresis originate from 4 types of polypeptide chains: the so-called αs1-, αs2-, ß-and k-casein. Variation in the prosthetic group content, genetic polymorphism and limited proteolytic cleavage by plasmin account for the great heterogeneity of whole casein. Genetic protein variants were used as markers for studying the mode of inheritance of the cognate alleles. The non-random assortment of prevalent casein alleles in both the bovine population and the progeny of heterozygous bulls provided strong evidence of an autosomal cluster of the αs1-, αs2-, ß- and k-casein loci (reviewed in [3, 4]). Furthermore, sequence analysis of variants differing in their phosphate contents gave strong support to the proposal that phosphorylation of caseins involves the enzymatic recognition of the tripeptide sequence SER/Thr-X-A, where A is an acidic residue, essentially Glu or SerP [3, 5], and X any amino acid residue. On the basis of internal homologies occurring within αs1-casein, it was suggested as far back as 1971 [6] that the cognate gene evolved through intragenic duplications and further study of the αs2-casein [7] led to the same suggestion. The ubiquity of the major phosphorylation site in the calcium-sensitive αs1-, αs2- and ß-casein gave some hint of a possible common origin, which was strongly supported by the striking homology of their conserved signal peptides [8, 9]. Furthermore, inter-species comparison of casein sequences showed the rapid evolutionary rate of the mature proteins [10, 11] but the striking conservation of the signal peptide [9].

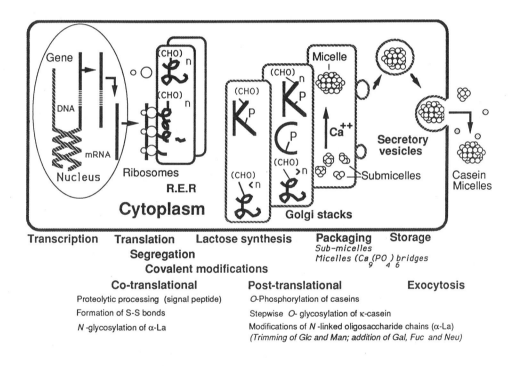

Fig. 17-2 Schematic representation of the biosynthesis, covalent modifications and intracellular transport of milk proteins. L, K and C symbolize α-lactalbumin, k-casein and the calcium-sensitive αs1-, αs2- and ß-casein, respectively.

Milk protein synthesis is under complex multihormonal control and the effects of lactogenic peptide and steroid hormones on cultured mammary cells are illustrated in

Fig. 17-3. In outline, milk protein gene expression is induced by prolactin and amplified at the transcriptional and posttranscriptional level by glucocorticoids in the presence of insulin, progesterone having an antagonistic effect. This schematic representation reflects our superficial knowledge of the intricate regulating mechanisms operating at the molecular level. Moreover, the regulation process appears to be different according to the animal species and even the type of milk protein gene (for more details, the reader is referred to reviews devoted to the topic [12, 17].

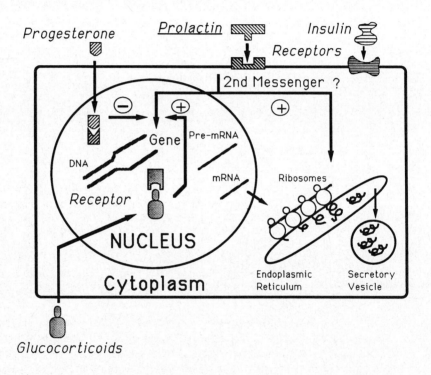

Fig. 17-3 Schematic representation of the hormonal control of milk protein gene expression in mammary epithelial cells. Stimulating (+), inhibitory (-) activity.

The mammary gland, which synthesizes and secretes in large amounts (up to one kg per day in the bovine species) a few specific proteins under the control of peptide and steroid hormones, is a nice model for studying the expression and regulation of genes. At least ten species, including dairy farm animals, are being investigated in this respect, using the current methods of molecular biology. Furthermore, these studies have many practical applications, some of them already under development. For example, milk protein cDNAs

```
          NON CODING 5' SIGNAL PEPTIDE                                              MATURE PROTEIN

αs1-Cn                -15                    -10                    -5              -1  +1              +5
Cow      CTT GACAACC ATG AAA CTT CTT ATC CTC ACC TGT CTT GTG GCT GTT GCT CTT GCC AGG CCT AAA CAT CCT ATC AAG CAC
Sheep    --- ------- --- --- --- --C --- --T --- --- --- --- --- --- --- --- --- --- --- --- --- --- --- ---
Rat      --- AG----- --- --- --- --- --- --- --C --- --- -C- --- --- --T CT- --- -G- GC- -A- CGT -GA A-T
G.P      --C AG----- --- --G --- --- --- --C --G --- --- TC- --- G-G --- --T- --G --- TT- --C T-- -G- ---
Rabbit   ---G---C--- --- --G --- --C --- --- --T --C --- --- --- --- AC- --- --- --- --A- --- TT- -A- T-A GGA ---

Amino acid      M   K  F/L  L   I   L   T   C   L   V   A  V/A  A  L/V  A  R/L P/H K/R H/A P/H IRF KRG H/N
                                                           S/T          M         F       L

αs2-Cn                -15                    -10                    -5              -1  +1              +5
Cow      ACAAAGCAAAC ATG AAG TTC TTC ATC TTT ACC TGC CTT TTG GCT GTT GCC CTT GCA AAG AAT ACG ATG GAA CAT GTC TCC
Sheep    -TC---T---- --- --- --- --- --T --- --- --- --- --C --- --- --- --- --- --- C-- -A- --- --- --- ---
Rat      -TC---T--C- --- --- --- --- --C --- --- --G G-- --- -C- --T --G --T C-- --- --- --- --- --- --- ---
G.P      --TC--T--C- --- --- --- C-- --- --C --- --- --C --- --- --G --T --C --- --C C-C -A- TCA --G --A CAG ---
Mouse    --TC-TT--T- --- --- --- A-- --T C-G --T --- --- --- --C --- --T --- --- C-G -G- --- --G --A TA- AT-

Amino acid      M   K  F/L F/I  I  F/L  T   C   L  L/V  A  V/A  A   L   A  K/N N/H T/K M/V E/K HDQ VKQ SPI
                                                                          Q   A/R  S                   Y

β-Cn                  -15                    -10                    -5             -i  +1              +5
Cow      AATTGAGAGCC ATG AAG GTC CTC ATC CTT GCC TGC CTG GTG GCT CTG GCC CTT GCA AGA GAG CTG GAA GAA CTC AAT GTA
Sheep    ---C------- --- --- --- --- --- --- --- --- --- --- --- --- --- --- --- --- --- --- --- --- --- ---
Rat      -C----C---- --- --- --- T-- --- --- --- --- --T --- --A --T --T --- --- --G --- AA- --T -C- T-- -C- --G
Rabbit   -C---GC--T- --- --- --- --T --- --- --- --- --C --T --- --G --- AA- --- C-- --- -G- --T
Mouse    -C----C---- --- --- --- T-- --- --C --- --- --T --- --C --T --- --- --- ACT AC- TTT ACT GTA TCC

Amino acid      M   K   V  L/F  I   L   A   C   L   V   A   L   A   L   A   R   E  L/Q E/D E/A LFT NTS V/S
Consensus      ARYC ATG AAR                                                     K/T  T  Q/F               V

                      -15                    -10                    -5             -1
Consensus       M   K  LFV LFI  I  L/F T/A  C   L  L/V A  VAS  A  L/V  A
                                                         T/L

κ-Cn               -21        ATG         -15                    -10                  -5              -1  +1
Cow      GGAAAGGTGCA ATG ATG AAG AGT TTT TTC CTA GTT GTG ACT ATC CTG GCA TTA ACC CTG CCA TTT TTG GGT GCC CAG GAG
Sheep    ----------- --- --- --- --- --- --- --- --- --- --- --A --- --- --- --- --- --- --- --- --- --- --- ---
Rat      -TG-------- --- --- --G- -A- --- A-- G-- --- A-- -A- --- --A --- C-- --T --- --C --- --- -C- --A G-- -T-
Mouse    AT--------- --- --- --G- -A- --- A-- G-- --G A-- -A- --T --- --- --- --T --- --C --- --- -C- --A G-- ATA

Amino acid      M   M  K/R S/N  F  F/I L/V  V  V/M T/N  I   L   A   L   T   L   P   F   L  G/A  A  Q/E EVI

β-Lg               -18            -15                    -10                  -5              -1  +1              +5
Sheep    CAGCTGCAGCC ATG AAG TGC CTC CTG CTT GCC CTG GGC CTG GCC CTC GCC TGT GGC GTC CAG GCC ATC ATC GTC ACC CAG
Amino acid      M   K   C   L   L   L   A   L   G   L   A   L   A   C   G   V   Q   A   I   I   V   T   Q
(pig)           -   R   -   -   -   -   T   -   -   -   -   -   T   -   -   -   -   V   E   -   -   P

α-La               -19            -15                    -10                  -5              -1  +1
Cow      GGGTCACCAAA ATG ATG TCC TTT GTC TCT CTG CTC CTA GTA GGC ATC CTA TTC CAT GCC ACC CAG GCT GAA CAG TTA ACA
Sheep    ----A------ --- --- --- --- --- --- --- --- --G --- --- --- --- --- --- --- --- --- --- --- --A --- ---
Goat     ----A------ --- --- --- --- --- --- --- --- --G --- --- --- --- --- --- --- --- --- --- --- --A --- ---
Rat      -A-C-GG---- --- --- CGT --- --T C-- --- T-- --- --CG T-T --T TCG C-G C-- --- TTT --A --C AC- G-- --T ---
G.P      CA-CAG----- --- --- --- T-- C-- --- T-G --G --G --- --- --G --T C-- --- GTG --- --C A-G --A C-T --C
Man      ----AG----- --- -G- -T- --- --- C-- T-- --G --G --- --- --G --- --C --- -T- -T- --C A-G --A --C ---

Amino acid      M  M/R SRF  F  V/F S/P  L  L/F  L  V/A G/C  I  L/S F/L H/P  A  TFV Q/L  A  ETK Q/E L/F  T

WAP                -19            -15                    -10                  -5              -1  +1
Rat      CCGCCGACACC ATG CGC TGT TCG ATC AGC CTC GTT CTT GGC CTG CTG GCC CTG GAG GTA GCC CTT GCT CGG AAC CTA CAG
Mouse    A-A---GT--- --- --T --C CTC --- --- --T --- --- --- --- --- --G --- --C --- -A- --- --- G--
Rabbit   -ACCT-CC--- --- --- --- CTC --- --G -CC --C --- --C --- --- --- --CG --- --C --- -T- GC- -CC A--

Amino acid      M   R   C  S/L  I   S   L  V/A  L   G   L   L   A   L   E  V/A  A   L   A  RQL N/A L/P QEK
```

Fig. 17-4 Inter-species comparison of nucleotide sequences of the major milk proteins cDNAs, in the region encoding the signal peptide. Dashes represent nucleotides identical to those of the bovine cDNAs taken as reference. Italicized numbers refer to the codons of the signal peptide (-) and the mature protein (+). The one-letter symbols of amino acids are written in italics. Consensus refers to the nucleotide sequence encompassing the initiation codon ATG of calcium-sensitive casein cDNAs (R and Y designate nucleotides containing purine and pyrimidine bases respectively).

or gene fragments are currently used as probes to search for and identify by RFLP, alleles associated with specific dairy properties [18, 19], allowing the selection at birth of animals with the best genotype for a given trait. Genes such as those encoding ß-lactoglobulin and WAP were engineered to direct the production of a functional human protein (α-antitrypsin, blood clotting factor IX or tissue plasminogen activator) in the lactating mammary gland of transgenic animals [20, 23].

The large amount of the milk specific protein mRNAs in the lactating mammary gland has facilitated their study and the nucleotide sequences of a score of mRNAs and a dozen genes coding for the major milk proteins are already known.

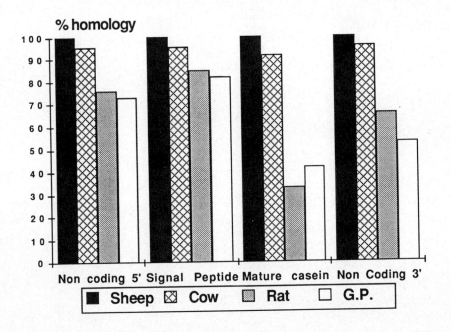

Fig. 17-5 Inter-species nucleotide sequence homology of different regions of αs1-casein cDNAs. Calculation of the percent of homology is based upon alignments of sequences referenced in the text. Interspecies comparison of milk protein cDNAs other than α-lactalbumin (data not shown) gave similar diagrams.

17.3 General Features of Milk Protein mRNAs

The rate of synthesis of milk proteins is primarily dependent upon the concentration of the relevant mRNAs. Their level increases steadily during late pregnancy to reach a peak at mid-lactation (up to 100,000 copies per cell, accounting for 60 % of total mRNA in the rat species). The differential accumulation of mRNAs is the result of a coordinated mechanism involving both an increase of the transcription rate (2-4 fold) and an efficient stabilization of mRNAs whose half-lives may increase up to 25-fold [12, 24].

Sizes of milk protein mRNAs range from 549 nucleotides for rabbit WAP to 1349 nucleotides for rat α-casein, excluding the poly(A) tail. The 5' and 3' untranslated regions amount to 28 to 44 % of the mRNA. The polyadenylation recognition signal AAUAAA occurs 12 to 19 nucleotides upstream from the poly(A) tail.

Inter-species comparison of homologous mRNAs confirmed the high rate of evolution of the open reading frames whose conserved domains are those encoding the major phosphorylation sites of the calcium-sensitive caseins and the signal peptide (Fig. 17-4). The synonymous substitutions at the third position of codons relevant to the signal peptide of calcium-sensitive caseins contrast with the random substitutions at all three positions of the codons relevant to the mature protein, which reflects a high selection pressure.

The faster evolutionary rate of the coding frame, as compared to the 5' and 3' flanking untranslated regions (Fig. 17-5), suggests that the latter may contain particular structures involved in the high rate of translation and/or the stability of the mRNA.

17.4   General Features of Milk Protein Genes

cDNAs were used as probes for somatic cell and *in situ* chromosomal hybridization studies to demonstrate directly the physical linkage of casein genes. They were localized on chromosomes 5 and 12 in the mouse [25] and the rabbit [26] species, respectively. The genes encoding milk whey proteins might be located on different chromosomes, as the mouse WAP locus is likely on chromosome 11 [25]. Human α-lactalbumin is on chromosome 12 [27].

The overall organization of a casein gene encoding the rat ɣ-casein (αs2-like), was first reported in 1982 [28], and the partial sequence of the rat ß-casein gene was published in 1985 [29]. Sequences of genes encoding whey proteins were first reported in 1984 for rat α-lactalbumin [30] and mouse an rat WAPs [31], and in 1988 for ovine ß-lactoglobulin [32, 33].

All milk protein genes sequenced so far have the canonical organization described for eukaryotic genes (Fig. 17-6) with the polyadenylation recognition signal, the consensus splice junction sequences (Fig. 17-7) and the CAP site, as well as the TATA box (occurring 19-30 nucleotides upstream from the transcription unit) and possible CAAT boxes characteristic of tissue-specific genes. But their specific activation requires the occurrence of particular and still unidentified sequences responsible for stage-tissue specificity, induction, amplification and inhibition of expression. The transcription units with sizes ranging from 2 kb for bovine α-lactalbumin to at least 17 kb for bovine αs1-casein, contain several introns: 3 for the α-lactalbumin and WAP genes but at least 8 for the genes encoding calcium-sensitive caseins. Their intervening sequences of different sizes, from 92 to 3000 bp, separate many small exons. Consequently, mature casein transcripts account for less than 15.5 % of the transcription unit (in the case of rat ß-casein, but only 6.5 % for bovine k-casein) in contrast with 36 % for the bovine α-lactalbumin transcript.

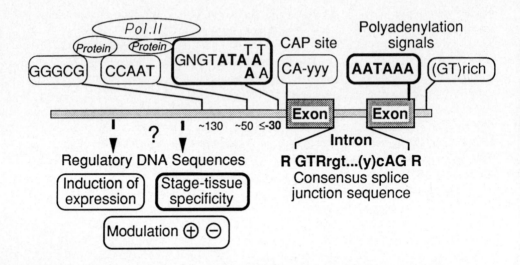

Fig. 17-6 Schematic representation of a functional milk protein gene. Bold-faced boxes refer to strong signals of tissue-specific genes [87]. A lower case letter refers to the nucleotide most frequently found at this position. R/r and y refer to nucleotides containing purine and pyrimidine, respectively. Pol.II = RNA polymerase II.

Most of these genes contain repeated sequences of different types interspersed throughout introns or/and flanking regions of the transcription unit. Our present knowledge of the function of milk protein genes is still superficial as the induction and modulation of expression have not been investigated in detail at the molecular level. But the recent isolation and sequence characterization of several genes allow now such experimental work.

The functional analysis of milk protein genes includes several steps:

1) Computer search for consensus sequences reported to be recognized by effectors such as hormone-receptor complexes and regulatory nuclear proteins. Potential recognition sites for glucocorticoids, progesterone and nuclear factor CTF/NF-1 were thus identified essentially in the 5' flanking region of milk protein genes.

2) Computer search for structural motifs either common to different milk protein genes or conserved during the evolution of a given gene (intra- and inter-species comparison). Such studies allowed Rosen's group [29, 34] to identify several structural patterns shared by the 5' flanking regions of the rat genes encoding the three calcium-sensitive caseins, and found later in their homologous counterparts from other species (Figs. 17-8, 17-9). Similarly, other teams [35, 36] reported the common occurrence of a highly conserved stretch of about 30 nucleotides localized a hundred or so nucleotides upstream from the transcriptional start site of three α-lactalbumin and five casein genes (Fig. 17-9). The so-called 'milk box' might be involved in the hormonal control or tissue-specific expression of these genes. Such investigations have some merit in indicating which regions of a milk

protein gene might be specifically of functional importance, but their real significance is still controversial for lack of experimental data. For example, there is no obvious milk box in ovine ß-lactoglobulin and that of bovine α-lactalbumin is partially deleted.

3) Structural modifications of the gene and study of the relevant functional alterations in transfected cells and transgenic animals. As described further, some progress has been made in this respect, since expression of WAP constructs with 2.5 kb of 5' flanking region and ß-lactoglobulin, α-lactalbumin and ß-casein constructs with less than 1 kb, was stage- and mammary tissue-specific.

4) Identification of gene regions bound to specific ligands, using foot-printing or gel retardation assays. At the moment, the only available data are those relevant to the WAP and the α-lactalbumin genes [37, 38].

As we shall see now, experimental data are still very scarce and much progress remains to be done in the field.

```
             EXON  INTRON                    EXON              EXON  INTRON                    EXON

    CONSENSUS (.r)  GTRrgt.//.yyyyyyyyyyyyNcAG (r.)   CONSENSUS (.r)  GTRrgt.//.yyyyyyyyyyyyNcAG (r.)

β Lg  Sheep (AG)  GTTCGA.//.AAGGTCCCTCTCCAG (GT)  β Cn  Sheep (AG)  GTAAGA.//.TCTTCCCATTCACAG (GA)
            (TG)  GTGGGC.//.GCAGCTGTCTTTCAG (GG)        Rat   --    ------    ---CTT--------- --
            (TG)  GTGAGT.//.CTTGACCGCGTCCAG (CC)        Rabbit --    ------              ?
            (GG)  GTGGGT.//.GTCCCTGCCCCATAG (TC)        Sheep (GA)  GTAAAT.//.TTTTTCTCTCTTTAG (CA)
            (GG)  GTGAGC.//.CTCCCTCCCCCACAG (GG)        Rat   AG    --GTG-    -G---GCTTT-AC-- A-
            (GG)  GTAAGC.//.CCATGTCCATTTCAG (GG)        Sheep (AG)  GTAAGA.//.ATTTTTCCTTTATAG (AC)
                                                        Rat   --    ------    ------TG----C-- --
WAP   Rat   (AG)  GTAGGC.//......TTTTGTACAG (TT)        Sheep (AG)  GTAAGA.//.TTTATTTTTCCAAAG (GA)
      Mouse --    ---A--    ....------G--- --            Rat   --    -----G    C--TC---CTTG--- --
      Rat   (CA)  GTAAGT.//.CCATCTCTTCCCCAG (TT)        Sheep (AG)  GTAAAA.//.TTTTTTTTTAACCAG (AA)
      Mouse --    ------    -------- ------ --           Rat   --    ---TTT    G---A---CTGG--- --
      Rat   (AA)  GTATGC.//.CTGTATCTCCTACAG (AA)        Sheep (AG)  GTAATT.//.TTTCCTTCTTTCCAG (GA)
      Mouse -G    ------    -CA-C--C------- TG           Rat   --    ----AC    --CTT--G----T-- --
                                                        Sheep (TT)  GTAAGT.//.TTTTTAATTTTTAAG (GT)
α La  Cow   (AT)  GTGAGT.//.TCGTCTTTCTTTCAG (GG)        Rat   C-    ------    A-A--TGCA------ --
      Rat   --    --A---    -TC-G-C--CC---- --           Sheep (TG)  ?   .//.        ?       (AA)
      G.P.  --    ------    --A-A--T--C---- --           Rat   --    GTAAGT.//.ATCTTTCTTTCACAG --
      Man   --    ------    --A------CC---- TG
      Cow   (CA)  GTGAGT.//.AGAGGCTCCTGCCAG (AG)  α Cn  Rat   (AG)  GTATTG.//.TCCTTTTCAACACAG (AT)
      Rat   --    ------    CTACA---- --T--- --           Cow   --    -----A              ?
      G.P.  --    --A---    TACAA---T--T--- --            Rabbit --    ------              ?
      Man   --    ------    TGTAA--------- --             Rat   (CT)  GTGAGT.//.TCTCTTACTTTTCAG (AG)
      Cow   (TG)  GTGAGT.//.CTTCTTCATGATCAG (GT)        Rat   (AA)  GTAAGT.//.ATTTATTCCTTGCAG (AC)
      Rat   --    ------    T----C-G------- -A            Rat   (AG)  GTGAGC.//.TTTTAATTCCCATAG (GA)
      G.P.  --    ------    T----C---T-C--- --            Rat   (AG)  GTAAGT.//.        ?
      Man   --    ----A-    -----C---T----- --

κ Cn  Cow   (AG)  GTAATC.//.TTTAATTTAATTTAG (CT)  γ Cn  Rat   (AG)  GTAATG.//.        ?
            (TG)  GTGAGT.//.TCTGAATCTCAACAG (GG)
            (TA)  GTAAGT.//.TTTATTTTTTTACAG (AG)
            (AG)  GTAAAC.//.CTGTCATTTTTATAG (GT)
```

Fig. 17-7 Known splice junction sequences of milk protein genes. Dashes represent nucleotides identical to those of the sequence taken as reference. In the consensus sequence, a lower-case letter refers to the nucleotide most frequently found at that position. R/r and y refer to purine and pyrimidine, respectively.

```
            -100                      αs1-Cn RAT                                              -24
TTGCAACAA  TTCTTAGAATT  TATG.....TGTCTGAAACA  AAACCACAAAATTAGCAT  TTCACTGCTCAGCAAGTTTAAATAGCTGTGG
            -100                      αs1-Cn RABBIT                                            -24
...GAA     TTCTTAGAATT  TAAA......GGTTGATCTG   AAACCACAAAATTAGCAT  TTTACTGATCACTGGGTTTAAATACTTGTG
            -100                      αs1-Cn COW                                               -22
GTGAGAGAA  TTCTTAGAATT  TAAA......GTTAAACCTG   AAACCACAAAATTAGCAT  TTTACTAATCAGTAGGTTTAAATAGCTTGG
            -100                      αs2-Cn RAT                                               -22
GATATTGCT  TTCTTAGAATT  CGAAT.....TAGGTATTTG   AAACCACAGAATTAGCAT  ATGATGCTAGAACCTGGTTTAAATAGTGCG
            -100                      β-Cn SHEEP                                               -22
CATGCAGAT  TTCTAAGAATT  CAAAT.....GTTTTATTTC   AAACCACAAAATTAGCAT  GCCATTAAATACTATATATAAAACAGCCAC
            -100                      β-Cn COW                                                 -22
CATGCAGAT  TTCTAGGAATT  CAAATCCA..GTTTTATTTC   AAACCACAAAATTAGCAT  GCCATTAAATACTATATATAAACAACCACA
            -100                      β-Cn RABBIT                                              -22
CATATGGAG  TTCTTGGAATT  GAAA.........TTTTTTC   AAACCACAAAATTAGCAT  GTCATTAAATGCAGTATATAAGCATCCCCA
            -100                      β-Cn RAT                                                 -26
CATGTGAAC  TTCTTGGAATT  AAGGAA......TATCTTAC   GAACCAC AAATTAGCAT  GTCATTAAGTATGGTATATATACAGTCACA
            -100                      κ-Cn COW                                                 -25
.TTTCCTTTCTGTCCATCTTCCTATTGGTGCAATGTAAAAGGAAGATAAATCTCATGACGCAAGACACTAACACCCTTTAATTAGTCTCTGGT
```

Fig. 17-8 Specific structural motifs (boxed) occurring upstream from the TATA box (underlined) of the genes encoding the calcium-sensitive caseins αs1, αs2 and ß. k-casein and whey proteins lack these sequences.

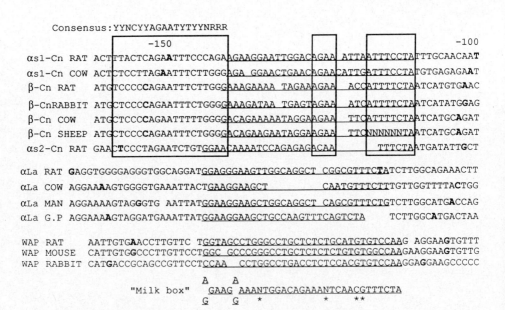

Fig. 17-9 Structural motifs (bold-faced boxes) common to the genes encoding the calcium-sensitive caseins αs1, αs2 and ß, and homologous sequences (underlined) shared with the α-La [35, 36] and WAP genes [38]. Sequences were aligned to get the best inter-species homology for a given gene. Bold-faced nucleotides occur at positions -150 and -100 with respect to the site of transcription initiation. According to *in vitro* assays [38], the so-called "milk box" consensus sequence [36] is the target of (a) mammary nuclear protein(s) in the α-La and WAP genes. * indicates a possible gap in the consensus sequence.

17.5 ß-Casein cDNAs and Genes

ß-casein, first sequenced in 1972 [39], is the longest casein in the bovine species with 209 amino acid residues including 5 phosphoserine. Comparison with its ovine counterpart [11] showed 20 amino acid substitutions and one deletion, which is considerable since bovine and ovine species are phylogenetically closely related. The high rate of evolution of this casein was confirmed later by comparing sequences of ß-casein cDNAs from the rat [40], bovine [41-43], ovine (Provot et al., unpublished work), mouse [44] and rabbit [45] species. Sizes of cDNAs range from 1072 for the rat to 1120 bp for the mouse, about 40 % of which are untranslated.

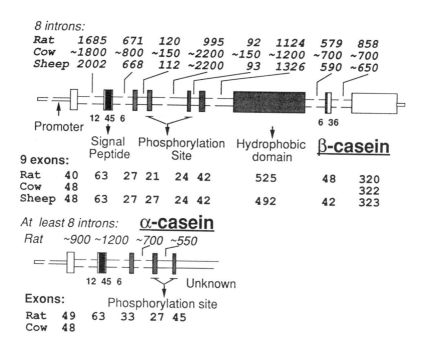

Fig. 17-10 Organization of the ß- and αs1-casein gene (adapted from [34]). Exons are represented by large boxes: black and dark-grey regions encode the signal peptide and the mature protein, respectively. Sizes of introns (upper numbers) and exons (lower numbers) are indicated.

The upper drawing refers to the partially sequenced ovine ß-casein gene (Provot et al., unpublished work): 2 and 0.3 kb from introns 4 and 8, respectively, are undetermined. In the rat [29], internal stretches from introns 1, 2, 4, 6 and 8, accounting for 2.1 kb, were not sequenced. In the cow, a restriction map comprising the positions of the exons and the sequences of exons I and IX are known [46]. A short 319 bp sequence upstream from the first exon of the rabbit ß-casein [47] has been reported.

The lower drawing refers to the rat α-casein (αs1-like casein) gene [34]. The 5' flanking region and the first exon were sequenced in the cow [34] and rabbit [47] species.

The rat ß-casein gene was the first casein gene to be almost entirely sequenced [29], leading to interesting observations. The 7.2 kb transcription unit contains 9 exons ranging in size from 21 to 525 bp (Fig. 17-10). The 5' untranslated region of the mRNA is divided into exons I and II, the latter encoding additionally the signal peptide and the first two amino acids of the mature protein. The major phosphorylation site of the casein is encoded by the 3' and 5' ends of exons IV and V, respectively, and the 525 bp of exon VII constitute 75 % of the coding frame. Interestingly, all splicing in the coding frame occurs between codons. On the basis of the distribution and size of exons and introns, and the sequence homology existing between some exons, Jones et al., [29] suggested that the ß-casein gene evolved through intragenic duplications: exon III might have generated exon IV and the set exon III-intron 3-exon IV further duplicated to give the set exon V-intron 5-exon VI, or reciprocally. Furthermore, comparison of the 5' flanking region with that of the αs2-like casein gene showed the occurrence of three conserved motifs in the 200 bp sequence flanking the transcription units, including a 18 nucleotide stretch neighbouring the TATA box (Figs. 17-8, 17-9).

Most of the sequences of the ovine (Provot et al., unpublished work) ß-casein gene, as well as a few parts of its bovine [46] and rabbit counterparts [47] are now known. The overall organization of the largest ovine and bovine transcription units are quite similar (Fig. 17-10) to the rat one and the difference in sizes is primarily due to the larger sizes of introns 1 and 4. The sequence of the domain encoding the ovine major phosphorylation site, although different from its rats counterpart (Fig. 17-11), also arises from a RNA-splicing event. This is advantageous as there is a selection pressure for the conservation of both the splicing sequence and the encoded phosphorylation recognition site. Ovine intron 4 contains several repeated sequences not yet identified, also found in the 5' flanking region. In the four species, the structural motifs first identified in the rat occur in the well-conserved sequence upstream from the transcription unit (Fig. 17-8).

Expression of the rat ß-casein gene was studied in transfected mammary epithelial cells and transgenic mice [48]. Expression without apparent hormonal regulation was first reported [49] for the transcription unit flanked by 2.3 and 0.4 kb on the 5' and 3' sides. However, later experiments [50] carried out with a series of 5'-deleted chimeric ß-casein-chloramphenicol acetyltransferase constructs led to the conclusion that cis-acting sequences mediating the synergic effect of prolactin and glucocorticoids occur between positions -285 and -170 from the CAP site. A ß-casein gene with 5' and 3' flanking regions of 3.5 and 3.0 kb, respectively, was specifically but poorly expressed in the mammary gland of transgenic mice with a yield representing at most 1 % of the level of endogenous ß-casein mRNA. These results suggest that distal sequences not present in the transgene may be needed for a good expression of the ß-casein gene.

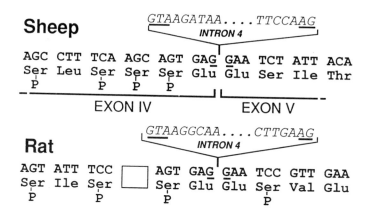

Fig. 17-11 The RNA stretch encoding the major phosphorylation site of bovine and rat ß-casein results from a splicing event. (Adapted from [29]). Phosphorylation requires the enzymatic recognition of the tripeptide sequence Ser-X-A, where A is an acidic amino acid residue [6].

17.6 αs1-Casein cDNAs and Genes

Bovine αs1-casein, the first sequenced casein, is a multiphosphorylated polypeptide chain of 199 amino acid whose intra-chain homology suggested that the relevant gene evolved from a primitive ancestor through duplication events [6]. Ovine [51], bovine [52, 53] and rabbit [54] αs1-casein cDNAs and their rat α-casein [55] and guinea-pig B-casein [56] cDNA counterparts were completely sequenced. Their size is over 1 kb and reaches 1349 nucleotides in the rat α-casein cDNA whose coding frame contains an insertion of 9.5 repeated elements of 18 nucleotides each. Inter-species comparison of the cDNAs showed that the coding frame of αs1-casein is the most divergent of the milk protein cDNAs (Fig. 17-5).

The coding frame is split into many exons and the estimated sizes of the rat [34] and bovine [57] αs1-casein genes are 10-15 and 20 kb, respectively. The partial organization of the first five exons comprising only 16 % of the rat mRNA [34] (Fig. 17-10) and fragmentary sequence data relevant to the 5' flanking region, exon I and intron 1 of the bovine [34] and rabbit [47] αs1-casein genes were reported. The 5' untranslated region of the rat αs1-casein mRNA is divided into exons I and II, the latter encoding additionally the signal peptide and the first two amino acids of the mature protein, as in the ß-casein gene. The major phosphorylation site of the casein is encoded by the 3' and 5' ends of exons IV and V, respectively.

The most conserved region is the 5' flanking region which contains in particular the 18 nucleotide motif specific to the calcium-sensitive casein gene family [34], occurring 17 nucleotides upstream from the TTTAAA box (Figs. 17-8, 17-9).

17.7   αs2-Casein cDNAs and Genes

αs2-casein is the most phosphorylated of the calcium-sensitive caseins with up to 13 phosphate groups in the bovine polypeptide chain made up of 207 amino acid residues. The two known bovine genetic variants A and D differ by an internal deletion of 9 amino acid residues.

The striking sequence homology first observed between the N- and C-terminal halves of the bovine mature polypeptide chain [7] suggested that the αs2-casein gene probably originated from the duplication of a primitive gene.

Ovine [58] and bovine [41] αs2-casein cDNAs, their rat γ-casein [55], guinea-pig A-casein [59] and mouse ε-casein [60] counterparts as well as a few stretches of the rat gene are known [28]. The cDNAs are quite different both in length (about 700 nucleotides in the mouse versus 1036 in the guinea-pig) and nucleotide sequences. In the ovine and caprine species, αs2-casein occurs as two non-allelic forms transcribed from 4 types of mRNAs resulting from the combination of two deletions affecting the 5' untranslated region and the coding frame (27 missing nucleotides), respectively ([58], Boisnard et al., unpublished work). This mRNA heterogeneity is likely to arise from an incomplete splicing of a unique pre-mRNA although the existence of at least two functional αs2-casein genes duplicated before the divergence of the ovine and caprine species is a plausible alternative.

The overall organization of the rat 15-kb αs2-like casein gene is known. It contains at least 9 exons, and many repeated sequences occur within or near the transcription unit. Although only fragmentary sequence data (including 680 bp upstream from the 44 untranslated nucleotides of exon I, the 5' end of intron 1 and the 63 bp exon II which encodes in particular the 15 amino acid signal peptide) are available, the 5' flanking regions of the transcription units of αs1-, ß- and αs2-like casein genes are clearly related as they share the same TATA box and neighbouring sequence motifs (Fig. 17-8). Another clue of the evolutionary relationship is the homology between the signal peptides (Fig. 17-4).

17.8   k-Casein cDNAs and Genes

k-casein, a polypeptide chain of 169 amino acid residues in the bovine species, is essential to micelle formation and stabilization, and its limited proteolytic cleavage triggers coagulation of milk. However, inter-species comparison of the C-terminal third of the polypeptide chain [10] released during coagulation, showed a high rate of evolution, confirmed later at the nucleotide level. k-casein cDNA sequences of rat [61], bovine [52, 62], ovine (Furet et al., unpublished work) and mouse [63] species are known and the structure of the k-casein gene was investigated partially in the bovine species [64]. The gene, which is at least 13 kb in length, is 15 times larger than the relevant 850 nucleotide-long mature bovine mRNA. The k-casein gene contains 4 introns ranging from 1847 to about 5800 bp and 84% of the coding frame are contained in the penultimate exon (Fig. 17-12). Intron 2 and to a lesser extent intron 3 contain many *Alu*-type repeats of the A family. One repeat of the C family also occurs in intron II.

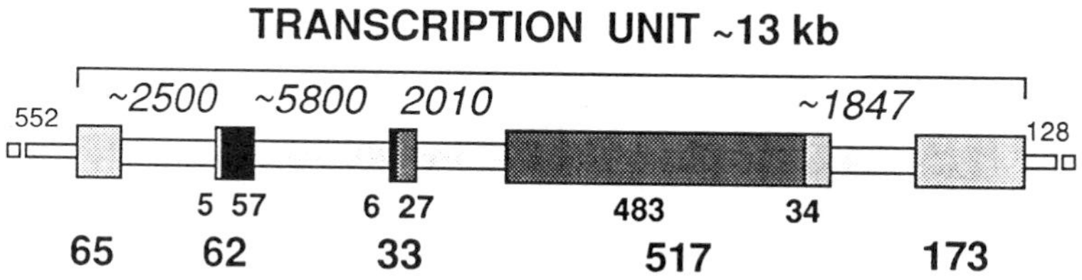

TRANSCRIPTION UNIT ~13 kb

Fig. 17-12 Organization of the bovine k-casein gene. Same legend as in Fig. 17-10. The bovine k-casein gene [64] encompasses 13 kb. Undetermined internal sequences of introns 1 (2.2 / 2.5 kb) and 2 (2.3 / 5.8 kb) account for 35% of the transcription unit.

The k-casein gene is the only milk protein gene whose signal peptide, different in both length and amino acid sequence from the consensus signal of the calcium-sensitive casein family (Fig. 17-4), is encoded by two exons. Furthermore, the 5' flanking region of the transcription unit does not contain the fore-mentioned invariant structural motif shared by the other caseins (Fig. 17-8). The k-casein gene, quite different structurally from the calcium-sensitive casein genes, is likely evolutionary related to the fibrinogen gene family which encodes proteins functionally similar to k-casein as their limited proteolytic cleavage triggers the clotting of blood. The most striking homology observed between the proteins and later the cDNAs, involves a nucleotide stretch corresponding to the 5' end of k-casein exon IV and the 3' end of γ-fibrinogen exon II, respectively [64].

Expression of k-casein and calcium-sensitive casein genes may be modulated in a different manner, at least in the rat species. In prolactin-deprived mammary explants, only transcription of the k-casein gene is stimulated in the presence of insulin, corticosterone and aldosterone [65].

17.9   Whey Acidic Protein (WAP) cDNAs and Genes

The Whey Acidic Protein (WAP) has been uncovered quite recently. There phosphorylated forms of an abundant (≈5 mg/ml) previously undescribed whey protein were isolated from rat milk and partially analyzed in 1978 [66]. The related unphosphorylated major murine whey protein designated "Whey Acidic Protein" [67] was biochemically characterized in 1981. This novel protein whose biological function is unknown, was described later in the rabbit [68] and camel [69] species and may also exist in the hamster species [70]. Ruminants' milk lacks this protein but the bovine colostral phosphoglycoprotein M-1 shares some similarity with the WAP in terms of amino acid composition and molecular weight (quoted in [67]).

The four WAP sequenced so far either at the protein (camel [69]) or at the cDNA level (rat [70, 71], mouse [70, 72] and rabbit [68]) share some structural homology with the 'four disulfide core' family of proteins such as neurophysins (the hypothalamic carriers for oxytocin and vasopressin), wheat germ agglutinin, red sea turtle protease inhibitor and snake venom toxins. Depending on the species, mature WAP contains between 108 and 118 amino acid residues, including 12 to 16 cysteine residues equally distributed into two structurally related domains.

The 5' and 3' untranslated regions account for 30% of the mRNA length which ranges from 549 in the rabbit to 576 nucleotides in the rat. With the exception of the sequence encoding the 19 amino acid signal peptide, the coding frame is less conserved than the latter regions. Obviously, the WAP is a protein which evolved rapidly.

Rat and mouse WAP genes [31] have a similar organization (Fig. 17-13): they contain 4 exons and the difference in the sizes of the transcription units (2.8/3.3 kb) is primarily due to the larger size of the mouse third intron. The two fore-mentioned cysteine domains are encoded by exons II and III, respectively, which suggests that the WAP gene evolved by intragenic duplication of one of these exons.

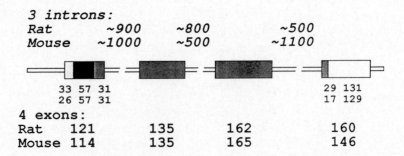

Fig. 17-13 Organization of the mouse and rat WAP genes. Drawing based on the data reported in [31]. Same legend as in Fig. 17-10. Only the 5' and 3' ends of the introns were sequenced.

WAP gene expression depends on prolactin and glucocorticoids [47, 73] and the transcript accounts for 15% of total poly(A)+mRNA in active rat lactating cells. Accordingly, several potential glucocorticoid receptor binding sites occur in the 5' flanking region of the transcription unit. In contrast, the WAP mRNA is barely detected in mammotrophic hormone-dependent rat tumors expressing the other milk protein mRNAs, which is correlated with the hypermethylation of the WAP gene [71]. Rat, mouse [31] and rabbit [47] 5' flanking regions of the WAP transcription unit share some homologous stretches, one of them spanning the unusual putative TATA (TTTAAAT) box occurring twenty or so nucleotides upstream from the transcription unit.

The 2.5 kb 5' flanking region of the murine WAP transcription unit contains sequence(s) responsible for its tissue-specific and temporal expression: hybrid genes constructed from oncogene (Ha-*ras* and c-*myc* [74, 75]) and tissue Plasminogen Activator [22, 23] transcription units fused to the relevant 2.5 kb WAP fragment, were expressed specifically in the mammary gland, although at a reduced rate. Accordingly, *in vitro* exonuclease III or DNAase protection and DNA-protein mobility shift assays [37] showed the preferential binding of nuclear proteins from rat lactating mammary cells to sites located in the promoter region -170/-20 of the WAP gene (Figs. 17-8, 17-9).

17.10 ß-Lactoglobulin (ß-Lg) cDNAs and Genes

ß-lactoglobulin (ß-Lg), the major whey protein of Ruminant's milk ($\approx$3 g/l in cow's milk), also found in several other species including pig, horse, dog and dolphin, is apparently lacking in rodents and human milk. Mature bovine ß-Lg is a 162 amino acid protein whose structure and physico-chemical properties have been extensively studied and its signal peptide is likely to contain 18 amino acid residues as shown for ovine and pig pre-ß-Lg [9]. The protein is able to bind miscellaneous hydrophobic molecules but its biological role is still unknown. However, amino acid sequence comparisons showed homology with several other secretory proteins such as mouse urinary protein (MUP), placental protein 14, and retinol-binding protein (RBP) [76, 77]. ß-Lg whose tertiary structure resembles that of RBP may be involved in vitamin A transport as specific receptors for the ß-Lg-retinol complex were found in the intestine of neonate calves [77].

The ovine ß-Lg cDNA [78] and its partially sequenced bovine counterpart [79, 80] share homology with the known cDNAs encoding the fore-mentioned proteins [76, 77] such as human PP14 cDNA. The overall 62% homology with the ovine cDNA is higher in the coding frame especially in the signal peptide domain. The ovine ß-Lg mRNA contains 785 nucleotides, excluding the poly(a) tail, of which 30% are untranslated.

The transcription unit of the ovine ß-Lg gene consists of 4662 nucleotides divided into 7 exons [32, 33] and its organization resembles that of the genes encoding the MUP and RBP (Fig. 17-14). *Alu*-like sequences occur 474 nucleotides downstream from the last exon.

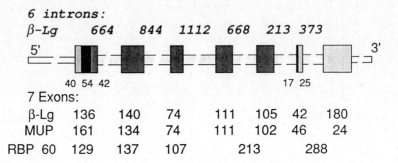

Fig. 17-14 Organization of the ovine ß-Lactoglobulin gene and two related genes encoding rat major urinary protein (MUP) and retinol binding protein (RBP), respectively. Adapted from [32]). Drawing based on the data reported in [32, 33]. Same legend as in Fig. 17-10. The 5' untranslated region of the RBP gene is divided into two exons, and the last two exons are equivalent to the pairs IV-V and VI-VII of the ß-Lg gene.

The ovine ß-Lg gene is already expressed at mid-pregnancy and ß-Lg mRNA concentration in total cellular RNA has increased about 20-fold at day 20 of lactation [78]. At this stage, ß-Lg mRNA represents about 5% of total poly(A)+RNA. In the ovine species, the accumulation rates of ß-Lg and casein mRNAs are asynchronous, suggesting a differential hormonal control on both transcription of the relevant genes and mRNA stabilization [78].

The ovine ß-Lg gene is highly and specifically expressed in the mammary gland of transgenic mice [81]: 15 mg of ß-Lg/ml of milk for a transgene with 3 kb 5' and 1 kb 3' flanking regions, respectively; 2 mg/ml for a transgene with only 0.8 kb upstream from the transcription unit. ß-Lg-human blood coagulation factor IX or α-antitrypsin fusion genes were also specifically expressed in the mammary gland of transgenic ewe [20].

17.11 α-Lactalbumin (α-La) cDNAs and Genes

α-La is a calcium metalloprotein occurring in all lactose-containing milks. It promotes the synthesis of lactose by interacting with and modifying the substrate specificity of UDP-galactosyltransferase (EC 2.4.1.22) which then uses glucose as acceptor instead of N-acetylglucosamine residues of glycoproteins. The structures, physico-chemical and biological properties of α-La have been extensively studied. The amino acid sequence determined either directly or deduced from the cDNA coding frame is known in 9 species and several amino acid residues structurally or functionally important have been identified (Fig. 17-15). Mature α-La which shares significant structural homology with lysozyme is made up of 123 amino acid residues in most species and the 19 amino acid signal peptide is the most divergent among those of the milk preproteins. This may reflect a lower selection pressure on the α-La signal peptide presumably correlated to the lower content of this protein in milk with respect to caseins.

α-La cDNAs have been sequenced in the ovine [82], caprine [83], bovine [84, 85], guinea-pig and human [86] species. A typical α-La mRNA such as the bovine one contains 725 nucleotides, excluding the poly(A) tail, of which 41% are untranslated. In contrast with the other milk protein cDNAs, inter-species comparison of α-La cDNAs showed a better homology in the coding frame than in the 5' and 3' untranslated regions, with a similar rate of divergence of the regions encoding the signal peptide and the mature protein [82].

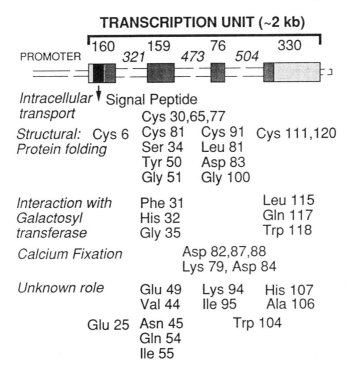

Fig. 17-15 Localization of the codons encoding the signal peptide and invariant amino acid residues of bovine pre-α-lactalbumin. The drawing is based on the data reported in [88]. Same legend as in Fig. 17-10. Numbers refer to positions of amino acids in the mature polypeptide chain.

The complete nucleotide sequences of the α-La gene from four species, rat [30], bovine [84], human [35] and guinea-pig [36], are known. The transcription unit of the αLa gene (Fig. 17-16) with its 4 exons is very similar to that of lysozyme, corroborating the previous hypothesis of a common ancestral origin. Functional domains of pre-α-La are encoded either by a unique exon (exon I for the signal peptide and exon III for the calcium binding loop) or by several exons (the domain responsible for the interaction with the UDP-galactosyltransferase is encoded by exons II and IV, and amino acid residues structurally important for polypeptide chain folding are transcribed from codons occurring within II, III and IV). According to RFLP patterns and analytical data of genomic clones (Soulier et al., unpublished work), several α-La-related sequences are present in the genome of domestic ruminants. Presumed bovine (Fig. 17-17) and ovine α-La pseudogenes were shown to contain a nucleotide stretch sharing homology (81 and 73% respectively) with the bovine α-La gene downstream from the end of exon II.

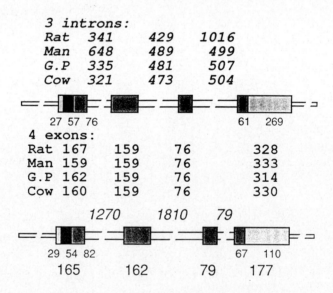

Fig. 17-16 Organization of the α-lactalbumin gene from four species. Comparison with the evolutionary related chicken white-egg lysozyme gene (lower drawing). Based on the sequence data referenced in the text. Same legend as in Fig. 17-10. The length of the transcription units is 2 kb for the rat, cow and guinea-pig α-La genes, 2.3 kb for the human α-La, and 3.7 kb for the lysozyme gene.

The α-La gene is already expressed during the development of the guinea-pig mammary gland which produces α-La but not caseins during pregnancy. On the contrary, in the rat species, there is a concomitant increase of expression of the casein genes and repression of the α-La gene until parturition [87]. However, in both species, the α-La gene activity is at its maximum during lactation. The differential expression of α-La and casein genes [14] reflects presumably a difference in the mode of regulation by prolactin, glucocorticoids and insulin. The bovine α-La gene is well expressed in the mammary gland of transgenic mice. One line carrying the α-La transcription unit flanked by 0.7 and 0.3 kb at the 5' and 3' ends, respectively, produces a milk containing about 0.5 mg bovine α-La per ml as compared to 1.2 mg/ml in cow's milk (Vilotte et al., unpublished work). The expression is tissue-specific since α-La mRNA was only detected in the mammary tissue. According to *in vitro* DNAase I footprinting assays [38], the 'milk box' of the rat α-La promoter contains a sequence recognized by (a) nuclear protein(s) apparently specific of the mammary gland and already present in virgin animals. Interestingly, this specific sequence corresponds to the deleted part of the bovine 'milk box' (Fig. 17-9).

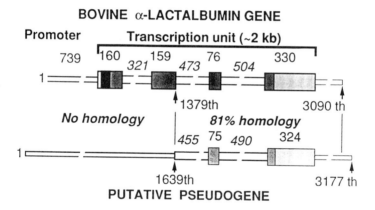

Fig. 17-17 Organization of a presumed bovine α-lactalbumin pseudogene. (Soulier et al., unpublished work). Same legend as in Fig. 17-10. The nucleotide stretch 1639-3177 of the genomic DNA fragment analyzed shares 81% homology with the region of the gene downstream from exon II.

17.12 Conclusion

Elucidation of the nucleotide sequences of several milk protein genes from different species was a prerequisite to a better understanding of their evolution and of the mechanisms involved in their expression and regulation at the molecular level. As predicted from the amino acid sequence data, the most conserved regions of the rapidly evolving coding frames are those encoding the signal peptide of all milk proteins, and the major phosphorylation site(s) of calcium-sensitive caseins whose common origin is substantiated by similar organization of the relevant genes.

At present, our structural knowledge is still insufficient since only a few genes were wholly or largely sequenced in more than one species: 4 for α-la, 2 for the WAP and the ß-casein. Fortunately, more sequence data will be available in the near future, allowing a more detailed and accurate analysis of conserved sequences which might play an important role. Nevertheless, analysis of sequence data already obtained showed the occurrence of structural motifs specific to the genes encoding calcium-sensitive caseins, as well as the identification of a consensus sequence ('milk box') shared by most but not all milk protein genes. However, their biological significance is still unknown.

Rapid progress in the field of milk protein gene expression and regulation will depend in part on the development of both permanent mammary cell lines and efficient cell-free transcription systems able to mimic what occurs *in vivo*. Although the application of transgenic mice for analyzing gene expression has some value, it is difficult to compare different transgenic animals as the copy number of integrated genes and their chromosomal location differ. Recent studies of *trans*-acting nuclear factors regulating the expression of milk protein genes are promising, and these experiments should provide useful information about the fine regulation of gene expression at the molecular level.

In practice, in the field of RFLP studies, the very short sizes of most exons make difficult the detection of the relevant genomic fragments with a cDNA probe in the standard conditions. The knowledge of the gene sequence will allow i) to choose and use as efficient probes, for RFLP studies, DNA genomic fragments devoid of repetitive sequences; ii) to develop allelic specific oligonucleotide probes recognizing *in vitro* amplified mutated DNA stretches, thus permitting the rapid identification of alleles of economic interest. Similarly, the knowledge of the sequences of ruminant α-la pseudogenes will allow the accurate chromosomal assignment of the authentic cognate gene by using a probe specific for the region of the gene deleted in the pseudogenes.

17.13 References

[1] Migliore-Samour, D., and Jollès, P., *Experientia 1988, 44,* 188-193.

[2] Swaisgood, H.E., In: *Developments in Dairy Chemistry - 1. Protein:* Fox, P.F. (ed.) London & New York: Applied Science Publishers, *1982*; Vol. 1, pp. 1-59.

[3] Mercier, J.C., Grosclaude, F., and Ribadeau-Dumas, B., *Milchwissenschaft 1972, 27,* 402-408.

[4] Grosclaude, F., In: *Proc. 16th Int. Conf. Anim. Blood Grps. Biochem. Polymorphisms:* International Society for Animal Blood Group Research., (ed.) Leningrad International Society for Animal Blood Group Research, *1979*; Vol. 1, pp. 54-92.

[5] Mercier, J.C., *Biochimie 1981, 63,* 1-17.

[6] Mercier, J.C., Grosclaude, F., and Ribadeau-Dumas, B., *Eur. J. Biochem., 1971, 23,* 41-51.

[7] Brignon, G., Ribadeau-Dumas, B., Mercier, J.C., Pélissier, J.P., and Das, B.C., *FEBS Lett., 1977, 76,* 274-279.

[8] Gaye, P., Gautron, J.P., Mercier, J.C., and Hazé, G., *Biochem. Biophys. Res. Commun., 1977, 79,* 903-911.

[9] Mercier, J.C., and Gaye, P., in: *Biochemistry of Lactation:* Mepham, T.B. (ed.) Amsterdam & New York: Elsevier, *1983*; pp. 177-227.

[10] Mercier, J.C., Chobert, J.M., and Addeo, F., *FEBS Lett., 1976, 72,* 208-214.

[11] Richardon, B.C., and Mercier, J.C., *Eur. J. Biochem., 1979, 99,* 285-297.

[12] Rosen, J.M., Supowit, S.C., Gupta, P., Yu-Lee, L.Y., and Hobbs, A.A., In: *Hormones and Breast Cancer:* Cold Spring Harbor Laboratory, *1981*; pp. 397-424.

[13] Rosen, J.M., Hobbs, A.A., Johnson, M.L., Rodgers, J.R., and Yu-Lee, L.Y., In: *Gene Regulation:* O'Malley, B.W., Fox, C. (eds.) New York: Academic Press, *1982*, pp. 275-290.

[14] Burditt, L.J., Parker, D., Craig, R.K., Getova, T., and Campbell, P.N., *Biochem. J., 1981, 194,* 999-1006.

[15] Rosen, J.M., Rodgers, J.R., Couch, C.H., Bisbee, C.A., David-Inouye, Y., Campbell, S.M., and Yu-Lee, L.Y., *Annals N.Y. Acad. Sci., 1986, 478,* 63-76.

[16] Choi, Y.J., Keller, W.L., Berg, J.E., Park, C.S., and MacKinlay, A.G., *J. Dairy Sci., 1988, 71,* 2898-2903.

[17] Houbedine, L.M., Djiane, J., Dusauter-Fourt, I., Martel, P., Kelly, P.A., Devinoy, E., and Servely, J.L., *J. Dairy Sci., 1985, 68,* 489-500.

[18] Rando, A., DiGregorio, P., and Masina, P., *Animal Genetics, 1988, 19,* 51-54.

[19] Levéziel, H., Méténier, L., Mahé, M.F., Choplain, J., Furet, J.P., Paboeuf, G., Mercier, J.C., and Grosclaude, F., *Génét. Sél. Evol., 1988, 20,* 247-254.

[20] Simons, J.P., Wilmut, I., Clark, J.A., Archibald, A.L., Bishop, J.O., and Lathe, R., *Biotechnology 1988, 6,* 179-183.

[21] Wilmut, I., Clark, J., and Simons, J.P., *New Scientist, 1988,* 56-59.

[22] Gordon, K., Lee, E., Vitale, J.A., Smith, A.E., Westphal, H., and Hennighausen, L., *Biotechnology, 1987, 5,* 1183-1187.

[23] Pittius, C.W., Hennighausen, L., Lee, E., Westphal, H., Nicols, E., Vitale, J., and Gordon, K., *Proc. Natl. Acad. Sci. USA, 1988, 85,* 5874-5878.

[24] Guyette, W.A., Matusik, R.J., and Rosen, J.M., *Cell, 1979, 17,* 1013-1023.

[25] Gupta, P., Rosen, J.M., D'Eustachio, P., and Ruddle, F.H., *J. Cell Biol., 1982, 93,* 199-204.

[26] Gellin, J., Echard, G., Yerle, M., Dalens, M., Chevalet, C., and Gillois, M., *Cytogenet. Cell. Genet., 1985, 39,* 220-223.

[27] Davies, M.S., West, L.F., Davis, M.B., Povey, S., and Craig, R.K., *Ann. Hum. Genet., 1987, 51,* 183-188.

[28] Yu-Lee, L.Y., and Rosen, J.M., *J. Biol. Chem., 1983, 258,* 10794-10804.

[29] Jones, W.K., Yu-Lee, L.Y., Clift, S.M., Brown, T.L., and Rosen, J.M., *J. Biol. Chem., 1985, 260,* 7042-7050.

[30] Qasba, P.K., and Safaya, S.K., *Nature, 1984, 608,* 377-380.

[31] Campbell, S.M., Rosen, J.M., Hennighausen, L.G., Strech-Jurk, U., and Sippel, A.E., *Nucleic Acids Res., 1984, 12,* 8685-8697.

[32] Ali, S., and Clark, A.J., *J. Mol. Biol.*, *1988*, *199*, 415-426.

[33] Harris, S., Ali, S., Anderson, S., Archibald, A.L., and Clark, A.J., *Nucleic Acids Res.*, *1988*, *16*, 10379-10380.

[34] Yu-Lee, L.Y., Richter-Mann, L., Couch, C.H., Stewart, A.F., MacKinlay, A.G., and Rosen, J.M., *Nucleic Acids Res.*, *1986*, *14*, 1883-1902.

[35] Hall, L., Emery, D.C., Davies, M.S., Parker, D., and Craig, R.K., *Biochem. J.*, *1987*, *242*, 735-742.

[36] Laird, J.E., Jack, L., Hall, L., Boulton, A., Parker, D., and Craig, R.K., *Biochem. J.*, *1988*, *254*, 85-94.

[37] Lubon, H., and Hennighausen, L., *Nucleic Acids Res.*, *1987*, *15*, 2103-2121.

[38] Lubon, H., and Hennighausen, L., *Biochem. J.*, *1988*, *256*, 391-396.

[39] Ribadeau-Dumas, B., Brignon, G., Grosclaude, F., and Mercier, J.C., *Eur. J. Biochem.*, *1972*, *25*, 505-514.

[40] Blackburn, D.E., Hobbs, A.A., and Rosen, J.M., *Nucleic Acids Res.*, *1982*, *10*, 2295-2306.

[41] Stewart, A.F., Bonsing, J., Beattie, C.W., Shah, F., Willis, I.M., and MacKinlay, A.G., *Mol. Biol. Evol.*, *1987*, *4*, 231-241.

[42] Jimenez-Flores, R., Kang, Y.C., and Richardson, T., *Biochem. Biophys. Res. Commun.*, *1987*, *142*, 617-621.

[43] Bayev, A.A., Smirnov, I.K., and Gorodetsky, S.I., *Genetika*, *1987*, *21*, 255-264.

[44] Yoshimura, M., Banerjee, M.R., and Oka, T., *Nucleic Acids Res.*, *1986*, *14*, 8224.

[45] Schaerer, E., Devinoy, E., Kraehenbuhl, J.P., and Houdebine, L.M., *Nucleic Acids Res.*, *1988*, *16*, 11814-11814.

[46] Gorodetsky, S.L., Tkach, T.M., and Kapelinskaya, T.V., *Gene*, *1988*, *66*, 87-96.

[47] Devinoy, E., Hubert, C., Jolivet, G., Thepot, D., Clergue, N., Desaleux, M., Dion, M., Servely, J.L., and Houdebine, L.M., *Reprod. Nutr. Develop.*, *1988*, *28*, 1145-1164.

[48] Yu-Lee, K.F., Demayo, F.J., Atiee, S.H., and Rosen, J.M., *Nucleic Acids Res.*, *1988*, *16*, 1027-1041.

[49] David-Inouye, Y., Couch, C.H., and Rosen, J.M., *Annals N.Y. Acad. Sci.*, *1986*, *478*, 274-277.

[50] Doppler, W., Groner, B., and Ball, R.K., *Proc. Natl. Acad. Sci. USA*, *1989*, *86*, 104-108.

[51] Mercier, J.C., Gaye, P., Soulier, S., Hue-Delahaie, D., and Vilotte, J.L., *Biochemie*, *1985*, *67*, 959-971.

[52] Stewart, A.F., Willis, I.M., and MacKinlay, A.G., *Nucleic Acids Res.*, *1984*, *12*, 3895-3907.

[53] Nagao, M., Maki, M., Sasaki, R., and Chiba, H., *Agric. Biol. Chem.*, *1984*, *48*, 1663-1667.

[54] Devinoy, E., Schaerer, E., Jolivet, G., Fontaine, M.L., Kraehenbuhl, J.P., and Houdebine, L.M., *Nucleic Acids Res.*, *1988*, *16*, 11813.

[55] Hobbs, A.A., and Rosen, J.M., *Nucleic Acids Res.*, *1984*, *10*, 8079-8098.

[56] Hall, L., Laird, J.E., and Craig, R.K., *Biochem. J.*, *1984*, *222*, 561-570.

[57] Bonsing, J., and MacKinlay, A.G., *J. Dairy Res.*, *1987*, *54*, 447-461.

[58] Boisnard, M., and Pétrissant, G., *Biochemie*, *1985*, *67*, 1043-1051.

[59] Hall, L., Laird, J.E., Pascall, J.C., and Craig, R.K., *Eur. J. Biochem.,* **1984,** *138,* 585-589.

[60] Hennighausen, L.G., Steudle, A., and Sippel, A.E., *Eur. J. Biochem.,* **1982,** *126,* 569-572.

[61] Nakhasi, H.L., Grantham, F.H., and Gullino, P.M., *J. Biol. Chem.,* **1984,** *259,* 14894-14898.

[62] Gorodetsky, S.I., and Kaledin, A.S., *Genetika,* **1987,** 596-604.

[63] Thompson, M.D., Dave, J.R., and Nakhasi, H.L., *DNA,* **1985,** *4,* 263-271.

[64] Alexander, L.J., Stewart, A.F., MacKinlay, A.G., Kapelinskaya, T.V., Tkach, T.M., and Gorodetsky, S.I., *Eur. J. Biochem.,* **1988,** *178,* 395-401.

[65] Vonderhaar, B.K., and Nakhasi, H.L., *Endocrinology,* **1986,** *119,* 1178-1184.

[66] McKenzie, R.M., and Larson, B.L., *J. Dairy Sci.,* **1978,** *61,* 723-728.

[67] Piletz, J.E., Heinlen, M., and Ganschow, R.E., *J. Biol. Chem.,* **1981,** *256,* 11509-11516.

[68] Devinoy, E., Hubert, C., Schaerer, E., Houdebine, L.M., and Kraehenbuhl, J.P., *Nucleic Acids Res.,* **1988,** *16,* 8180-8180.

[69] Beg, O.U., Bahr-Lindström, H., Zaidi, Z.H., and Jörnvall, H., *Eur. J. Biochem.,* **1986,** *159,* 195-201.

[70] Hennighausen, L.G., Sippel, A.E., Hobbs, A.A., and Rosen, J.M., *Nucleic Acids Res.,* **1982,** *10,* 3733-3744.

[71] Dandekar, A.M., Robinson, E.A., Appella, E., and Qasba, P.K., *Proc. Natl. Acad. Sci. USA,* **1982,** *79,* 3987-3991.

[72] Hennighausen, L.G., and Sippel, A.E., *Nucleic Acids Res.,* **1982,** *10,* 2677-2684.

[73] Hobbs, A.A., Richards, D.A., Kessler, D.J., and Rosen, J.M., *J. Biol. Chem.,* **1982,** *257,* 3598-3605.

[74] Groner, B., Schonenberger, C.A., and Andres, A.C., *TIG,* **1987,** *3,* 306-308.

[75] Schoenenberger, C.A., Andres, A.C., Groner, B., Van der Valk, M., Lemeur, M., and Gerlinger, P., *EMBO J.,* **1988,** *7,* 169-175.

[76] Julkunen, M., Seppälä, M., and Jänne, O.A., *Proc. Natl. Acad. Sci. USA,* **1988,** *85,* 8845-8849.

[77] Papiz, M.Z., Sawyer, L., Eliopoulos, E.E., North, A.C.T., Findlay, J.B.C., Sivaprasadarao, R., Jones, T.A., Newcomer, M.E., and Kraulis, P.J., *Nature,* **1986,** *324,* 383-385.

[78] Gaye, P., Hue-Delahaie, D., Mercier, J.C., Soulier, S., Vilotte, J.L., and Furet, J.P., *Biochimie,* **1986,** *68,* 1097-1107.

[79] Jamieson, A.C., Vandeyar, M.A., Kang, Y.C., Kinsella, J.E., and Batt, C.A., *Gene,* **1987,** *61,* 85-90.

[80] Ivanov, V.N., Judinkova, E.S., and Gorodetsky, S.I., *Biol. Chem. Hoppe-Seyler,* **1988,** *369,* 425-429.

[81] Simons, J.P., McClenaghabn, M., and Clark, J., *Nature,* **1987,** *328,* 530-532.

[82] Gaye, P., Hue-Delahaie, D., Mercier, J.C., Soulier, S., Vilotte, J.L., and Furet, J.P., *Biochimie,* **1987,** *69,* 601-608.

[83] Kumagai, I., Tamaki, E., Kakinuma, S., and Miura, K.I., *J. Biochem.,* **1987,** *101,* 511-517.

[84] Vilotte, J.L., Soulier, S., Mercier, J.C., Gaye, P., Hue-Delahaie, D., and Furet, J.P., *Biochimie, 1987, 69,* 609-620.

[85] Hurley, W.L., and Schuler, L.A., *Gene, 1987, 61,* 119-122.

[86] Hall, L., Craig, R.K., Edbrooke, M.R., and Campbell, P.N., *Nucleic Acids Res., 1982, 10,* 3503-3515.

[87] Nakhasi, H.L., and Qasba, P.K., *J. Biol. Chem., 1979, 254,* 6016-6025.

[88] Bucher, P., and Trifonov, E.N., *Nucleic Acids Res., 1986, 14,* 10009-10026.

[89] Shewale, J.G., Sinha, S.K., and Brew, K., *J. Biol. Chem., 1984, 259,* 4947-4956.

18. Studies on Mitochondrial DNA in Farm Animals

W. Hecht

18.1 Introduction

Mitochondria have always been one of the most attractive objects in biochemistry and related fields of research. This is hardly astonishing, if we look at the function of these organelles, which is a prerequisite for the life of almost all eucaryotes: the synthesis of energy equivalents utilizable by the cell.

The strikingly fascinating feature of mitochondria for the geneticist is their genetic apparatus, particularly the mitochondrial DNA (mtDNA).

The discovery of restriction endonucleases and the advent of recombinant DNA technology have greatly prompted studies on mtDNA from a variety of species. Relatively simple isolation procedures, small genome size, ubiquitous abundance in eucaryotes, and the unisexual mode of inheritance have predestinated mtDNA as an object in evolutionary, population genetic, and molecular studies.

The purpose of this contribution is to give a concise review on topics concerning the mitochondrial genome of farm animals. It is strictly confined to the following species: horse, donkey, cow, pig, sheep, goat, rabbit, and chicken.

For details of methods currently used for the analysis of mtDNA, the reader is referred to a collection, recently published by Hauswirth et al. [1].

18.2 Some Properties of mtDNA

18.2.1 Maternal Inheritance and Copy Number in Different Cell Types

Hutchison et al. [2] were the first to demonstrate maternal inheritance of mtDNA in mammals. They analyzed the transmission of mtDNA from parents to offspring in reciprocal crosses of horse and donkey. Fragment patterns of purified mtDNA produced by the restriction endonuclease Hae III were different in both species. Thus the transmission of mtDNA types from parents to hybrid offspring could be monitored. In all cases examined, mule mtDNA resembled the horse pattern, whereas hinny mtDNA fragments were identical to those of the donkey. These results are compatible with the hypothesis that mtDNA is maternally inherited, they do however not rule out the possibility that paternal mtDNA contributes to the mtDNA pool of a zygote at all.

Two mechanisms underlying the observed maternal inheritance of mtDNA may be discussed. The first one is a stochastic model for the selection of mtDNA types, both maternal and paternal for replication in the cells of the developing embryo. In the bovine oocyte, the copy number of mtDNA molecules has been quantitated as 2.6×10^5, which is one hundred times the value found in somatic cells [3]. In contrast, the number of mitochondria in the bull sperm midpiece has been estimated to be 72 [4]. If the number of mtDNA molecules in the bull sperm mitochondria is reduced to one per organelle, as in the mouse [5], the ratio of maternal to paternal mtDNA copies in the zygote would be approximately 3600 : 1. A ratio, wide enough to explain, why the paternal contribution to the mtDNA pool could remain undetected by experimental procedures unless paternal mtDNA is selectively favored.

A second explanation for maternal inheritance relies on the assumption, that specific processes inhibit replication of sperm derived mtDNA in the oocyte or early embryo. In the ram, Bartoov and Fisher [6] presented evidence, that sperm mtDNA differs from mtDNA from somatic cells in several physical properties such as base composition and contour length.

To date however, there is no experimental proof of the hypothesis, that the described alterations specifically eliminate sperm mtDNA during embryogenesis.

18.2.2 Size of mtDNA in Farm Animals

In some of the species, kept as farm animals, the size of mtDNA molecules has been determined. Methods used for sizing include electron microscopy, restriction fragment analysis and nucleotide sequencing; the last method yielding undoubtedly the exactest results. Available size data are listed in Table 18-1.

The size of almost all farm animals' mtDNA equals approximately 16 kilobases except the rabbit, whose mtDNA is one kilobase larger. It is yet unknown, which parts of the mitochondrial genome of the rabbit contribute to this increase in total size.

Table 18-1 Size of mtDNA in farm animals.

Species	Size (Kilobases)	Method (a)	Reference	
Cattle	16.338	SA	Andersson et al.	[7]
Chicken	16.43	RFA	Wakana et al.	[8]
Pig	16.35	RFA	Watanabe et al.	[9]
Rabbit	17.3	EM	Brown	[10]
Rabbit	17.3	RFA	Ennafaa et al.	[11]
Sheep and	15.8 - 16.4	RFA	Upholt and Dawid	[12]
Goat	15.8	EM	"	

(a) SA, sequence analysis; RFA, restriction fragment analysis; EM, electron microscopy.

18.2.3 The D-Loop

The term D-loop was introduced to mark a region of animal mtDNA, where replication is believed to be initiated. The newly synthesized daughter strand displaces the parental heavy strand and thus a loop structure is observed upon electron microscopic inspection of such molecules. Consistently this region is termed displacement- or D-loop. For general reviews on replication and transcription of the animal mitochondrial genome see Clayton [13] and Clayton [14].

Control elements for transcription and replication of the mitochondrial genome have been mapped to the bovine D-loop by S1 nuclease analysis and primer extension techniques [15]. The 5' termini of D-loop structures are heterogenous in bovine mtDNA and map at two distinctive positions [15]. Furthermore, King and Low [16] demonstrated, that the frequency of origins of heavy strand replication at both positions is dependent on the growth state of the bovine cells, they examined. This finding is of special importance in the light of considerations put forward by Clayton [13]. He stated, that the D-loop is a prominent candidate for an element in mtDNA capable of interacting with chromosomal DNA.

18.2.4 Rate of Nucleotide Substitution in mtDNA

The rate of base substitutions in mtDNA seems to be more rapid than in genes encoded by the nuclear genome [10]. From data on sheep and goat mtDNA Upholt and Dawid [12] estimated that the nucleotide substitution rate in mtDNA is about 10^{-8} per year, which is approximately 5 - 10 times higher than in nuclear genes. A similar value can be calculated from data on pig mtDNA presented by Watanabe et al. [17]. In their report nucleotide sequence divergence between typical mtDNA from european and asian pigs was calculated as 1.75%. Taking 0.8 - 0.9 x 10⁶ years as time of divergence between these genealogical groups of pigs [17], the rate of nucleotide substitution per year is approximately 2 x 10⁻⁸ per year.

Table 18-2 Mapped restriction enzyme cleavage sites in mtDNA of farm animals.

Restriction enzyme	Cattle (a)	Chicken (b)		Goat	Sheep	Pig	Rabbit
AvaI	3	5					4
AvaII		4					
BamHI	3	2				5	2
BclI							6
BbeI						0	
BglII	2					3/1	
BstEII	1						
ClaI							2
DraI						7	
EcoRI	3	2	0	5/4	3	3	2
EcoRV						1	1
HaeII		3					
HhaI	5						
HincII		8					
HindIII	3	4	4	3/4	3	4	7
HpaI	6	3					4
KpnI	2	2				1	
MluI						1	
PstI	2					3	2
PvuII		4	4			2	2
SacI	1					2	3
SalI	1	3	3			1	
SmaI						0	
SstII		2					
StuI						8/10	
XbaI	6	1	1			3	2
XhoI	1	2	2			1	
XmnI							4
Reference	[20]	[8, 19]		[12]		[9, 17]	[11]

Numbers indicate frequency of cleavage by the respective enzyme, numbers separated by slashes refer to polymorphic sites.
(a) A complete restriction map can be deduced from the nucleotide sequence data reported by Andersson et al. [7],
(b) First column refers to first author cited.

18.2.5 Physical Mapping, Gene Maps, and Genomic Organization

Restriction maps, i.e. the physical mapping of the relative positions of cleavage sites of one or more restriction endonucleases, have been established for most of the species discussed in this survey. Table 18-2 presents information on these cleavage maps in different species relative to restriction enzymes mapped and the number of cleavage sites per genome. Alignment of the physical maps with specified genes, encoded by mtDNA has been reported for several species. Hauswirth et al. [18] have mapped the genes for small and large ribosomal rRNA and the D-loop on the physical map of bovine mtDNA. Andersson et al. [7] have presented a complete mitochondrial gene map for the bovine species, derived from the complete sequence of this genome. Except for some minor modifications this map coincides exactly with the human map with regard to gene content and gene order [7].

The gene content of a typical mammalian mtDNA consists of 13 genes, transcribed into mRNA, 22 tRNA genes, 2 rRNA genes and the noncoding region already mentioned, the D-loop [14].

Glaus et al. [19] have mapped both the mitochondrial rRNA transcripts and the D-loop in chicken mtDNA. Relating the pig restriction map to the respective gene map was accomplished by partial sequencing of cloned mtDNA fragments [17]. The rabbit physical map has been oriented by hybridization analyses, using cloned mtDNA sequences with known gene content from Xenopus [11]. In sheep and goat, the position of the D-loop has been determined by electron microscopic analysis of heteroduplexes [12].

Assuming that the gene content and gene order is evolutionary conserved in all species listed in Table 18-2, their complete gene maps are known, or, the other way round, all mapping data are in good agreement with the hypothesis, that the genetic maps are conserved throughout these species.

18.3   Heterogeneity of mtDNA

18.3.1   Intra- and Interspecies Comparisons

Potter et al. [21] studied the Hae III restriction patterns of mtDNAs from several mammalian species. Of special interest are their analyses of mtDNA cleavage patterns from species capable of interbreeding. The horse and donkey patterns and the cow and buffalo patterns had approximately 50 % of comigrating bands. Among ten horses five different restriction patterns were found, but these differed only slightly from each other and all variant patterns were detected in ponies. The authors point out, that no heterogeneity was detected within any single individual, when analyzing several tissues.

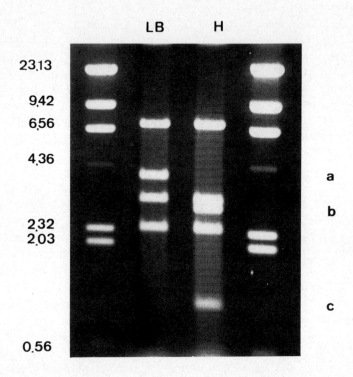

Fig. 18-1 Restriction fragment length polymorphism in pig mtDNA after HincII digestion.
LB = Belgium Landrace; H = Hampshire.
Fragment sizes of marker in kilobases on the left. Variable mtDNA fragments b and c in
Hampshire sum up to variable fragment a in Belgium Landrace. The Hampshire pattern is
identical to those found in other european breeds (Hecht et al., unpublished data).

Based on similar comparisons of three sheep and two goat mtDNAs, Upholt and
Dawid [12] calculated the sequence divergence between three individual sheep as 0.5-1%,
between two goats as 2 %, and between sheep and goat as 6-11 %.

Watanabe et al. [9, 17] analyzed mtDNA restriction pattern from pigs both of asian
and european descendence. Their findings suggest, that the patterns produced by restriction
enzymes BglII, SacI, StuI, and TaqI clearly discriminate between these groups. According
to these studies, the breeds Hampshire, "Landrace" (not specified), Duroc, and Pitman
Moore make up the european group, while Japanese wild boars, Taiwan native breeds, and

the Ohmini and Göttinger Minipigs represent the asian type. Large White is exhibiting both types of mtDNA and so has maternal lineages originating in both groups. This view is further supported by analyses of restriction patterns, observed using restriction enzymes HinfI and HaeIII in a sample of 41 and 12 animals respectively. In this study however (Hecht et al., unpublished data) the samples derived from pigs of the Hampshire breed exhibited patterns identical to the asian type, as judged by comparison to Göttinger Minipig mtDNA. This indicates, that this breed, just as the Large White, represents maternal lines both of european and asian origin. Among european breeds of pigs only two variable cleavage sites have been detected: one in an animal from the Duroc breed, using restriction enzyme EcoRV [9] the other in the Belgium Landrace, using HincII (Hecht et al., unpublished data). This variant restriction pattern is depicted in Fig. 18-1.

Table 18-3 Restriction fragment length variation in mtDNA of farm animals.

Species	Restriction-enzyme	Number of animals examined	different patterns	Reference	
Cattle	Hind III	13	2	Watanabe et al.	[22]
	Taq I	11	2		
	Msp I	11	2		
	Hha I	40	2	Hertner (personal communication)	
	Hae III	18	8	Laipis et al.	[23]
Chicken	Msp I	16	2	Wakana et al.	[8]
	Eco R I	?	3	Glaus et al.	[19]
				Wakana et al.	[8]
Goat	Eco R I	2	2	Upholt and Dawid	[12]
	Hind III	2	2		
	Hae III	2	2		
Horse	Hae III	10	5	Potter et al.	[21]
Rabbit	Eco R I	11	2	Ennafaa et al.	[11]
Sheep	Hae III	3	3	Upholt and Dawid	[12]
Pig	Bgl II	34	2	Watanabe et al. [9, 17]	
	Sca I	34	2		
	Stu I	34	2		
	Eco R V	34	2		
	Taq I	10	2		
	Hinc II	41	2	Hecht et al.	
	Hinf I	41	2	(unpublished data)	
	Hae III	12	2		

Low levels of intraspecific sequence heterogeneity, found in chicken, have been attributed to an evolutionary bottleneck, possibly caused by the domestication of this species [8, 19]. It is remarkable, that in the bovine species almost all restriction fragment length polymorphisms reported to date, occurred within breeds, possibly indicating less stringent evolutionary situations in the breeding history of this species.

A survey of all polymorphic mtDNA restriction patterns reported in farm animals is given in Table 18-3.

18.3.2   Heterogeneity of mtDNA in a Maternal Lineage

One of the most exciting studies on mtDNA variation in farm animals are those reported by Laipis and Hauswirth [24], Laipis et al. [23], Hauswirth and Laipis [25] and Olivo et al. [26]. In brief, these authors describe experiments on mtDNAs from maternally related Holstein cows. Among the members of this family two phenotypes of restriction fragment length patterns produced by endonuclease HaeIII were detected and the variable site mapped to unidentified reading frame 5 [25]. In an extended study, using nucleotide sequencing data, Olivo et al. [26] presented evidence, that in one of the HaeIII phenotypes mentioned above, there is microheterogeneity constituting three types of base sequences in the D-loop region.

18.3.3   Heterogeneity of mtDNA within Individual

In general, mtDNA of a single mammal is thought to be homogeneous with respect to primary structure and in this way forms a molecular clone [21]. In rare cases exceptions have been observed, where different mtDNA types coexist in an individual, a condition often referred to as heteroplasmy. In mouse [27] and man [28] extreme cases have been reported, exhibiting severely deleted molecules along with typically sized ones.

Among farm animals, heteroplasmy has been described in mtDNA from cow and rabbit. Hauswirth et al. [29] analyzed the nucleotide sequence of the bovine D-loop region adjacent to the tRNA gene for phenylalanine. A stretch of cytosine residues immediately neighbouring this gene varied in length from 9 - 19 bases in independent clones, derived from the same tissue of single individuals. Interestingly, this homopolymer region is the preferred target of bovine heart mitochondrial endonuclease, an enzyme possibly involved in mtDNA replication [30].

Analyzing gel separated rabbit mtDNA fragments, Ennafaa et al. [11] discovered, that certain fragments appeared as fuzzy bands. These fragments contain the D-loop region and vary in size by approximately one hundred basepairs. Ennafaa et al. [11] interpret this observation as a result of general heteroplasmy in all rabbits examined.

18.4 Concluding Remarks

Profound knowledge has accumulated about structure and function of mitochondrial DNA. Up to now, bovine mtDNA is the best characterized organelle genome among farm animals. Techniques such as the polymerase chain reaction [31] and the development of transfer methods for in vitro manipulated mtDNA will further stimulate research on the function of this unique molecule [13]. A first clue, that variation in quantitative traits may be correlated to differences in physiological parameters of mitochondrial metabolic pathways comes from statistical analyses [32, 33]. It might be a promising approach to find out, whether structural differences in mtDNA contribute to phenotypic variation in such traits.

Acknowledgement

I am grateful to U. Hertner (Institute for Animal Breeding and Genetics, University of Göttingen) for providing unpublished data and mtDNA samples from Göttinger Minipigs.

18.5 References

[1] Hauswirth, W.W., Lim, L.O., Dujon, B., and Turner, G., In: *Mitochondria a Practical Approach:* Darley-Usmar, Rickwood, D., Wilson, M.T. (eds.) Oxford: IRL Press, *1987*; pp. 171-244.

[2] Hutchison III, C.A., Newbold, J.E., Potter, S.S., and Edgell, M.H., *Nature 1974, 251,* 536-538.

[3] Michaels, G.S., Hauswirth, W.W., and Laipis, P.J., *Develop. Biol. 1982, 94,* 246-251.

[4] Bahr, G.F., and Engler, W.F., *Expl. Cell Res. 1970, 60,* 398-340.

[5] Hecht, N.B., Liem, H., Kleene, K.C., Distel, R.J., and Ho, S., *Develop. Biol. 1984, 102,* 452-461.

[6] Bartoov, B., and Fisher, J., *Int. J. Androl. 1980, 3,* 594-601.

[7] Andersson, S., de Bruijn, M.H.L., Coulson, A.R., Eperon, I.C., Sanger, F., and Young, I.G., *J. Mol. Biol. 1982, 156,* 683-717.

[8] Wakana, S., Watanabe, T., Hayashi, Y., and Tomita, T., *Anim. Genet. 1986, 17,* 159-168.

[9] Watanabe, T., Hayashi, Y., Ogasawara, N., and Tomoita, T., *Biochem. Genet. 1985, 23,* 105-113.

[10] Brown, M., *Ann. N.Y. Acad. Sci. 1981, 361,* 119-134.

[11] Ennafaa, H., Monnerot, M., El Gaaied, A., and Mounolou, J.C., *Génét. Sél. Evol. 1987, 19,* 279-288.

[12] Upholt, W.B., and Dawid, I.B., *Cell 1977, 11,* 571-583.

[13] Clayton, D.A., *Cell 1982, 28,* 693-705.

[14] Clayton, D.A., *Ann. Rev. Biochem.* **1984**, *53*, 573-594.

[15] King, T.C., and Low, R.L., *J. Biol. Chem.* **1987**, *262*, 6204-6213.

[16] King, T.C., and Low, R.L., *J. Biol. Chem.* **1987**, *262*, 6214-6220.

[17] Watanabe, T., Hayashi, Y., Kimura, J., Yasuda, Y., Saitou, N., Tomita, T., and Ogasawara, N., *Biochem. Genet.* **1986**, *24*, 385-396.

[18] Hauswirth, W.W., Laipis, P.J. Gilman, M.E., O'Brien, T.W., Michaels, G.S., and Rayfield, M.A., *Gene* **1980**, *8* , 193-209.

[19] Glaus, K.R., Zassenhaus, H.P., Fechheimer, N.S., and Perlman, P.S., In: *The Organization and Expression of the Mitochondrial Genome:* Kroon, A.M., and Saccone, C. (eds.) Elsevier: North Holland Biomedical Press, **1980**, 131-135.

[20] Laipis, P.J., Hauswirth, W.W., O'Brien, T.W., and Michaels, G.S., *Biochim. Biophys. Acta* **1979**, *565*, 22-32.

[21] Potter, S.S., Newbold, J.E., Hutchison III, C.A., and Edgell, M.H., *Proc. Nat. Acad. Sci. USA* **1975**, *72*, 4496-4500.

[22] Watanabe, T., Hayashi, Y., Semba, R., and Ogasawara, N., *Biochem. Genet.* **1985**, *23*, 947-957.

[23] Laipis, P.J., Wilcox, C.J., and Hauswirth, W.W., *J. Dairy Sci.* **1982**, *65*, 1655-1662.

[24] Laipis, P.J., and Hauswirth, W.W., In: *The Organization and Expression of the Mitochondrial Genome:* Kroon, A.M., Saccone, C.(eds.) Elsevier: North Holland Biomedical Press, **1980**; pp.125-130.

[25] Hauswirth, W.W., and Laipis, P.J., *Proc. Natl. Acad. Sci. USA* **1982**, *79* , 4686-4690.

[26] Olivo, P.D., Van de Walle, M.J., Laipis, P.J., and Hauswirth, W.W., *Nature* **1983**, *306*, 400-402.

[27] Boursot, P., Yonekawa. H., and Bonhomme, F., *Mol. Biol. Evol.* **1987**, *4*, 46-55.

[28] Holt, I.J., Harding, A.E., and Morgan-Hughes, J.A., *Nature* **1988**, *331*, 717-719.

[29] Hauswirth, W.W., Van De Walle, M.J., Laipis, P.J., and Olivo, P.D., *Cell* **1984**, *37*,1001-1007.

[30] Low, R.L., Cummings, O.W., and King, T.C., *J. Biol. Chem.* **1987**, *262*, 16164-16170.

[31] Saiki, R.K., Scharf, S., Faloona, F., Mullis, K.B., Horn, G.T., Erlich, H.A., and Arnheim, N., *Science*, **1985**, *230*, 1350-1354.

[32] Brown, D.R., DeNise, S.K., and McDaniel, R.G., *J. Anim. Sci.* **1988**, *66*, 1347-1354.

[33] Dzapo, V. and Wassmuth, R., *Z. Tierzüchtg. Züchtungsbiol.* **1983**, *100*, 280-295.

19. Patentability of Genetic Inventions in Animals

R. Moufang

19.1 Introduction

Whilst initially developed in the area of lower organisms, modern genetic technologies are increasingly being applied to more complex biological entities. During the last few years, major breakthroughs have been achieved in plant biotechnology [1], as well as in animal production [2]. Methods of manipulating embryos and transferring genes have been refined and now constitute standard procedures used for a variety of purposes [3]. Possible applications of transgenic animal technology [4] comprise models for oncogenesis and diseases [5], immunological research, marking of chromosomal regions, and direct use of transgenic lines as farm animals [6]. In addition, during the past 40 years traditional breeding techniques also have been considerably improved, a development resulting in important genetic gains in a number of farm animals [7].

To keep society in step with modern developments of genetics, the legal system has to respond to newly raised issues in an adequate manner. As to emerging technologies in general, the interests of society appear to be twofold: On the one hand, possible risks and undesirable applications have to be avoided or minimized. On the other hand, great efforts should be made to take possession of and fully use the positive potential of a new technology. Law provides society with different instruments to accomplish this twofold task: Whereas consistent legal frameworks have been developed or are currently being developed in order to counteract potential dangers of new technologies, other legal fields - among which the patent law has the most prominent place - serve the purposes of stimulating innovation and of fostering progress. To recognize this "legal division of labour" leads to the following important insight: An instrument conceived with the aim of promoting technology, i.e. patent law, can hardly be shaped in such a way that it acts as an efficient safeguard against abuses or dangers of new technologies. The same is true vice versa.

When addressing the legal aspects of animal genome research, one might consequently suggest the following principle: On the one hand, possible safety risks, e.g. caused by the release of transgenic animals into the environment or by the production of food by newly engineered animals, [8] should be minimized by adequate regulations of the competent authorities. Furthermore, animal welfare statutes might have to be strengthened in order to prevent certain undesirable developments [9]. On the other hand, patent law has to be adapted as well, in order to encourage innovation in animal breeding and genome research. It should be realized by politicians and lawyers that an increasing need for protection of major achievements emerges in this field [10].

The issue of patents on animals has, however, stimulated a heated and somewhat confusing debate in the United States. In Europe similar discussions have started recently [11]. In the following, an attempt will be made to render the current controversy more rational by presenting a picture of the relevant aspects that deserve further exploration and discussion. Animal breeders and genome researchers merit a clear answer to their question as to what achievements in their working field can be patented. Unfortunately, in this area the law is far from being unequivocal and well settled. The following picture must therefore be considered as a tentative one.

After outlining some general features of patent law (section 2), the historical perspective will be briefly reviewed (section 3) in order to prepare the ground for the intricate and complex legal subtleties that surround the question whether animal inventions constitute patentable subject matter under European and German patent law (section 4). A short survey of solutions developed by foreign patent systems will equally be given (section 5). Thereafter, further topical problems in the field of patent law which might be of interest for animal breeders and genome researchers will be discussed (section 6). Finally, the results of the analysis are summarized (section 7).

19.2   General Features of Patent Law

The granting of patents for new inventions as a way of encouraging innovation is a long-established instrument of promoting technical progress and industrial development [12]. Strong support can be found for the assertion that patent law fulfills this task in an efficient manner [13].

Although most industrialized countries - with the exception of the socialist countries - have similar patent law structures, there are considerable differences in detail. This means that a particular invention may be patentable under the laws of one country but not under the laws of another country [14]. The legal rights conferred to the patentee are limited to the territory of the country that has granted the patent (principle of territoriality). Consequently, an inventor needs to obtain a patent in each country where he wishes to have protection [15]. The obvious disadvantages of this situation have been the driving force for different attempts at achieving international harmonization of patent law. So far, the most successful and most important of these initiatives has been the European Patent Convention (EPC) of 5 October 1973 which went into force in October 1977 and was ratified by most EC member countries and further European states (Austria, Liechtenstein, Sweden, Switzerland). Patent applications filed with the European Patent Office (EPO) in Munich are dealt with in a uniform application procedure and , if successful , lead to

European patents which are valid in all EPC member states designated by the applicant [16].

Notwithstanding the great success of the European patent system, the national route remains equally open. Consequently, an inventor who wishes protection for his invention e.g. in the Federal Republic of Germany may also file an application with the German Patent Office. As to their substantive patent law, the EPC member states have harmonized their legislation to a wide extent [17]. The situation under German and European patent law [18] is therefore almost identical. Under both systems, the prerequisites of a successful patent application are the following: In order to constitute *patentable subject matter,* the application must be concerned with an invention in the legal sense and may not fall under special exclusionary provisions [19]. In addition, the application has to fulfil the *criteria of patentability,* i.e. novelty, inventive step, and industrial applicability [20]; furthermore, the inventor must *disclose* the invention in a manner sufficiently clear and complete for it to be carried out by a person skilled in the art [21]. Both under European and German patent law, the application will be carefully examined by the Patent Office in order to find out whether a patent can be granted. Although this examination may last several years, the patent application will in any case already be published after the expiry of a period of eighteen months from the date of filing or from the date of priority [22]. Subsequent to the publication of the application, the files relating to it may be inspected on request by any third parties [23]. This is in sharp contrast to U.S. patent law which allows public access to the information contained in the patent application only if and when a patent is granted.

The rapid development of modern biotechnology has raised numerous questions in the field of industrial property law. In order to analyze and improve the current state of patent law vis-à-vis biotechnological inventions, several in-depth-studies have been completed, both in the national and in the international context [24]. Furthermore, legislative action has been taken or is currently under consideration. The recently published EC Proposal for a Council Directive on the Legal Protection of Biotechnological Inventions [25] can probably be considered to be the most important of these initiatives. The proposed directive is intended to harmonize possible solutions in the construction of ambiguous legal provisions and to strengthen patent protection. It covers microbiological achievements and plant biotechnology, as well as the results of animal breeding and genome research [26].

19.3 Animal Inventions and Patent Law: The Historical Perspective

The roots of the present controversy over "animal patents" can be traced down to the beginning of this century, since the question whether and to what extent innovative achievements in the field of animal breeding constitute patentable inventions has been discussed in Germany for nearly 100 years. The eminent legal scholar *Josef Kohler* already stated in 1900 that new methods for feeding oysters, for artificially hatching, or for artificially breeding silkworms, etc., should be patentable whereas, in his opinion, "general instructions concerning the best mixture of races" could not fall within the domain of patent protection [27]. In 1913, Quade vigorously pleaded for legal protection of new animal breeds and proved that the definition of natural forces used by the patent office at that time had already been extended to embrace many "vegetative" capacities of animals [28]. The Imperial Patent Office (Kaiserliches Patentamt), however, was not willing to grant patents for new achievements in animal breeding. In a decision of 1914, its Appeal Division pronounced an obiter dictum according to which patents were not available for processes in the production of animals [29]. Eight years later, a similar position was taken by the revocation division of the Reichspatentamt [30].

Whilst patent law began to develop a much more favorable attitude towards microbiological and plant breeding inventions as early as the first half of this century [31], animal breeding inventions continued to be viewed as nonpatentable subject matter for a rather long time [32]. In the 1960s, the German Federal Patent Court (Bundespatentgericht) refused several patent applications in the field of animal breeding by arguing that such achievements would not constitute patentable subject matter [33]. In 1969, however, these decisions, which had constructed the patent statute in an extremely narrow way, were overruled by the Federal Supreme Court (BGH) in its famous "Red Dove" landmark decision [34]. The Court made it clear that achievements in the field of biology were not excluded from patent protection in principle, and held that the term "invention" had to be regarded as a basic concept in a field of law having as its most important task encompassing the patentable results of the most recent states of science and of research [35]. Consequently, the term "invention" was redefined as "a teaching to methodically utilize controllable natural forces to achieve a causal, perceivable result". The Court distinguished three kinds of biological inventions by stating that (a) on the one hand, the course of biological events can be affected with means other than animate matter; (b) in reverse, inanimate matter can be influenced by biological means; (c) finally, the means as well as the final result may lie within the field of biology. All of these groups of biological inventions were held to be patentable subject matter, the last group also embracing the achievements of animal breeding.

Notwithstanding such new and encouraging dicta, in this case the final outcome for the applicant was negative, since the Court maintained that every invention had to fulfil the so-called repeatability requirement, and stated that this requirement had not been met by the applying animal breeder. With good reasons, one could therefore view the Red Dove decision as of somewhat dubious value, since the affirmed openness of patent law for animal breeding was deeply undermined by the difficulties inherent in traditional breeding techniques to meet the repeatability requirement [36].

While under German case law - with the "Red Dove" decision -a tendency was at last emerging that seemed more favorable to the achievements of animal breeders, the European patent law harmonization process [37] had already begun. During this process which laid the foundations of the actual legal situation in most West European countries the protection needs of animal breeders were completely neglected. When in 1963 the Strasbourg Convention [38] was concluded, the signatories were permitted not to grant patents for plant or animal varieties or essentially biological processes for the production of plants or animals [39]. The travaux préparatoires of the Strasbourg Convention do not shed much light on the reasons which led to this exclusionary provision, especially insofar as the field of animal husbandry is concerned [40]. It can be stated, however, that neither economic arguments nor ethical considerations [41] have played any major role. Apparently, the true reasons were rather simple: The legal experts did not want that the great task of harmonizing widely divergent patent systems in Europe to be further complicated by lengthy and controversial debates about issues of - as they erroneously may have thought - minor importance. Consequently, they excluded certain biological inventions from the unification project [42].

The decision of 1963 to sacrifice certain biological inventions to promote European patent law harmonization is reflected both in the EPC and in the national patent laws of its signatories, with one fundamental difference: Whereas the Strasbourg Convention has left it to its contracting states to decide whether they would grant patents or not for the mentioned biological inventions, the EPC as well as - with slight modifications - German patent law nowadays explicitly exclude patents for these kinds of inventions [43], Art. 53 EPC reads as follows:

"European patents shall not be granted in respect of: ... b) plant or animal varieties or essentially biological processes for the production of plants or animals; this provision does not apply to microbiological processes or the products thereof."

Without any doubt, this provision particularly discriminates against the achievements of animal breeders, not only compared to microbiological inventions, but also to plant varieties for which a special protection system is frequently available. Whereas certain authors are trying to justify this discrimination as being a corollary of the exclusion of inventions contrary to "order public" and morality [44], a clear majority opinion that can be deduced from legal literature deplores and argues against the great injustice done to animal breeders [45].

19.4 Patentability of Animal Inventions Under Current European and German Law

Without any doubt, the interpretation of Art. 53 lit. b EPC is a formidable task: The language used in this poorly drafted provision is far from being clear, especially when taking into account the new developments of biology. Nevertheless, this task has to be accomplished today, since the European patent system itself can only be changed by lengthy legal procedures (including diplomatic conferences, ratification procedures, etc.).

A careful analysis of Art. 53 lit. b EPC will show that this provision by no means excludes all inventions in the field of animal genetic engineering, but only some of them. It is, however, extremely difficult to draw an exact demarcation line. In this context, most commentators plead for observing the following guidelines: As a general rule, exclusionary provisions have to be construed narrowly [46]. In addition, due attention has to be given to the protection needs of modern biotechnology [47]. It will be seen in the following that also the proposed EC directive was developed along these guidelines.

19.4.1 The Exclusion of Animal Varieties

Art. 53 lit. b EPC - as well as § 2 No. 2 of the German Patent Act (Patentgesetz = PatG) - excludes "animal varieties" from patentability, this provision not being applicable to "products of microbiological processes". Unfortunately, there is no unanimous definition of the term "animal variety" among scientists. According to Webster's Dictionary (2nd edition), a variety is "a group of animals or plants related by descent but distinguished from other similar groups only by characters considered too inconstant or too trivial to entitle it to recognition as a species, or whose distinguishing characters are dependent on breeding controlled by man for their perpetuation; often, any group of lower rank than a species" [48]. The dictionary further remarks that "there is a prevailing tendency to abandon the word variety on account of its indefiniteness in meaning" [49]. In the third edition of Webster's dictionary, variety is defined as "any of various intraspecific groups of plants or animals" [50]. Referring to words sometimes used in a more or less synonymous way (subspecies, race, breed, strain, stock), the dictionary further states: "These words show variable uses according to the period of scientific writing in which they appear and have been used to designate closely related groups of plants or animals narrower in scope than a species. Variety and subspecies often apply to a group distinguished from others in a general class by characteristics too minor to constitute criteria of a species. Sometimes variety designates a group produced by human research and control" [51].

The interpretation of Art. 53 lit. b EPC is further complicated by the fact that, at this very point, the three official languages of the Convention differ considerably: Whereas the French version ("races animals") approaches the English text, the German term "Tierarten" is equivalent to "species". Whether such linguistic confusion has to be explained by an unawareness of the respective scientific meaning or whether other reasons could be found [52], is an idle question. In any case, strong arguments support the conclusion that, under European law, the English and French version should prevail [53]. More difficult to answer is the question whether under German (national) patent law the term Tierarten should also be construed as meaning "Tierrassen" [54].

A very controversial issue is the problem whether the exclusion of animal varieties means that *animals in general* are non-patentable subject matter, or whether it only means, in contrast, that *animals in form of a variety* can not be patented. Following the interpretative guidelines mentioned above [55], the second alternative seems preferable. When taking this route, the exclusionary provision is not rendered totally superfluous by making it a mere matter of claim drafting skill whether an application will be successful or not: If the subject matter of an invention concerns the creation/production of one single variety, it falls under Art. 53 lit. b phrase 1 EPC. Since the provision also embraces all individual organisms that represent the newly created variety, it is irrelevant in such a case whether the patent application avoids any reference to the word variety, race, subspecies etc. and "only" claims individual animals or embryos [56]. If, however, an invention enables the skilled person to produce animals - all of which express e.g. the same transferred gene whilst belonging to completely different varieties or even different species -, patent claims directed to such animals do not fall under the exclusionary provision and should be granted [57].

In its very recent Lubrizol decision which, however, concerned the field of plant breeding, the EPO has taken a similar stance [58]: Starting from the assumption that the term "plant variety" means a multiplicity of plants which are largely the same in their characteristics (i.e. 'homogeneity) and remain the same within specific tolerances after every propagation or after every propagation cycle (i.e. "stability"), the Board concluded that only possession of both these criteria, homogeneity and stability, would be a prerequisite for a "plant variety". Since the claimed hybrid seeds or plants, considered as a whole generation population, are not stable, the Board did not consider them as a "variety" [59].

No valid argument seems conceivable that could preclude this reasoning to be transferred from the field of plant breeding to the area of animal breeding [60]. This view is supported by Art. 3 (I) (1) of the proposed EC Directive which reads as follows: "Micro-organisms, *biological classifications other than* plant or *animal varieties as well as parts of* plant and *animal varieties* other than propagating material ... shall be considered patentable subject matter" [61].

Animal parts that do not suffice to establish a variety should equally be patentable [62]. This is certainly valid for genome sections like genes, promoters, enhancer sequences, transposons, plasmids, and chromosomes. Somatic cell lines should equally be patentable [63], at least if they have been derived from sexually reproducing animals for which no cloning system currently exists. Patent applications for animal sperm and non-fertilized egg cells constitute a borderline case [64].

It has also been argued that the breeder of a new animal variety might successfully claim new products derived from these animals, e.g. lamb meat of high nutritional value or mink furs of a certain color [65]. This argument merits further analysis: On the one hand, food products can generally be protected under European and German patent law [66]. On the other hand, such claims would not reflect any more the real core of the invention which resides in the breeding of the new animal variety, but would focus on the "periphery of the invention".

In all cases in which - according to the foregoing analysis -the subject matter of a patent application has to be considered as an animal variety, these applications are in principle embraced by the exclusionary provision of Art. 53 lit. b EPC. But, as it is explicitly stated in phrase 2, this provision does not apply to the products of microbiological processes. It has therefore been argued that animal varieties produced by genetic engineering techniques might be regarded as patentable. Since, in modern animal genome technology, single genes are frequently cloned and multiplied in microorganisms, recombined with a vector (transposon, plasmid, virus) and microinjected into animal embryonic cells or otherwise transferred to host animals, it has been suggested that the resulting animals and their offspring constitute the products of a microbiological process and that, consequently, the new animal variety is patentable [67]. However, legal uncertainties remain also to this extent, since it is extremely difficult to draw an exact borderline between microbiology and "macrobiology" [68].

19.4.2 The Exclusion of Essentially Biological Processes for the Production of Animals

Besides applying for product protection, an inventor can generally also draft process claims. In this respect, animal breeders will have to surmount a further obstacle erected by Art. 53 lit. b EPC, since this provision explicitly excludes "essentially biological processes for the production of animals" from patentability [69]. In the recent past, this provision has been increasingly criticized for not being clear at all and for drawing a completely arbitrary demarcation line between patentable and non-patentable subject matter [70].

The exclusionary provision certainly does not embrace processes which focus on certain mechanical devices to be used in the course of animal production, e.g. a method for preventing piglets from suffocating under the dam in a brooding pen [71]. With regard to patent applications directed to genetic processes, however, the scope of the provision is currently discussed among patent experts in a highly controversial manner. Three different opinions will be summarized in the following:

Firstly, one may assert that the whole field of genetics belongs to the realm of biology and that, consequently, all genetic processes and methods are - as being essentially biological processes - excluded [72].

Secondly, one might suggest that those animal genetic inventions are not essentially biological in which an important part of the process relies on physical or chemical means (e.g. mutagenic processes making use of physical or chemical mutagens, microinjection methods etc.). According to this view, a single case analysis is required under which even modern genetic technologies *can* be excluded from patentability *if* the biological parts of the process are not outweighed by the physical or chemical steps [73].

Thirdly, one can argue as follows: The legislator saw an antinomy between "technical" processes (in which intervention of man occurs) and "biological" processes (in the sense of natural, uncontrollable processes). Therefore, the term "essentially biological" should be construed as "essentially natural and uncontrollable" or "essentially without human intervention" [74].

The latter interpretation, however, would deprive the exclusionary provision of any independent meaning and reduce it to a mere concretization of the repeatability requirement that generally prevails in patent law. Consequently, only traditional breeding techniques (methods of crossing, inter-breeding, or selectively crossing) could then be considered to be essentially biological. In the field of plant breeding inventions, the Lubrizol decision of the EPO [75] went even a step further: Whereas the Examining Division had held the claimed process unpatentable since "it had never been disputed that at least the classical breeders' methods were considered as essentially biological", the Board of Appeal espoused a somewhat different point of view and set the decision of the Examining Division aside: Under the analysis of the Board, a "fundamental alteration of the character of a known process for the production of plants" is required in order to avoid the verdict "essentially biological" [76]. "The necessity for human intervention alone is not yet a sufficient criterion" since something "beyond a trivial level" has to be contributed [77]. "In the present case, which presents a multistep process, each single step as such may be characterized as biological in a scientific sense. However, instead of the traditional approach ... the specific arrangement of the steps ... provides a process with a reversed sequence... This arrangement of steps is decisive for the invention and permits the desired control of the special result in spite of the fact that at least one of the parents is heterozygous. The facts ... clearly indicate that the claimed processes ... represent an essential modification of known biological and classical breeders' processes, and the efficiency and high yield associated with the product in the present case show important technological character" [78].

These dicta indicate that the Lubrizol decision can be summarized as an attempt to narrow the exclusion of "essentially biological processes for the production of plants and animals" by regarding the provision simply as a mere concretization of the patentability criterion of inventive step.

A further argument could be advanced to support the patentability of processes employed in modern animal genome technology: Since the exclusionary provision does not apply to microbiological processes, one might assert that numerous methods in genetic engineering have a microbiological character [79]. This argument was already presented in the parallel context of the exclusion of animal varieties [80].

19.5   The International Perspective

19.5.1   Development and Current Status of U.S. Patent Law

In the United States, discussions about animal patents have - quite similarly to the situation under German patent law - a rather long tradition. As early as 1906, arguments were made to support the assertion that patent law embraces new animal breeds [81]. The Plant Patent Act of 1930 which introduced a special protection scheme for new plant varieties was initially considered to be only a part of much broader legislation granting patent-like protection to all originators of plants and animals and products thereof, such as fruits, roots, eggs, leaves, and seeds [82]. In 1966, the Patent, Trademark and Copyright Section of the American Bar Association adopted a resolution asking for appropriate protection of achievements in the field of animal husbandry [83]. In spite of these discussions in academic and professional circles, the principal question, i.e. whether animals constitute patentable subject matter, was addressed neither by court decisions nor by the legislator and remained unresolved for a long period. In 1975, an application that concerned newly developed dwarf chickens was dealt with by the Court of Customs and Patent Appeals, after it had been rejected by the Board of Appeals of the Patent and Trademark Office (PTO) [84]. The appellants claimed to have discovered a dwarfism gene in chickens which, when employed in a controlled breeding method, produces dwarf hens which, when mated with "normal" cocks, lay eggs which hatch into chicks that mature into normal size heavy meat fowl of good eating characteristics. Whereas the rejection of the PTO was based on the opinion that the claims were directed to non-statutory subject matter, the CCPA maintained that the invention was not described in a precise and uncontradictory way [85]. Thus, the fundamental question was finally left unanswered by the Court.

In the recent past, however, U.S. patent law developed in a direction more favorable to biological inventions. This tendency is reflected by three important legal cases: In its famous landmark decision "Diamond v. Chakrabarty" of 1980 [86], the U.S. Supreme Court held a bacterial strain in which a plasmid from another strain had been inserted to be patentable subject matter. In its view, the legislator had clearly recognized that the relevant distinction was not between living and inanimate things, but between products of nature, whether living or not, and human-made inventions. Reviewing the legislative history, the Court held that Congress intended statutory subject matter to "include anything under the sun that is made by man" [87]. In 1985, the Board of Appeals and Interferences of the PTO decided that patents on new plants are equally available under general patent law and that this form of exclusionary right is not preempted by the fact that the legislator introduced special protection schemes for plants in 1930 (Plant Patent Act) and 1970 (Plant Variety Protection Act) [88]. The next step "up the chain of life" [89] was completed in April 1987. In its decision "Ex parte Allen" [90], the Board of Appeals and Interferences of the PTO expressly stated that it considered animals to be patentable subject matter. However, the application that claimed polyploid oysters was rejected as not complying with the patentability requirement of non-obviousness.

In order to implement the new developments of patent case law into the PTO rules, the U.S. Commissioner of Patents and Trademarks made the following announcement on 7 April 1987:

"The Patent and Trademark Office now considers non-naturally occurring non-human multicellular living organisms, including animals, to be patentable subject matter within the scope of 35 USC 101" [91]. This statement has raised a bitter controversy engaged by animal rights movements and by several farmers' unions. Several bills were introduced in Congress with the purpose of banning animal patents [92] or mandating a moratorium on the grant of such exclusionary rights [93]. A series of hearings were held in the second half of 1987, but no bill actually passed both houses of Congress and became law [94].

On 13 April 1988, Harvard University was granted a patent covering (but not limited to) a new breed of genetically altered mice [95]. This patent has rekindled debate in Congress over a moratorium on patents for higher life forms [96]. But only one bill, the "Transgenic Animal Patent Reform Act" (introduced by Congressman Kastenmeier) passed the relevant committees [97] and was approved by the House of Representatives by voice vote on 13 September 1988. The bill accepted the patentability of animals, but limited the scope of protection by granting certain exemptions to farmers who reproduce animals on their farm [98]. Since the Senate, however, took no action prior to the end of the 100th Congress in October 1988, the bill had to be reintroduced into the new Congress [99].

Furthermore, a law suit was filed on 28 July 1988 by animal rights groups and an organization of farmers, challenging the PTO's decision to issue patents on multicellular animals. The plaintiffs charged that the PTO decision was made without the required public input or proper congressional authorization [100]. In January 1989, the U.S. District Court for the Northern District of California held that the PTO did not exceed its statutory authority when promulgating the new rule. The lawsuit was consequently dismissed [101].

19.5.2 Overview of the Legal Situation in Other Countries

An international analysis reveals that in most countries legal uncertainty reigns as to the patentability of animal inventions [102]. Specific exclusionary provisions shaped similarly to Art. 53 lit. b EPC can be found in numerous patent statutes and lead to comparable problems of interpretation [103]. On the other hand, some national patent systems appear to be more favorable towards the protection of the results of modern animal husbandry [104]. Especially in several socialist countries, new achievements of animal breeding can be protected either by patents or by so-called inventor's certificates. This situation prevails in Bulgaria [105], Hungary [106], Rumania [107], and in the Soviet Union [108].

19.6   Further Patent Law Issues of Topical Interest

There are a number of further patent law issues of which animal genome researchers and breeders who seek protection of their achievements should be aware. In the following, some of them will be succinctly reviewed.

Patent law contains a *safeguard against* the patenting of *inventions* the publication or exploitation of which would be *contrary to "order public" or morality* [109]. It has been argued that patenting animals should be considered to be unethical [110]. Several arguments are employed to support this assertion. None of them, however, appears to be persuasive [111]: This is true, in particular, as to the argument that morality would forbid animal patents because animals are living creatures which should not be made the objects of exclusionary rights. For millenniums, society has allowed animals to be killed for food. For millenniums, law has allowed individual animals to be owned by humans [112]. If the above-mentioned argument were right, every farmer who owns livestock would act in an immoral way. According to another argument, encouragement of the development of transgenic animals by patenting is wrong because it would lead to inappropriate animal suffering. Even if one accepts the idea that certain developments of modern animal breeding techniques could cause unnecessary harm to animals and are therefore undesirable, the argument is much too sweeping since animal breeding in general is certainly a commonly accepted activity from which society has reaped and will continue to reap great benefits.

Furthermore, one should never forget that a patent is essentially a "negative" right which excludes others from using the invention, but does not "automatically" confer to the patentee the "positive" right to use the invention. Each patentee has to respect the whole legal system, including e.g. animal welfare statutes. Consequently, the exploitation of a patented animal invention will be unlawful, if, in a specific case, exploitation is prohibited by another body of law. Since patent law works at an early stage in which not all consequences of possible concrete applications of an innovation can be adequately appreciated, the above-mentioned exclusionary provisions foreclose the patentability only in cases in which the purpose of an invention or all of its possible applications are unethical. Of course, if the only aim of an invention is to make animals suffer, it will be clearly unpatentable [113]. Apart from such exceptional cases, ethical considerations speak in favor of animal patents since the lack of sufficient protection constitutes an unjustifiable discrimination of animal breeders and animal genome researchers vis-à-vis inventors in other fields of technology [114].

The dichotomy between *unpatentable discoveries* and patentable inventions [115] should also not foreclose the patentability of the achievements of animal genome research. Genes that have been isolated from nature can generally be patented although, in such a case, the scope of protection does not include the use of the gene in its natural environment [116].

The *patentability criteria* of novelty, inventive step and industrial applicability do not constitute insurmountable difficulties for animal genome inventions. Researchers should be reminded, however, that under European patent law the *novelty requirement* is absolute and does not grant a general grace period [117]. From this it follows that also previous publications and communications of the inventor himself destroy the novelty of a subsequent patent application, a consequence frequently overlooked by researchers working in the public sector. The patentability requirement of industrial applicability is generally rather easy to fulfil. "Industry" also includes agriculture [118]. There is, however, an important exception:

Diagnostic, surgical and therapeutical methods practiced on the human or animal body are not regarded as susceptible of industrial application [119].

Inventions must be *disclosed*. Because of inherent difficulties in describing biological entities in an unequivocal way, it has been accepted by patent authorities for many years that, when an applicant claims a microbiological invention, cultures of the microorganisms can be deposited in recognized depositories [120]. An international treaty was even concluded in 1977 in order to make multiple deposits superfluous [121]. In this context, it has to be reminded that current patent law practice treats animal cell lines like microorganisms [122]. Whether the deposit of animal propagating material or of individual animals will, however, be accepted by the patent authorities is still uncertain [123].

Concerning the *scope of protection* of animal patents, numerous controversial questions exist for which traditional patent case law does not provide very much guidance [124]. In this respect, the proposed EC Directive will be of particular value [125]. Whether certain impacts of animal patents on agriculture should be considered as undesirable and, consequently, be avoided by specific legislative action (e.g. creating farmers' privileges for sexual reproduction of patented animals on the farm), is a political question which might have to be answered soon [126], but which does not touch the fundamental issue of patentability of animal inventions itself.

19.7 Conclusions

Despite the provision of Art. 53 lit. b EPC, patent law is not a closed book for animal breeders and animal genome researchers. On the contrary, a great part of modern animal breeding achievements can be patented. This is a valid observation, especially insofar as particular animal gene sequences, vectors, modified animal viruses and somatic cell lines are concerned. Transgenic animals should also be patentable if they do not constitute a single variety. The usual patent law requirements (novelty, inventive step, industrial applicability, enabling disclosure) must, however, always be fulfilled. Moral reasons against patenting of animals inventions are not well-founded. If certain aspects of the impact on agriculture are considered as politically undesirable, specific legislative action could be taken without unduly hampering technological progress in the field of animal breeding by general exclusionary provisions.

19.8 References

[1] Cf. for details Caplan, A., et al., *Science 1983, 222,* 815-821; Gasser, C.S., Fraley, R.T., *Science 1989, 244,* 1293-1299; Horsch, R.B., et al., *Science 1984, 223,* 496-498; Marx, J.L., *Science 1986, 230,* 1148-1150; id., *Science 1988, 240,* 145; Ream, L.W., and Gordon, M.P., *Science 1982, 218,* 854-859; Roberts, L., *Science 1988, 241,* 1290.

[2] Cf. Capecchi, M.R., *Science 1989, 244,* 1288-1292; Hammer, R.E., et al., *Nature 1985, 315,* 680-683; Jaenisch, R., *Science 1988, 240,* 1468-1474, with numerous references; Maelicke, A., *Nachr. Chem. Tech. Lab. 1985, 33,* 716-718; Marx, J.L., *Science 1988, 242,* 33; Palmiter, R.D., et al., *Science 1983, 222,* 809-814; Pursel, V.G., et al., *Science 1989, 244,* 1281-1288; Rubin, G.M., and Spradling, A.C., *Science 1982, 218,* 348-353; Spradling, A.C., and Rubin, G.M., *Science 1982, 218,* 341-347.

[3] Jaenisch, loc.cit. (supra note 2), at 1473. The author (at p. 1468 et seq.) lists three major methods for generating transgenic animals: microinjection of DNA into pronucleus, retrovirus infection, and in-vitro-establishment of embryonic stem cells later injected into host blastocysts.

[4] Cf. for details Jaenisch, loc. cit. (supra note 2), at 1470 et seq.

[5] Cf. Leonard, J.M., et al., *Science 1988, 242,* 1665-1670; Marx, J.L., *Science 1988, 242,* 1638. A good example for a model for oncogenesis is the invention disclosed in the U.S. patent No. 4 736 866, see infra (note 95).

[6] Cf. *Commercial Biotechnology: An International Analysis,* Congress of the United States, Office of Technology Assessment (ed.), Washington, D.C.: U.S. Government Printing Office, *1984,* p. 168 et seq.; Hansel, W., Prospective Developments in Agriculture, in: *Animal Patents: The Legal, Economic and Social Issues,* Lesser, W. (ed.), London: Macmillan Press, *1989*; Sun, M., *Science 1988, 240,* 136.

[7] Cf. *Impacts of Applied Genetics: Micro-Organisms, Plants, and Animals,* Congress of the United States, Office of Technology Assessment (ed.), Washington, D.C.: U.S. Government Printing Office, *1981,* p. 165 et seq.; Kräußlich, H., *Gewerblicher Rechtsschutz und Urheberrecht, Internationaler Teil* (GRUR Int.) *1987,* 340-344.

[8] Cf. for details Jones, D.D., *Food Drug Cosmetic Law Journal 1985, 40,* 477-493; id., *Food Drug Cosmetic Law Journal 1988, 43,* 351-368.

[9] Cf. Sojka, K., *Monatsschrift für deutsches Recht 1988,* 632-637.

[10] Three main factors can be made responsible for this situation:
a) for a long time, traditional breeding activities have been controlled by administrative statutes and regulations leading in an indirect way to certain protection against competition. In the perspective of the European Common Market, this regulatory practice comes under strong pressure and has to undergo major changes. A good example is the Decision of the European Court of Justice "Re Artificial Insemination: E.C. Commission v. France (Case 161/82)" of 28 June 1983, *Common Market Law Reports 1984, 2,* 296-320, although the action of the Commission was finally dismissed.
b) The investment costs in modern genome research are heavy and considerably exceed those of traditional breeding.
c) Once they are reached, achievements in modern genome research might be rather easily exploited by third parties.
Cf. for details Kräußlich, loc.cit. (supra note 7), at 344; Looser, J., *Gewerblicher Rechtsschutz in der Tierzüchtung,* Diplomarbeit Stuttgart-Hohenheim, *1984,* p. 144 et seq.; id., *Gewerblicher Rechtsschutz und Urheberrecht* (GRUR) *1986,* 27-30; von Pechmann, E., *GRUR Int. 1987,* 344-349.

[11] Cf. for example the "Information Release on Commissions's Proposal to Patent Life in Europe" by the International Coalition for Development Action & North-South Coalition (ICDA) of 12 July 1988. See also Zell, R.A., *Bild der Wissenschaft 1989, No. 3,* p. 36-47.

[12] Cf. for details Adler, R.G., *Science 1984, 224,* 457, at 458; Beier, F.K., Crespi, R.S., Straus, J., *Biotechnology and Patent Protection: An International Review,* Paris: OECD-Publication, *1985,* p. 15.

[13] Cf. Beier, F.K., 11 *International Review of Industrial Property and Copyright Law 1980 (IIC),* 563-584, with numerous references. Nevertheless this basic assertion is still disputed. For an economic analysis of the patent law see e.g. Firestone, O. J., *Economic Implications of Patents,* Ottawa: University of Ottawa Press, *1971*; Taylor, C.T., Silberston, Z.A., *The Economic Impact of the Patent System,* Cambridge: University Press, *1973.*

[14] Significant differences between the various patent systems exist particularly in the field of "animal inventions", see infra (Section 5).

[15] Cf. Saliwanchik, R., *Patenting Biotechnological Inventions: A Guide for Scientists,* Madison, Wisconsin: Science Tech Publishers, *1988*, p. 9.

[16] The scope of protection that follows from a European patent is, however, - with some exceptions - determined by the national laws of the designated states. Therefore, a European patent has to be viewed legally as a multitude of national patents (so-called "Bündelpatent"). In contrast to that, the Community Patent Convention (CPC) - concluded in 1975, but not yet in force - will lead to patents which have the same legal effects in the whole EC territory.

[17] A further international treaty, i.e. the Strasbourg Convention on Unification of Certain Points of Substantive Law on Patents for Invention, concluded in *1963*, served as an additional vehicle for this harmonization.

[18] Here and in the following, the term "European patent law" refers to the EPC.

[19] Cf. Art. 52-53 EPC; §§ 1-2 German Patent Act (PatG).

[20] Cf. Art. 52 (I), 54-57 EPC; §§ 1 (I), 3-5 PatG.

[21] Cf. Art. 83 EPC, § 35 (II) PatG.

[22] Cf. Art. 93 (1) EPC; § 32 PatG. "Priority" refers to the right of the inventor to claim the filing date of an application filed in country A also in the context of a subsequent application in country B unless a period of 12 months - starting with the first application - has elapsed. This right internationally follows from Art. 4 Paris Convention for the Protection of Industrial Property of 1883. Cf. for details regarding this multilateral treaty Beier, F.K., *IIC 1984, 15,* 1-20.

[23] See for a critical analysis of the "early publication system" Beier, F.K., *GRUR Int. 1989,* 1-14, at 11 et seq., *IIC 1989, 20,* 407; Straus, J., and Moufang, R., *Patent- und eigentumsrechtliche Aspekte der Hinterlegung und Freigabe von biologischem Material für Patentierungszwecke,* Baden-Baden, Nomos Verlagsgesellschaft, *1989.*

[24] Cf. Beier et al., op.cit. (supra note 12); Bent, S.A., Schwaab, R.L., Conlin, D.G., Jeffery, D.D., *Intellectual Property Rights in Biotechnology Worldwide,* New York: Stockton Press, *1987*; Cooper, I.P., *Biotechnology and the Law,* New York: since *1982* (looseleaf); Moufang, R., *Genetische Erfindungen im gewerblichen Rechtsschutz,* Cologne/etc.: Carl Heymanns Verlag, *1988.* Further reference should be made to the work of the Committee of Experts on Biotechnological Inventions established by the World Intellectual Property Organization (WIPO) in Geneva, particularly to WIPO Document BIG/281 "Industrial Property Protection of Biotechnological Inventions: An Analysis of Certain Basis Issues - Prepared by Straus, J." = Straus, J., *Gewerblicher Rechtsschutz für biotechnologische Erfindungen,* Cologne/etc.: Carl Heymanns Verlag, *1987.*

[25] Cf. *Official Journal of the European Communities* of 13 January *1989,* C 10/03; *GRUR Int. 1989,* 52-56 = *IIC 1989, 20,* 55-63.

[26] For details see infra (sections 4 and 6); cf. also Whaite, R., Jones, N., *EIPR 1989, 11,* 145-153.

[27] Kohler, J., *Handbuch des deutschen Patentrechts,* Mannheim: J. Bensheimer Verlag, *1900*, p. 439.

[28] Quade, F., *GRUR 1913*, 2-4. The controversy is also reflected in the references cited by Kisch, W., *Handbuch des deutschen Patentrechts*, Mannheim/etc.: J. Bensheimer Verlag, *1923*, p. 25 (note 15).

[29] Reprinted in: *Blatt für Patent-, Muster- und Zeichenwesen* (Bl.f.PMZ) *1914*, 257; similar words can also be found in a later decision, cf. *Bl.f.PMZ 1915*, 31.

[30] The decision is reprinted in: *Bl.f.PMZ 1924*, 6 et seq.

[31] Patents on new plant varieties and seed have been granted in Germany as early as 1934. Cf. the decisions of the Appeal Division of the Reichspatentamt, *Mitteilungen der deutschen Patentanwälte (Mitt.) 1936, 94-95*, 286; see also Benjamin, R., *Journal of the Patent Office Society (JPOS) 1936, 18*, 462-463, 886-887. After the Second World War several plant patents have been granted by the reestablished German Patent Office, cf. Wuesthoff, F., *GRUR 1957*, 49-56; Moufang, op. cit. (supra note 24), at 95 et seq.; Decision of the Federal Supreme Court (BGH) of 6 July 1962, *GRUR 1962*, 577-580 - "Rosenzüchtung".

[32] An exception was the German Patent No. 875 739 granted for a sex selection process in the course of which hormones were injected into fertilized poultry eggs. Cf. Wuesthoff, F., *GRUR 1977*, 404-411, at 405.

[33] Decision of 30 July 1965, *Entscheidungen des Bundespatentgerichts*, Vol. 8, p. 121-136; Decision of 5 May 1967 (unpublished); Decision of 31 January 1968, *Entscheidungen des Bundespatentgerichts*, Vol. 10, p. 1-4. For details cf. Moufang, op.cit. (supra note 24), p. 87 et seq.

[34] Ex parte Schreiner, *IIC 1970, 1*, 136-142 = *GRUR 1969*, 672-676 - "Rote Taube" with notes by Heydt, L.

[35] Cf. *IIC 1970, 1*, 136, at 137.

[36] Cf. for more details Beier, F.K., *IIC 1972, 3*, 423-450, at 428 et seq.

[37] Cf. supra (section 2).

[38] Cf. supra (note 17).

[39] Cf. Art. 2 lit. b Strasbourg Convention.

[40] Regarding the field of plant breeding, the provision might have been influenced by the concept that special protection schemes would be better suited to the specific needs of traditional plant breeders. Based on this concept, the so-called plant variety protection system was developed nationally and - with the UPOV Convention of 1963 - internationally, although already in the 1930s patents for plants had been granted in different countries. The numerous problems caused by this bifurcation of the traditional path of industrial property are beyond the scope of the present study. But cf. for details Moufang, op.cit. (supra note 24), p. 185 et seq., 383 et seq.; Straus, J., *IIC 1984, 15*, 426-442.

[41] Contra: Burnier, D., *La notion de l'invention en droit européen des brevets*, Genève: Librairie Droz, *1981*, p. 124.

[42] Cf. Beier, F.K., and Straus, J., *Industrial Property* (Ind. Prop.) *1986*, 447-459, at 453 et seq.

[43] Cf. Beier and Straus, loc.cit. (supra note 42), at 454; Moufang, op.cit. (supra note 24), p. 101.

[44] Cf. Mousseron, J.M., *Traité des brevets,* Paris: Librairies Techniques, *1984,* p. 458; Mathély, P., *Le droit européen des brevets d'inventions,* Paris: Librairie du Journal des Notaires et des Avocats, *1978,* p. 148: "Pour les races animales, il s'agit là d'un problème d'ordre moral en considération du respect de la vie." See for a discussion of the moral objections infra (section 6).

[45] Cf. Beier and Straus, loc.cit. (supra note 42), at 457; Blum, R., *GRUR 1972,* 205-211, at 209; Bruchhausen, K., in: Benkard, G., *Patentgesetz - Gebrauchsmustergesetz,* 8th ed., Munich: C.H. Beck'sche Verlagsbuchhandlung, *1988,* § 2 PatG note 17; Byrne, N.J., *IIC 1985, 16,* 1-18, at 5; Hesse, H.G., *GRUR 1971,* 101-106; Moufang, op.cit. (supra note 24), p. 216 et seq.; v. Pechmann, loc.cit. (supra note 10); Trüstedt, W., *GRUR 1986,* 640-645, at 641; Warcoin, *Revue du droit de la propriété intellectuelle 1985, 1,* 2-5, at 4.

[46] Cf. Bruchhausen, loc.cit. (supra note 45), § 2 PatG note 12; this principle has also been recognized by the Boards of Appeal of the EPO and by the German Federal Supreme Court (BGH), see e.g. the EPO decision "Lubrizol" of 10 November 1988 (Technical Board of Appeal, Case No. T 320/87, still unpublished).

[47] Cf. Preamble of the proposed EC directive.

[48] *Webster's New International Dictionary of the English Language, Second Edition, Unabridged,* Springfield, Massachusetts: Merriam Publishers, *1952,* p. 2819.

[49] Ibidem; cf. also the following dicta which can be found in the "Lubrizol" decision, loc.cit. (supra note 46): "The term 'variety' is not defined in the EPC at all. There is further no generally recognised taxonomic definition of 'variety' as there is for 'species' or 'genus'."

[50] *Webster's Third New International Dictionary of the English Language,* Springfield, Massachusetts: Merriam Publishers, *1961,* p. 2534.

[51] Ibidem.

[52] It has been suggested that the German delegation had tried to avoid the term "Tier*rassen*" for historical reasons.

[53] See for details v. Pechmann, loc.cit. (supra note 10), at 347.

[54] Taking into account the need for a harmonized interpretation, the question should be answered in the affirmative. See, however, Bruchhausen, loc.cit. (supra note 45), § 2 PatG, note 12a, 17.

[55] Cf. supra (note 46).

[56] This is in line with the EPO decision "Propagating material/CIBA GEIGY" (Technical Board of Appeal), *Official Journal of the European Patent Office* (O. J. EPO) *1984,* 112, according to which "the legislator did not wish to afford patent protection under the EPC to plant varieties of this kind *whether in the form of propagating material or of the plant itself."* For the contrary view see, however, Byrne, N.J., IIC *1986,* 17, 324-330, at 328.

[57] Cf. for further details Moufang, op.cit. (supra note 24), p. 190 et seq., 214.; Straus, J., Animal Patents - The Development and Status of European Law, in: *Animal Patents: The Legal, Economic and Social Issues,* Lesser, W. (ed.), London: Macmillan Press, *1989.*

[58] Cf. supra (note 46). Whereas founded on the reasoning of the Ciba Geigy decision (cf. supra, note 56), the Lubrizol decision further developed the European patent case law since it concerned a "purely" genetic invention.

[59] Instead of arguing "lack of stability", the Board might have been better advised to found its decision on "lack of homogeneity" since the patent claims were directed to hybrid seeds belonging to the genus Brassica or even to hybrid seeds in general. Cf. e.g. claims 20-25 of the application under review (European patent application No. 81 303 287.7).

[60] Cf. Bent et al., op.cit. (supra note 24), p. 156. In its very recent decision "Onco Mouse / HARVARD" of July 14, 1989, an Examining Division of the EPO did not share this point of view and refused a patent for transgenic animals. Cf. *GRUR Int. 1989*, 499; the decision is being published in the November issue of *O. J. EPO*. Cf. also Straus, loc.cit. (supra note 57). 6).

[61] This corresponds to the legal situation in Switzerland where it follows from the newly amended Examination Guidelines (reprinted in: *GRUR Int. 1986*, 541) that product claims to whole animal species and genera are allowed. Cf. also Straus, loc.cit. (supra note 57).

[62] With regard to this issue, it can be questioned whether the language used by the proposed EC Directive ("parts of animal varieties other than propagating material") is wholly appropriate.

[63] Cf. also Straus, op.cit. (supra note 24), p. 73. Numerous patents on somatic animal cells and hybridomas have already been granted by the EPO. One example is the European patent 73 056 "Production of polypeptides in vertebrate cell culture" granted to Genentech. Reference can also be made to the EPO decision "Submitting culture deposit information/IDAHO" (Legal Board of Appeal of 30 November 1987, Case No. J 08/87, J. 09/87, published in an abridged version in *O. J. EPO 1989*, 9 et. seq.), dealing with an application which concerned the production of polypeptides in insect cells. Cf. also the UK Patents 1 015 262 and 1 300 391 cited by Byrne, loc.cit. (supra note 45), at 5.

[64] Animal sperm is held patentable by Bruchhausen, loc.cit. (supra note 45), § 2 PatG, note 17. It may be argued, however, that sperm and egg cells already represent the variety in question.

[65] Cf. v. Pechmann, loc.cit. (supra note 10), at 345; id., *GRUR 1987*, 475-481, at 480. Contra: Rogge, E., *GRUR 1988*, 653-659, at 657 et seq.

[66] As to the patentability of chemical, pharmaceutical and food products, Art. 167 EPC admits reservations of the Contracting States. This right is, however, strictly limited in time.

[67] Cf. Bent et al., op.cit. (supra note 24), p. 150, 156; Curry, J.R., The Patentability of Genetically Engineered Plants and Animals in the US and Europe, London: Intellectual Property Publishing Limited, *1987*, p. 31. The definitions contained in Art. 5, 6 and 19 of the proposed EC directive open the door even more in this direction:

Art. 5: "Microbiological processes shall be considered patentable subject matter. For purposes of this Directive, this term shall be taken to mean and to include a process (or processes) carried out with the use of or performed upon or resulting in a micro-organism."

Art. 6: "A process consisting of a succession of steps shall be regarded a microbiological process, if the essence of the invention is incorporated in one or more microbiological steps of the process."

Art. 19: "For the purposes of this Directive:

(a) the word 'microorganism', where used, shall be interpreted in its broadest sense as including all microbiological entities capable of replication, e.g. as comprising, *inter alia,* bacteria, fungi, viruses, mycoplasmae, rickettsiae, algae, protozoa, and cells; ...".

[68] Cf. for details Moufang, op.cit. (supra note 24), p. 200.

[69] Again, some additional linguistic confusion is created by the German text of this provision which employs the term "Züchtung" (breeding) instead of "Erzeugung" (production).

[70] Cf. Bent et al., op.cit. (supra note 24), p. 160 et seq.; Moufang, op.cit. (supra note 24), p. 193 et seq., with further references; Straus, op.cit. (supra note 24), p. 67 et seq.

[71] Cf. the patent application dealt with by the EPO decision of 24 November 1988 (Technical Board of Appeal, Case No. T 58/87, still unpublished): The opponent to the granted patent did not even raise an objection that the method were unpatentable because of being an essentially biological process for the production of animals. The decision itself is also silent on this point. One must therefore assume that the Board implicitly found the exclusionary provision inapplicable.

[72] This is, especially, the position of Oredsson, T., *Nordiskt Immateriellt Rättsskydd 1985,* 229-259, at 238.

[73] Cf. also Bruchhausen, loc.cit. (supra note 45), § 2 PatG, note 12; Moufang, op.cit. (supra note 24), p. 197 et seq. This solution respects the wording of Art. 53 lit. b and avoids the outdated antinomy "biological/technical".

[74] In this sense e.g. the EPO Guidelines for Examination, Part C IV, 3.4; Straus, op.cit. (supra note 24), at 75 et seq.; Teschemacher, R., Patentable Subject Matter Under the European Patent Convention (EPC) in the Field of Biotechnology, in: *Symposium on the Protection of Biotechnological Inventions,* Geneva: WIPO-Publication *1987,* p. 87-102, at 95. This position necessarily leads to the assumption that the question whether biological or chemical or physical entities and means are used does not play any decisive role. Cf. also Art. 7 of the proposed EC Directive:

"A process in which human intervention consists in more than selecting an available biological material and letting it perform an inherent biological function under natural conditions shall be considered patentable subject matter."

According to Byrne, loc.cit. (supra note 45), at 8, patents can be granted for techniques relating to freezing and storing sperm and embryos of domesticated animal breeds, in vitro fertilization, and the implantation of embryos. But see also infra (note 119).

[75] Cf. supra note 46. The process claims of the application were directed to processes for rapidly developing hybrids and commercially producing hybrid seeds in general or belonging to the genus Brassica.

[76] Lubrizol decision (unpublished), p. 10.

[77] Lubrizol decision (unpublished), p. 9.

[78] Lubrizol decision (unpublished), p. 10-11.

[79] Cf. Curry, op.cit. (supra note 67), p. 31; Straus, op.cit. (supra note 24), p. 73, referring to the transfer of DNA sequences cloned by rDNA technology via microinjection into pronuclei of fertilized animal eggs.

[80] Cf. supra (section 4.1, note 67 and 68). In this context, reference should be made again to Art. 5, 6 and 19 lit. a of the proposed EC Directive.

[81] Cf. Cooper, op.cit. (supra note 24), p. 6-4 et seq., referring to the testimony of Walker during a Congressional hearing on the socalled "horticultural patent" bill (H.R. 18851) which, however, did not become law.

[82] Cf. Dienner, J.A., *JPOS 1953, 35,* 286-295, at 290; Cooper, op.cit. (supra note 24), p. 6-5; Rossman, J., *JPOS 1935, 17,* 632-644, at 643; Thorne, H.C., *JPOS 1923, 6,* 23-28, at 27.

[83] 1966 Resolution 22; cf. Cooper, op.cit. (supra note 24), p. 6-4.6; Luckern, P.J., and Hesseltine, C.W., *American*
Patent Law Association Quarterly Journal 1979, 7, 236-277, at 259.

[84] In re Merat and Cochez, *United States Patent Quarterly* (USPQ) *1975,* 186, 471-476.

[85] Cf. the critical analysis by Cooper, op.cit. (supra note 24), p. 6-9 et seq. Cf. also Clark, P.T., *AIPLA Q. J. 1988/89, 16,* 442-456, et 446 et. seq..

[86] *USPQ 1980, 206,* 193-202; cf. also Wegner, H.C., *European Intellectual Property Review (EIPR) 1980, 2,* 304-307; Johnson, G., *University of Florida Law Review 1980, 32,* 820-828.

[87] *USPQ 1980, 206,* 193-202, at 197.

[88] Ex parte Hibberd, *USPQ 1985, 227,* 443-448 = *GRUR Int. 1986,* 570-574.

[89] Cf. Bernstein, H.H., Patenting the Microorganisms: In re Bergy, the First Step Up the Chain of Life, *George Mason University Law Review 1978, 2,* 265-286.

[90] *USPQ 2d 1987, 2,* 1425 et seq. = *GRUR Int. 1988,* 601-602.

[91] Cf. Quigg, D.J., *JPOS 1987, 69,* 328.

[92] For details see Brody, B.A., An Evaluation of the Ethical Arguments Commonly Raised Against the Patenting of Transgenic Animals, in: *Animal Patents: The Legal, Economic and Social Issues,* Lesser, W. (ed.), London: Macmillan Press, *1989.*

[93] Cf. Brody, loc.cit. (supra note 92), referring to the bill "H.R. 3119", introduced on 5 August 1987, by Congressman Rose and others; cf. also Raines, L.J., *Issues in Science and Technology 1988, 4,* 64-70; Bodewig, T., *GRUR Int. 1987,* 520, 631.

[94] Cf. Brody, loc.cit. (supra note 92); Markey, H.T. *II C 1989, 20,* 372; Raines, loc.cit. (supra note 93), at 64.

[95] U.S. patent No. 4 736 866, cf. Brody, loc.cit. (supra note 92).

[96] Cf. Booth, W., *Science 1988, 240,* 718.

[97] H.R. 4970; cf. *Patent, Trademark and Copyright Journal (PTCJ) 1988*, 36, 499; and *PTCJ 1988, 36*, 503 et seq...

[98] Cf. for a critical analysis of this bill the contributions of Lesser, W., and Milligan, R., in: *Animal Patents: The Legal, Economic and Social Issues*, Lesser, W. (ed.), London: Macmillan Press, *1989* (forthcoming).

[99] On 22 March *1989* as H. R. 1556; cf. *PTCJ 1989, 37*, 521; *PTCJ 1989, 38*, 555; cf. also O'Connor, K.W., Animal Patents: Congressional Perspectives, in: *Animal Patents: The Legal, Economic and Social Issues*, Lesser, W. (ed.), London: Macmillan Press, *1989*.

[100] Cf. *PTCJ 1988, 36*, 347.

[101] Cf. *PTCJ 1989, 37*, 295-296.

[102] Cf. WIPO Document BioT/CE/IV/2 of 24 June 1988, p. 23 et seq.

[103] Cf. for details Bent et al., op.cit. (supra note 24), p. 471 et seq.; Question 93, *Annuaire de l'Association Internationale pour la Protection de la Propriété Industrielle* (AIPPI-Annuaire) *1988* / I, p. 64-73; WIPO Document BioT/CE/IV/2, p. 23 et seq.

[104] Argentina, Australia and Japan belong to this group of countries, cf. Question 93, loc.cit. (supra note 103), p. 71; WIPO Document BioT/CE/IV/2, p. 26. Very recently, a Japonese patent has been granted for "swine with a shortened cornua uterus".

[105] Cf. for details Dietz, A., *GRUR Int. 1969*, 243-252, at 244; Looser, Diplomarbeit, op.cit. (supra note 10), p. 54; Question 93, loc.cit. (supra note 103), p. 71; WIPO Document BioT/CE/IV/2, p. 26.

[106] Cf. for details Bent et al., op.cit. (supra note 24), p. 497; Cooper, op.cit. (supra note 24), p. 6-14; Vida, A., GRUR Int. *1970*, 149-156, at 152 = IIC *1970*, 1, 225-234 at 229 et seq.; Vida, A., and Dietz, A., *Das ungarische Patentrecht*, Cologne/etc.: Carl Heymanns Verlag, *1976*, p. 26 et seq.; Question 93, loc.cit. (supra note 103), at 71; WIPO Document BioT/CE/IV/2, p. 26.

[107] Cf. for details Laudner, W., *Inventii si Inovatii 1986*, 146-147; cf. also *Inventii si Inovatii 1986*, 103-107; see also Eminescu, Y., Der Rechtsschutz der Erfindungen in Rumänien, in: *Das Patentrecht der südosteuropäischen Staaten*, Dietz, A. (ed.), Weinheim/etc.: Verlag Chemie, *1984*, p. 27-34, at 30.

[108] Cf. for details Lebedev, V. Ju., *Voprosy isobretatel'stva 1985*, No. 6, p. 26-30; id., *Voprosy isobretatel'stva 1987*, No. 4, p. 26-29; Tret'jakova, V.G., *Voprosy isobretatel'stva 1986*, No. 9, p. 20-24; Vinnicuk, D.T., et al., *Voprosy isobretatel'stva 1987, No. 1*, p. 25-28.

[109] Cf. Art. 53 lit. a EPC, § 2 No. 1 PatG. For details see Beier and Straus, loc.cit. (supra note 41), at 448 et seq.; Dolder, F., *Mitteilungen der deutschen Patentanwälte 1984*, 1-7; Moufang, op.cit. (supra note 24), p. 219 et seq.

[110] Cf. the references cited supra, note 44.

[111] This view is shared e.g. by the following authors: Brody, loc.cit. (supra note 92); Hoffmaster, B., *Intellectual Property Journal 1988, 4*, 1-24; Straus, loc.cit. (supra note 57).

[112] Cf. also Brody, loc.cit. (supra note 92).

[113] This point was made already by Kohler, J., *Deutsches Patentrecht,* Mannheim: Bensheimer Verlag, *1878,* p. 70.

[114] Cf. Brody, loc.cit. (supra note 92); see also the references cited supra (note 45).

[115] Cf. Art. 52 (II) lit. a EPC; § 1 (II) No. 1 PatG; for details cf. Beier, F.K., and Straus, J., *Der Schutz wissenschaftlicher Forschungsergebnisse,* Weinheim/etc.: Verlag Chemie, *1982,* p. 13 et seq.; Moufang, op.cit. (supra note 24), p. 159 et seq.

[116] For details cf. e.g. Moufang, op.cit. (supra note 24), p. 172 et seq.

[117] This is in sharp contrast to U.S. patent law. Cf. for details Loth, H.-F., *Neuheitsbegriff und Neuheitsschonfrist im Patentrecht,* Cologne etc.: Carl Heymanns Verlag, *1988;* Straus, J., *The significance of the novelty grace period for non-industrial research in the countries of the European Economic Community,* Luxembourg: Office for Official Publications of the European Communities, *1988.*

[118] Cf. Art. 57 EPC; § 5 (I) PatG.

[119] Cf. Art. 52 (IV) EPC; § 5 (II) PatG. This provision, however, does not apply to products for use in any of these methods. Cf. for further details the EPO decisions of 14 October 1987, "Pigs I/WELLCOME" (Technical Board of Appeal, Case No. T 116/85), *O. J. EPO 1989,* 13-24; of 15 October 1987, "Pigs II/DUPHAR" (Technical Board of Appeal, Case No. 19/86), *O. J. EPO 1989,* 24-29; and of 24 November 1988 (Technical Board of Appeal, Case No. T 58/87, still unpublished). See furthermore the British decision Occidental Petroleum Corporation's Application of 5 March 1984, *GRUR Int. 1985,* 120 - "Embryo-Transplantierung".

[120] Cf. for details Moufang, op.cit. (supra note 24), p. 307 et seq.; Straus and Moufang, op.cit. (supra note 23).

[121] Cf. Budapest Treaty on the International Recognition of the Deposit of Microorganisms for the Purposes of Patent Procedure; in January 1989, 22 States were parties to the treaty, cf. *Ind.Prop. 1989,* 16.

[122] Cf. EPO Decision of 30 November 1987, "Submitting culture deposit information/IDAHO" (Legal Board of Appeal, Case No. J 08/87 and J 09/87, published in an abridged version in *O. J. EPO 1989,* 9). The application under review concerned the production of polypeptides in insect cells. Cf. also Byrne, loc.cit. (supra note 45), at 5; Cooper, I.P., *Rutgers Journal of Computers, Technology and the Law 1980,* 8, 1-46, at 44 et seq.; Straus, loc.cit. (supra note 57).

[123] Cf. Straus, loc.cit. (supra note 57). See also v. Pechmann, loc.cit. (supra note 10), at 346; id., loc.cit. (supra note 65), at 480 et seq.; contra: Rogge, loc.cit. (supra note 65). For technical details cf. Foote, R.H., The Technology and Costs of Deposits, in: *Animal Patents: The Legal, Economic and Social Issues,* Lesser, W. (ed.), London: Macmillan Press, *1989.*

[124] Cf. for details Moufang, op.cit. (supra note 24), p. 367 et seq.; WIPO Document BioT/CE/IV/2, p. 29 et seq.

[125] Cf. Art. 10-13 of the proposed EC Directive.

[126] Cf. Milligan and Lesser, loc.cit. (supra note 98); Sorensen, A.A., Animal Patents: The Perspectives of Farmers, in: *Animal Patents: The Legal, Economic and Social Issues,* Lesser, W. (ed.), London: Macmillan Press, *1989.*

20. Application of Genome Analysis in Animal Breeding

H. Geldermann

20.1 Introduction

In recent years several traits were improved in farm animals. For example, milk production in W.Germany (Fig. 20-1) increased in steps which correlate with various influences. The development was forced by traditional breeding methods based on phenotypic assessment for performance i.e. production, reproduction and viability. The estimates of such complex phenotypes are the result of a combined action of numerous genes and environmental factors. Since animal breeders were not able to visualize the real genotype of individuals they selected on the basis of phenotypic values using complex biometrical procedures and applied them to populations. However, the network of effects masks the genotype, thus the complex phenotype of an individual allows only an imperfect estimate of the true genetic potential. The estimated parameters are inadequate to analyze genes responsible for the expression of economically relevant traits. Even where family data showed effects of major genes, e.g. the halothane susceptibility in pigs, the double muscling in cattle or the dwarfism in chicken, the primary gene products and their mechanisms of regulation remain largely unknown.

   Thus, in animal breeding the selection of superior genotypes is restricted to phenotypic correlates so that environmental effects prevent breeding progress. Moreover, selection for high production combined with insufficient information on the genetic and physiological background does not only result in desirable effects, but also, unwillingly, promotes the increase in metabolic disorders and reproductive problems and the decline in animal product quality. This situation limits further advances in performance, whereby - apart from economical and political factors - the following genetic and breeding influences are predominant:

- the degree of genotypic variability in a population,
- the insufficient knowledge on the biological basis of performance, and
- the characteristics of sexual reproduction that influence selection intensity and generation interval.

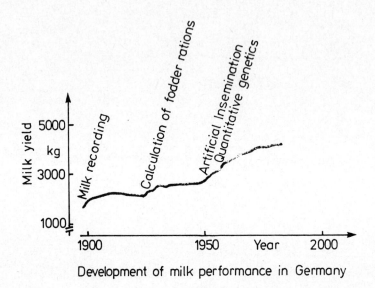

Development of milk performance in Germany

Fig. 20-1 Development of milk yield per cow in Germany and the impact of major influences.

Some of these problems were attenuated by technical advances during the last two decades. In farm animals biotechnology and with it analysis and manipulation of body functions [1, 2] had the largest impact. Special domains of biotechnical research have developed (Table 20-1).

   The first biotechnology used in animal breeding was artificial insemination. It improved estimation of breeding value, choice of mating and selection intensity. A further step, which changed the pattern of reproduction, was embryo transfer including superovulation and conservation of embryos by deep-freezing. In future, particular impulses are expected from methods related to molecular genetics.

20.2   Application of Molecular Genetics for Genome Analysis in Animal Breeding

The technology of molecular genetics is used for gene manipulation and genome analysis. Gene manipulation implies experimental manipulation of gene arrangement and/or information, whereas genome analysis is concerned with gene loci and genotypes. The approach of genome analysis may be divided into analysis of gene structure and function, gene mapping, and diagnosis of gene variants in populations.

Table 20-1 Methods in animal breeding.

Domain of Techniques		Examples
Quantitative Genetics		selection methods, mating systems
Biotechnology	Reproduction	artificial insemination, superovulation, embryo transfer, in vitro fertilization, cloning, androgenesis.
	Cytogenetics	diagnosis of chromosomes, chromosome mapping, sex diagnosis, induction of chimerism, combination and transfer of genomes, chromosomes or chromosomal segments.
	Biochemical genetics	analysis of gene products and their variants, gene mapping, control of identity, twins, parents etc.
	Molecular genetics	analysis of gene structure and function, identification of gene variants, gene mapping, gene conservation, modification of genes, gene transfer in micro-organisms, somatic cells or germline cells.

20.2.1   Analysis of Gene Structure and Function

For direct selection on the DNA level animal breeders need information on genes and haplotypes coding for economic relevant phenotypes; these genes are referred to as target or candidate genes.

20.2.1.1 State of Knowledge

Studies of target genes are directed towards
- detection, isolation and characterization of selected genes or chromosomal sections,
- analysis of gene expression,
- diagnosis of DNA sequences specific for regulation and differentiation.

In farm animal species only a few target genes are known. They include genes for disease resistance, reproduction, regulation of metabolism, structural composition of products, morphological features, sex differentiation and proteins of medical interest. However, up to now almost no information on DNA structure is available for these genes. Only for bovine milk protein genes [3, 4, 5], the MHC [6, 7] and some hormone genes [8, 9] have studies reached the genomic DNA level. Nevertheless, the number of genes described for farm animals grows rapidly because of numerous experimental activities. The amount of sequenced DNA shown in Table 20-2 may indicate the state of the art. A total of almost 500 DNA sequences with an overall length of 490 kb are known for farm animal species.

Molecular genetic studies did increase information not only on gene structure but also on basic physiological processes. As regulation of gene expression becomes traceable on the primary level, main components of production traits were elucidated for causal interactions between genotypes and trait values. For example the analysis of milk protein genes in connection with gene expression serves to study yield and quality of milk (Fig. 20-2).

Table 20-2 Sequenced DNA segments in farm animal species [10].

Species	Number of known DNA sequences	Length (kb)
Bovine	111	110.3
Sheep	13	8.5
Goat	23	16.2
Pig	7	12.7
Horse	1	1.3
Rabbit	83	89.1
Chicken	253	252.4
Summary	491	490.0

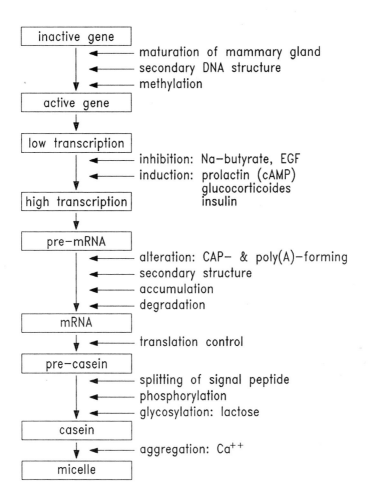

Fig. 20-2 Expression of casein genes.

20.2.1.2 Possible Applications

In farm animals, genes were investigated in order to detect target genes, to conserve gene material and to provide DNA sequences for further experiments.

Analysis of Genes Influencing Production Traits

Using DNA technology, information on gene expression and regulation opens the pathways from distinct gene variants to differences in trait formation. Extensive studies in human and laboratory animals gave prove to the fundamental success of such an approach. Thus, creation of new and profound knowledge of target genes for production traits is part of the present effort. From these studies, components of biological systems should be characterized and considered in breeding programs thereafter. For one aspect, criteria may be found to estimate the metabolic loading of an organism through production. Then, the minimizing of the loading in relation to a given level of production, may improve fertility and viability.

An example for analyzing genes interesting to breeders is the genetic variability in disease resistance. The efficiency of animal production, quantity as well as quality, can be improved significantly by a reduction of disease related losses [11]. Genetic variations, a prerequisite for successful selection, were found for viral, bacterial and parasitic infections [12, 13]. A large proportion of resistance variability among individuals results from genetically determined immunoresponsiveness. Rapid, strong and flexible host reaction is crucial to pathogen elimination. It depends on specific and non-specific mechanisms. Selection for factors influencing resistance changes the disposition for special diseases [14]. For example, the MHC influences disease resistance as well as fertility, and a large number of polymorphisms exist. It is a major objective to describe the general adaptability for each breeding individual. Presently this is limited by the lack in profound information on molecular mechanisms of resistance and by negative correlations among various immune response functions, e.g. phagocytosis, immune response gene expression, cell-mediated and humoral immunity. To solve these problems further molecular genetic studies are badly needed.

Gene Conservation

Genomic DNA can be isolated and stored from endangered breeds as well as from superior breeding individuals or strains (Figure 20-3). Such gene conservation by DNA banks - containing genomic DNA (e.g. in sperms) or cloned DNA fragments (e.g. in bacterial cells) - maintains genetic resources and enables complex research programs. Later, using DNA banks, specific genes can be isolated, cloned and analyzed. Subsequently, gene variants and haplotypes may be identified that are relevant for special breeding purposes in order to screen for them in other populations and/or in subsequent generations.

Providing DNA Sequences for Gene Transfer Experiments

Genes and gene variants have been studied to isolate DNA sequences for transfer into tissue cells or into cells of germplasm. To support these techniques, selection, modification and combination of appropriate DNA sequences are urgently needed. Open questions do not only relate to advantageous genes but also to their variants, tissue specifity and regulating effects. Actual projects and results are summarized [15, 16, 17], pointing to the role of genome research for promoting successful progress of gene transfer.

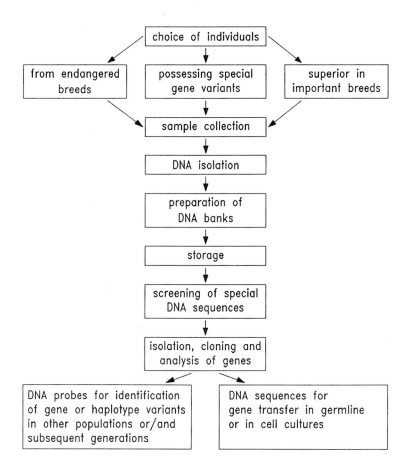

Fig. 20-3 Approaches of gene conservation in animal breeding.

Providing DNA Sequences for Gene Expression in Cell Cultures

Isolated DNA sequences from farm animal species were already used in cell culture systems for production of biologically relevant substances. Thereby, large amounts of gene products were obtained That are important for various experiments, e.g. to analyze mechanisms for metabolic regulation and their effects on production, reproduction and viability. Key elements of such physiological processes can be identified and may then be used as external agents for manipulation of body functions, e.g. to improve lactation or growth by application of external hormones. Well known examples were performed with recombinant growth hormones (e.g. bovine somatotropin, BST [9, 18]) produced in biotechnical processes. Possible practical application in animal production is a matter of controversy, but the scientific use of *ex corpora* produced, biologically active substances opens new areas to basic studies.

20.2.2   Detection and Characterization of Gene Variants in Populations

Conventional assessments of genetical variation deduced from the phenotype to the underlying gene variants. Thus, a phenotype could be associated with the presence of a distinct allele only for simple inherited traits, such as blood groups, proteins or enzymes. However, interindividual differences for production traits are usually of multifactorial origin. Therefore, in most cases effects could not be assigned to singular genes. It is thus of central importance to use cloned DNA sequences as probes for a direct identification of genotypes in populations.

20.2.2.1 State of Knowledge

For a direct assay of DNA variants different approaches are available (Fig. 20-4). Restriction Fragment Length Polymorphisms (RFLPs) reveals a main group of variants. As reviewed in Table 20-3, variable restriction fragments are produced by base alterations or by chromosomal rearrangements. A base alteration can change the restriction site for an enzyme and thereby initiate Restriction Site Variants (RSVs) [24]. Chromosomal rearrangements generate Variable Number of Tandem Repeats (VNTRs) in tandem-repetitive regions of DNA. These are dispersed in eukaryotic genomes and show substantial length polymorphisms ("hypervariability"). They arise from unequal exchanges that alter the number of short tandem repeat elements in a subset [21, 22]. A hybridization probe containing the tandemly repeated core sequence can detect many highly polymorphic loci simultaneously, and the visualized patterns are usually named DNA-fingerprints.

DNA variants were found for several eukaryotic species and in large numbers. Nevertheless, information is limited for farm animal species (Tab. 20-4). However, the progress is accelerating, since the major part of polymorphisms has been described within the last years.

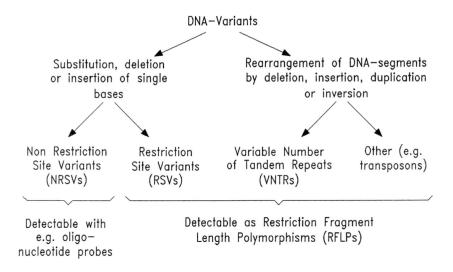

Fig. 20-4 Assay of DNA variants.

Table 20-3 Characteristics of Restriction Fragment Length Polymorphisms (RFLPs).

Name	Restriction Site Variants (RSVs)		Variable Number of Tandem Repeats (VNTRs)	
Type of DNA probe	locus specific DNA sequence (cDNA, genomic DNA)	set of DNA sequences linked in the genome	core sequence of VMTR loci (Minisatellites or Simple Tandem Repeats	DNA sequence specific for a single VNTR locus
Number of alleles or haplotypes	≤ 2	many	many	many
DNA loci marked per probe	1	several, linked loci	several, more or less unlinked	1
Degree of Heterozygosity	low	?	high	high
Choice of genes	yes	yes	no	yes
Sum of detectable DNA loci	>10,000	>10,000	>1,500	>1,500
References	[19]	[20]	[21, 22]	[23]

Table 20-4 Restriction Fragment Length Polymorphisms (RFLPs) in farm animal species [25].

Species	RSVs (Number of genes or gene complexes used)	VNTRs (Number of probes used)
Cattle	28	6
Sheep	5	1
Goat	5	1
Pig	9	4
Horse	7	5

The criteria of DNA polymorphisms which determine their application in genetics and breeding can be summarized:

- Restriction fragments separated by electrophoresis and oligonucleotide probes allow a direct monitoring of gene variants. Consequently the identification is not influenced by dominance or gene expression so that the gene variants can be registered almost independently from sex, age, tissue or environment.
- DNA markers are more or less stable. The somatic stability of most of the DNA variants enables the identification of genotypes from different tissue or semen samples, whereas the germline stability is crucial for studies of inheritance.
- DNA probes can be selected according to gene, chromosomal region or trait. Thus, in line with the actual objective, special DNA probes can be used or developed.
- Many gene variants can be identified. For breeding purposes, polymorphisms are important which have high degrees of variability, i.e. VNTRs or RSVs plotted with complex DNA probes (Table 20-3).

20.2.2.2 Possible Applications

Unfortunately, cost and time are still considerable for the test of an individual. Moreover, in farm animal species the limited number of known DNA variants (Table 20-4) restrict practical use. But, remarkable innovations are stimulated from the DNA variants and a large spectrum of contributions can be realized for animal breeding within a reasonable period of time.

Estimation of Breeding Values on the Basis of Genotypes (Genotype-Assisted Selection, GAS)

The utility of DNA variants in animal breeding is based on tight linkage between marker and target genes. A linkage permits to infer the presence of a desirable gene by analysis of the marker gene. Classical examples are genes linked to disease resistance genes [11, 12, 13, 14]. The use of several DNA probes may enable screening for many different disease resistance genes simultaneously without the need to inoculate the population.

DNA variants allow to analyze genes or chromosomal sections which cause part of the genotypic variance for a production trait [26]. However, the application of gene variants for judging economically relevant traits is limited by several restrictions [27, 28, 29]. If marker genes are taken which are only linked and therefore indirectly connected with the genes of the production trait of concern, then associations between marker and target genes are functions of distances, types of linkage and degrees of linkage disequilibria. Thus, reasonable associations are restricted to families and have to be tested for each generation. This implies considerable expenditures [28, 29]. However, the closer the linkage between marker and target genes and the more significant the judged target genes act on a production trait, the more evident the application of the marker genes in breeding.

Consequently, two approaches can be followed: One, as many marker gene loci as possible may be used (Marker Assisted Selection, MAS; [30, 31]), e.g. including the enormous variability of VNTRs. Two, marker genes can be selected which have already an a priori probability to be associated or closely linked with the production traits of concern. This holds true e.g. for milk protein genes relative to the amount and properties of milk proteins [27]. Genes or chromosomal regions were identified directly and variants contributed to economically relevant expressions of traits [30, 31]. Consequently, such genes cause part of the genotypic variance for the production trait and are therefore the target genes (or target chromosome regions); their use in breeding may be named as Target gene Assisted Selection (TAS).

Information on marker and target genes can be implemented in conventional breeding programs. For example, based on desired alleles or haplotypes, young sires may be preselected prior to their entry into further test programs. For dairy cattle progeny tests include only a small number of sires, but many young individuals are available with high estimated breeding values on the basis of phenotype and breeding value of ancestors and sibs. Then, among young individuals those can be preselected and used in the later steps of the test program that possess superior alleles, haplotypes and/or chromosomal sections from their proven sires and dams. Such a Genotype-Assisted Selection has several advantages, e.g. enlarged genetic gain without a change in generation intervals (Fig. 20-5), early preselection, independence from sex, use of effects through special alleles or allelic combinations. In future, a first selection step will be possible to screen embryos before transfer on the basis of genomic DNA from a few blastomeres, and identify those embryos which have the desired sex and the desired alleles.

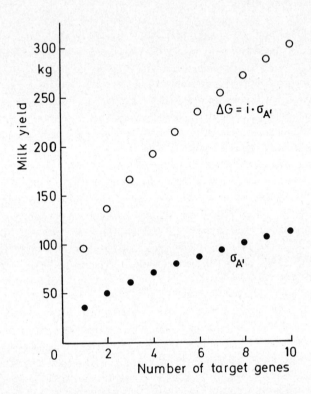

Fig. 20-5 Influence on selection gain, ΔG, by preselection according to genotypes of target genes for milk yield.

Assumptions for each target gene: additive gene effect $a = 50$ kg milk; dominance effect $d = 0$; allele frequency $p = q = 0.5$.

Selection intensity for sires $i = 2.7$.

Average effect of allele A^1 resp. A^2: $\alpha_1 = q[a + d(q - p)]$; $\alpha_2 = -p[a + d(p - q)]$.

Thus for individuals with different genotypes of the target gene, A^1A^1, A^1A^2 and A^2A^2, the breeding values (G) are $2\alpha_1$, $\alpha_1 + \alpha_2$ resp. $2\alpha_2$ [32]. Then, based on K target genes, the index of breeding value becomes

$$I = \sum_{j=1}^{K} G_j$$

and the variance of breeding values is

$$\sigma_A^2 = \sum_{j=1}^{K} \{2p_jq_j[a_j + d_j(q_j - p_j)]^2\}$$

Introgression of a New Character into a Highly Improved Population by Marker Gene Backcross Breeding (MBB)

Superior farm animal breeds have accumulated desirable genetic properties from several breeds. The breeding strategy for further improvement of such a breed tries to introduce only a distinct trait or trait value from the source of a "donor" breed. In a first step, this requires a cross between the superior donor breed and the breed to be improved ("recipient" breed). Segregating progeny, obtained by backcrossing on the recipient breed, then contains mosaics of chromosomal pieces derived from both parents. Apart from the desirable gene variants, backcross individuals may also carry many undesirable genes from the donor breed. Thus, on the basis of phenotypic values, the traditional backcross breeding selects in hybrid individuals the desirable characteristics of the recipient breed along with introduced traits from the donor breed. In several cycles the hybrid individuals are crossed back to the recipient breed so that at last the individuals obtained are genetically similar to the recipient breed, with the exception of gene variants for the added target character.

The introduction of marker genes can trace the transfer of tightly linked target genes. Thereby, segregating progeny are screened without influence of gene expression. This takes place prior to expression, i.e. at a young stage (before the trait is expressed) or for recessive genes. An example for the introgression of a gene with effects on important characteristics is given in Fig. 20-6. The use of marker genes forces the fixation of the desired allele(s) so that at the other gene loci alleles of the recipient breed are preserved. Since marker genes can monitor genes with effects on quantitative and on qualitative traits, the flow of genes is to be traced without limitation to the type of character. For breeding application consideration of the complete genome of individuals is necessary to yield information on the parental origin of alleles at all particular sites in the genome. Then, throughout the genome, the composition can be estimated for an individual's chromosomes in terms of its parents'. The information will permit deduction of the chromosome region(s) that should be included into the recipient breed. Examples in farm animal populations are the high incidence of multiple birth in some sheep breeds when compared to others, or horned and hornless cattle breeds. In such cases, where a single trait is of main interest, large differences in allele frequency of the breeds mentioned are expected at the relevant target genes, and the backcross individuals, when scored for marker genes, can be restricted to individuals with the desired expression of the production traits.

Backcrossing results in the transfer not only of desirable gene(s), but also of additional linked, undesirable genes ("linkage drag"). Recombinations in these chromosomal regions can be monitored by marker genes. It is possible to select individuals that have recombinations close to the gene of interest. The ability to select for desirable recombinants in a chromosomal region of interest is a function of recombination frequencies, of the number of markers usable in that region, and of the number of individuals analyzed. Therefore, manipulation of recombination, saturated gene maps, as well as high efficient screening methods will be of major interest.

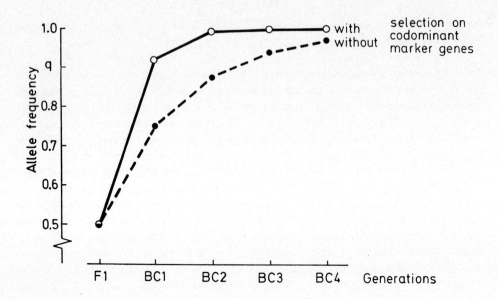

Fig. 20-6 Number of backcrosses for the nearly fixation of a dominant target gene with and without the use of codominant marker genes.

From a donor breed only one allele should become evident, whilst alleles of other gene loci are eliminated by prolonged backcrosses (BC) and selection. When a dominant allele causes the desired characteristic, linked codominant marker genes are helpful to monitor the introgression of the desired chromosome section, and individuals with inferior gene material can be culled. If selection is directed solely to the desirable trait it takes 6 backcross generations to reach closely the genotype of the recipient breed. Additional selection for codominant alleles of linked loci that predict the genotype with a probability of 80%, the genotype of the recipient breed is recovered as early as after 2 backcross generations.

Backcross breeding combined with DNA variants makes it possible to improve or to adjust existing breeds for highly specific characters [33]. Any gene, that can be detected by a DNA probe, can also be monitored for its transfer into other breeds. With many gene markers it will be possible to alter important properties of breeds, such as disease resistance, and several chromosomal sections can be considered simultaneously. The ability to adjust existing breeds in a few generations allows breeders a quicker response to market demands, as well as to unexpected environmental pressures.

Primitive breeds and wild species contain a tremendous amount of genetic variation. Breeders have generally been unable to utilize these resources. Even when fertile hybrids are obtained, there is no assurance that the gene(s) for the character(s) of interest can be successfully transferred into an economically important breed. Undesirable effects of linkage drag are of special disadvantage when breeding with exotic germplasm, and the method of traditional backcross breeding is much too time-consuming and ineffective. However, DNA markers can be used to select individuals with little undesirable donor DNA, whilst gene(s) controlling a character of interest from exotic germplasm can be located and the transfer of small amounts of superior foreign chromosomal DNA containing the desired genes can be traced into the recipient breed. In the long run, the use of DNA markers to access genes from exotic germplasm may prove to be a significant contribution for the introduction of new genes into the gene pool of important farm animal breeds. This may help to improve the genetic variation of relevant characters.

Identification of Chromosome Combinations which Generate Heterosis Effects

It is widely accepted that effects of heterosis depend on the relative amount of heterozygous genotypes in an individual. Therefore, using the F2-generation of a cross between populations, marker genes would enable to analyze the type and magnitude of genetic effects for marked chromosome regions. On the other hand, with the help of many marker genes a degree of heterozygosity may be calculated, which is representative for the genome.

The data on marker gene heterozygosity will then help to monitor influences of genetic drift. Populations as well as breeding lines may be selected for their ability to show hybrid effects in special cross breeding programs. Furthermore, the expected degrees of heterozygosity, estimated on the basis of many marker genes in the parental generation for an anticipated offspring, can be used to recommend matings within populations. This kind of prognosis may assist breeding for improved fertility and viability, for example in cattle.

Sex Diagnosis

Identifications of chromosomal combinations that cause special effects may also include sex diagnosis at the level of DNA-analysis. Using sex specific DNA probes diagnosis is possible at early embryo stages, and almost perfect prediction of the sex of an embryo has thus been achieved [34, 35]. The procedures involve micromanipulation to remove a few blastomeres and to hybridize their DNA with marked probes. Thereby, the sex determination depends on the presence or absence of special DNA sequences - in mammals those which are specific for the Y-chromosome. The hybridization techniques are independent of cell cycles and stages of embryonic development. Technical difficulties posed by the availability of a very small quantity of DNA from the early embryo could be overcome by the very sensitive polymerase chain reaction (PCR) technique [36, 37]. As a drawback, the diagnosis lasts several hours so that the remaining embryos have to be conserved deep frozen or in culture before transfer of typed embryos is possible.

The ability to determine the sex of an embryo may increase efficiency of breed improvement and thus stimulate commercial applications [38].

Identification of Recessive Alleles in Heterozygous Individuals

An increasing amount of information is available on the inheritance of pathogenetic disposition for many diseases. Certain alleles in host genomes cause a susceptibility for the development of a disease and the transmission of a disposition to the following generation. In many species conditions are often similar for genetic defects. Especially in mammalia a basic genetic homology has been observed. Thus the extensive experience for defect alleles in the human has considerable relevance for studies in other species, and sustains molecular genetic research in farm animals.

In farm animal species, biochemically defined defects have been detected to identify heterozygous individuals ("carrier") which possess only one recessive defect allele at the locus considered. They are therefore phenotypically not distinguishable from the clinically normal situation. For example, in cattle the carrier of a recessive lethal allele can be identified by a deficiency in the activity of uridine monophosphate synthetase (DUMPS; [39]). In contrast, the primary gene products of the mendelian inherited malignant hyperthermia in pigs still remains unknown. In this case, several marker genes and the sensitivity against halothane can be used to identify the parental haplotypes [40, 41]. Subsequently the genotype at the "halothane locus" can be predicted for an offspring with a probability of more than 90% [41]. Thus, marker genes are usable to distinguish individuals heterozygous for a recessive defect allele from those without that allele.

More advanced attempts apply DNA probes to label defect genes [42]. As soon as regular gene sequences were cloned, direct (= analysis of variants for the regarded gene) or indirect (= diagnosis according to variants of genes linked with the mutant gene) tests were developed for some mutants at the DNA level. For practical breeding purpose it is valuable that DNA probes identify the defect genes independent from gene expression and age of the individual to be tested. Thus, recessive defect alleles can be found in heterozygous individuals. The early detection of defect genes would help to cull those individuals for which a higher disposition to the disease is expected later in life. This type of selection is introducable to embryo tests combined with the typing for sex and desired genes. Such a more stringent selection may be used i.e. for male individuals which are used to produce numerous offsprings; from it, as shown in Fig. 20-7, the frequencies of defect alleles in populations will be reduced remarkably. The significance of controlling defect genes is described by Shook [11] and Patterson et al. [13].

Identification of Genetical Sources

The source of DNA can be discriminated with the help of special DNA probes. Thereby, semen samples, tissues, individuals, breeding lines or breeds can be identified. Especially the multilocus heterogeneity of VNTRs, known as DNA fingerprints, result in extreme diversity ("hypervariability") of patterns in individuals of the same population (Fig. 20-8). The VNTRs exceed a combination of all other presently available techniques by many orders of magnitude [21, 22, 43, 44] so that very precise identification is possible.

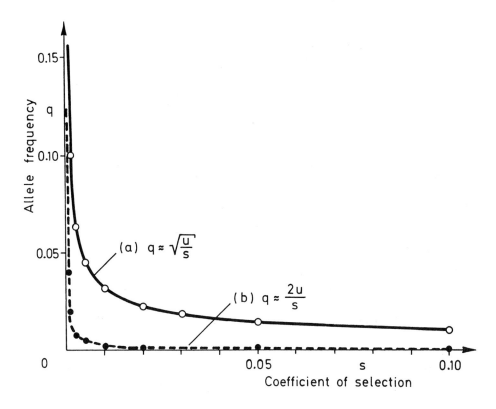

Fig. 20-7 Equilibria between mutation and selection in populations.
Assumed mutation rate $u = 10^{-5}$.
(a) Selection against a recessive allele by culling the
homozygous defect individuals only.
(b) Selection against a recessive allele as given for (a) with
an additional culling of sires which are heterozygous ("carriers").

These characteristics of marker gene based evaluation opens new perspectives for application, e.g. control of questionable semen samples, identification of newly bred strains or assignments of individuals to the gene pool of a breeding program. Moreover, germplasm resources (different breeds of a species, wild progenitor species) can be screened in order to identify variants of chromosomal segments with potential use for improvement of desired traits in commercial breeds, e.g. using Marker gene Backcross Breeding programs (see above).

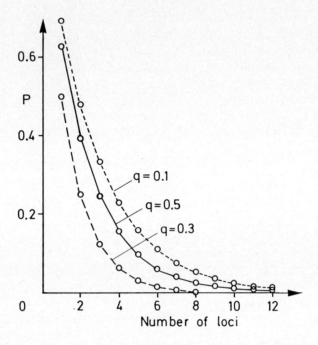

Fig. 20-8 Probability, P, of identical DNA fingerprints for two unrelated individuals in a population (according to [22]).
The symbols used are:
  f : number of all differentiated DNA fragments,
  x : average probability for a fragment occurring in an individual to be present also in a second randomly chosen individual of the same population. For a HARDY-WEINBERG-population, x is related to the allele frequency
  q : $x = q^2 + 2q(1 - q) = 2q - q^2$, and
f/x : number of gene loci.
The probability of two individuals to be identical for one fragment becomes
$$P = x^2 + (1 - x)^2$$
so that for f fragments of a DNA fingerprint follows
$$P = (1 - 2x + 2x^2)^{f/x}.$$

Control of Genetic Relationship

Control of genetic relationship includes tests for putative parents, analysis of population structures and dynamics, estimation of genetic distances between populations etc. DNA criteria ensure very precise statements and can be gained from various tissue and semen samples in addition to the presently used blood.

For example, parentage control helps to improve the estimation of breeding values from information of relatives and thus increases the genetic gain [44]. Thereby, DNA probes offer considerable advantages for parentage controls, since a single technique already yields much higher exclusion probabilities (Figure 20-9) than multiple tests with blood groups and protein variants (Table 20-5). Thus, DNA technology enables positive declarations of true parentships or the stringent association of offspring with one of several parents in question, e.g. after multiple inseminations.

Table 20-5 Approximate exclusion probabilities for parentage controls obtained from cattle.

Blood groups	70 - 90 %
Biochemical polymorphisms	40 - 60 %
DNA-Fingerprints	> 99.9 %

Apart from parentage control the new DNA variants allow to obtain information on the relationship between individuals of a population to monitor relevant genetic processes. For example, in modern breeding programs, populations of farm animals often have small effective size. The dispersive genetic drift may become more effective than the systematic selection forces so that valuable genotypic variation gets lost. Thus, a screening for marker gene variation in a population helps to maintain diversity of genotypes from one generation to the next.

Even data on relationships between breeds are meaningful for breeders. Populations with a large genetic distance contain different allele frequencies and alleles. They are thus qualified for cross-breeding programs and as resources for particular gene variants.

20.2.3 Gene Mapping

Research on genome structure includes the analysis of gene arrangements in the chromosomes, called gene mapping.

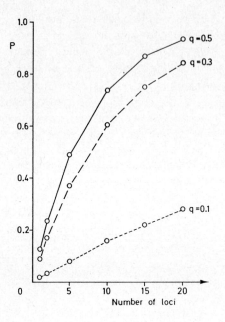

(a) Proving of one parent, the other is unknown.

$$P_j = 2p_j^2 q_j^2 \quad \text{for each locus } j$$

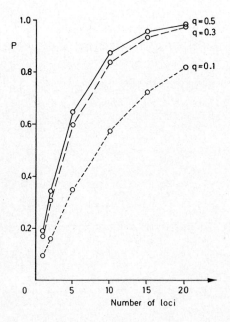

(b) Proving of one parent, the other is known.

$$P_j = p_j q_j (1 - p_j q_j) \quad \text{for each locus } j$$

Fig. 20-9 Probability, P, of exclusions in parentage controls [46].
Assumptions: 2 alleles per locus with codominant expression and identical frequencies (p and q) for all loci.

20.2.3.1 State of Knowledge

During the last decade, new methods of cell genetics and recombinant DNA technology did improve the efficiency of gene mapping. With the help of somatic cell hybrids and banding of chromosomes, single chromosomes were identified and genes could be assigned to chromosomes or chromosomal regions. The number of assigned genes grew rapidly after DNA sequences were used as probes for *in-situ*-hybridization of chromosomes [53]. Moreover, a mapping with RFLPs started, based on Restriction Site Variants (RSVs) or Variable Number of Tandem Repeats (VNTRs) (see Table 20-3). Combining molecular biological techniques with analysis of recombination, DNA sequences were used as probes to detect polymorphic loci and to pursue the segregation for homologous regions of the chromosomes from parents to the offspring generation [47]. Linkage relationships were tested in pedigrees by established methods [48, 49]. The DNA marker loci could be arranged into linkage groups. Since many different DNA sequences can be tested, genetic linkage maps can be constructed which contain a very large number of markers at close intervals. For the purpose of mapping, DNA-variants offer significant advantages: they are screened at the level of DNA and thus normally behave codominantly, have a high number of allelic variation and are more or less free of epistatic effects.

The progress of gene mapping in the human grew exponentially and contains more than 3500 mapped genes, most of them identified by DNA probes. For farm animal species only a small number of genes is mapped when compared to species for which close gene maps are available (Table 20-6). However, extensive data, presented in human and mouse, are helpful for further activities, because most methods for marker gene analysis are transferable to several species, and the marker genes show homologies for the genomic arrangement in different mammalian species [52, 54, 55, 56].

Table 20-6 Status of gene mapping in some mammalian species.

Species	Number of chromosomes (1N)	Groups of linkage or syntheny	Number of mapped genes	Reference
Human	23	24	ca. 3,650	[50, 51]
Mouse	20	21	ca. 1,250	[52]
Cattle	30	31	57	[52, 53]
Sheep	27	18	34	[52]
Pig	19	16	38	[52]
Horse	32	6	21	[52]

One prerequisite for further detailed linkage studies is a large number of genes which are already mapped and cover most of the genome. The number of families required for such gene mapping is a function of the family size, the information content of the marker genes, and the recombination distance between the loci in question. Generally, large families yield more information than smaller. For gene mapping, DNA markers can be selected according to their value of information. Botstein et al. [47] defined the Polymorphism Information Content (PIC) as

$$1 - \left(\sum_{i=1}^{k} p_i^2 \right) - \sum_{i=1}^{k-1} \sum_{j=i+1}^{k} 2p_i^2 p_j^2$$

with k   :   number of alleles and
     p_i, p_j :   frequency of the ith resp. jth allele.

Distances between the mapped genes desirable for an appropriate probability estimate of other genes by linkage analysis can be calculated with the LOD-scores [48]. As seen from Fig. 20-10, distances of up to about 0.2 Morgan are feasible with reasonable numbers of families. Thus, newly detected genes, even those affecting quantitative traits, can be mapped easier if a detailed gene map for the species is already at hand or if a large number of DNA loci is included at once.

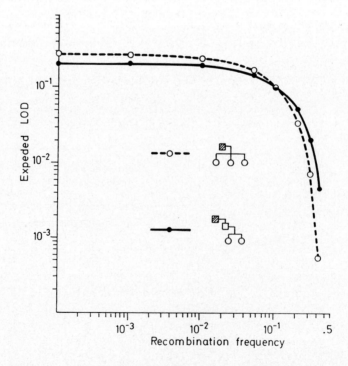

Fig. 20-10 Expected LOD scores as a function of recombination fraction [57].

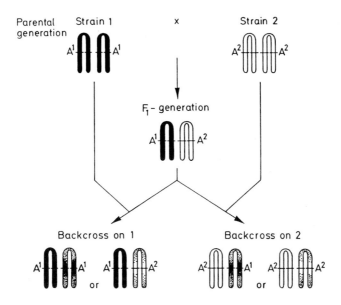

Fig. 20-11 Outline for a backcross experiment to produce reference families (according to [59]).

The linkage situation is given for a gene locus with the genotypes A^1A^1, A^1A^2 and A^2A^2. For the backcross generation the average arrangements of genes within the chromosome are shown for many individuals.

Examples for parental strains 1 and 2: More general mapping - in cattle, a European and a Zebu breed; in pigs, wild boar and a domesticated breed; in sheep, a merino breed and caraculs. Analysis of special genes - Zebu and N'Dama crosses established in Africa to study the inheritance of trypanosomiasis [62]; homozygous halothane positive and negative individuals of Landrace pigs; Boroola merinos and European merino breeds.

Because as many genetic markers as possible should be pursued within a distinct material of individuals or cells, families are important in which alleles of various loci are segregating. For this reason so-called reference families were established by human genetists [58]. A reference material is a collection of samples from members of informative families. Informative individuals have, relative to a target trait or gene, a maximal heterogeneity of trait values and/or related genotypes. The samples (DNA, tissues, fluids, bone marrow, cell lines etc.) are stored and available for several research projects. This type of informative material has been described to study marker gene effects on the values of quantitative traits in cross-bred generations of mice [59]. Analogous cross-bred generations were proposed for farm animal species as well. Hetzel [60] emphasized the importance of a collaborative effort between laboratories to maintain informative families for cattle. Moreover, Maijala [61] proposed reference families by collecting data, which do not only include marker genes but physiological and immunological traits and results from *in-situ*-hybridization of chromosomes and somatic cell techniques. To this end, individuals of extremely different breeds can be crossed, followed by a crossing of the F1 generation or a back-crossing on both parental breeds (Fig. 20-11). The generated individuals are maximally heterogenic both, for their genotypes and for their phenotypes that are responsible for differences between the original breeds or strains.

20.2.3.2   Possible Applications

Gene maps can serve as a tool to analyze the genetic variation of production traits on the level of genome [55]. For domestic animal species this general goal of gene mapping implies a large scale of important applications.

Map-Based Gene Cloning

Methods are well-established for isolation and cloning of a gene on the basis of its product. However, for many important genes the products are unknown in farm animals. Even for genes already identified by classical genetics, the mechanisms by which they act are largely unknown. One approach to clone genes without knowledge of their products is the transposon tagging [63]. But, a large number of transposon-mutagenized individuals must be screened which limits the value for farm animals. The other method for a product independent gene isolation is offered by a map-based cloning. Often referred to as reverse genetics [64], this approach is based on physical linkage to mapped RFLP markers. Once tightly-linked RFLP markers are known, the flanking gene regions are identified using libraries with overlapping clones. A move or "walk" along the chromosome from the cloned gene to the gene of interest has already been used to clone genes involved in hereditary diseases of human [65]. For reasonable chromosomal distances large DNA segments were cloned in cosmid or yeast chromosome vectors [66, 67]. Poustka and Lehrach [68, 69] described linking and jumping libraries containing linked terminal DNA sequences from large DNA fragments. The approach of reverse genetics, however, needs a method to identify the DNA segment that actually contains the gene(s) of interest and is only feasible for genes physically tightly linked with the marker [33].

Analysis of Relations between Known Marker Genes and Hypothetical Target Genes for Production Traits by Interval Gene Mapping (IGM)

As mentioned, for most of the genes relevant in animal breeding no primary products or functions are known. To study such genes, an Interval Gene Mapping (IGM) may serve as a first crucial step. For this purpose, informative family groups can be used where marker genes segregate and also genes influencing the variance of the production criteria in question. Thereby, the association of variation in production criteria with single marker genes allows an indirect identification of closely linked but unknown target genes. This implies, in principle, a resolving of genetic components of a more or less complex trait.

However, heritable multifactorial characters, often referred to as quantitative, are influenced by a combined action of several genes. For expression of such characters little is known on the number, chromosomal position and effects of genes. Therefore the problem is that only high numbers of DNA markers make it feasible to measure and map the effects of genes underlying quantitative traits (Quantitative Trait Loci, QTL [26]). To detect QTLs two individuals may be crossed which are genetically different for quantitative characters of interests. The progeny (F_2 or backcross generation) is then obtainable, segregating for these traits as well as for DNA markers. An outline of an analysis of effects on a quantitative trait assessable from gene markers in a F2-generation of breeding strains is given in Fig. 20-12. Associations between a segregating marker gene and the values of quantitative characters should then be due to linkage of the marker gene to one or several QTLs. The ability to detect QTLs by marker genes is a function of the magnitude of QTL effects, the number of offspring studied, and the recombination frequency between marker gene and QTL(s). Likelihood maps can be established to show chromosomal intervals and effects for within located QTLs, as described for the genome of tomato [33]. Linked markers, bounding an interval which may contain QTLs, will reduce the likelihood, that genotypes at the marker genes will differ from genotypes at within linked QTLs, to the square of the recombination frequency between the flanking markers. For a demonstration of effects and interactions of QTL(s) each interval between marker genes can be studied as a discrete entity , and at last a biophysical basis of complex traits will be available [33].

To dissect complex traits into their genetic components many mapped marker loci are necessary. As early as 1980 Botstein et al. [47] proposed to cover the whole genome of a species with polymorphic DNA markers. Each gene should then be judged by gene linkage. Assuming a genome size of 2500 cM and aiming at a probability of greater 95% for a linkage of 20 cM or less between marker genes and unknown genes, about 215 randomly distributed marker genes have to be used (Fig. 20-13a, [70]). The number of marker genes is reduced to about a third if their loci are regularly distributed (Fig. 20-13b). In some cases a few chromosome sections may contain the most important genes ("major genes") of the considered trait, e.g. the halothane gene region in pigs relative to meat quality and stress resistance. Then the number of genes necessary for breeding purposes is considerably reduced (Fig. 20-13c) because markers of flanking regions of target genes will be sufficient.

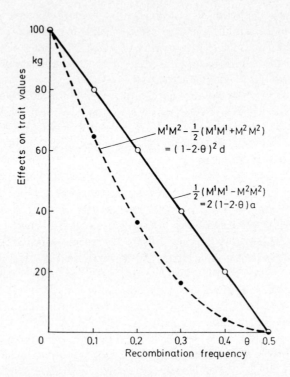

Fig. 20-12 Estimate of genetic effects on a quantitative trait by a segregation analysis in a F2-generation using the example of milk yield in cattle.

Assumptions: Parental sources with different alleles at a marker gene M and a target gene A ($M^1M^1A^1A^1$ and $M^2M^2A^2A^2$); effects of the gene locus A on milk yield: a = d = 100 kg.

For a target gene with the genotypes A^1A^1, A^1A^2 and A^2A^2, effects on the quantitative trait are designed as +a, d resp. -a [32]. Then, from a marker gene, with a recombination frequency Θ to the target gene and with the genotypes M^1M^1, M^1M^2, M^2M^2, the following effects on the quantitative trait are measured in the F_2-generation:

$(1-2\Theta)a+2\Theta(1-\Theta)d$; $((1-\Theta)^2+\Theta^2)d$; $-(1-2\Theta)a+2\Theta(1-\Theta)d$ resp..

The mean trait values are compared for individuals with different marker genotypes: $\frac{1}{2}(M^1M^1 - M^2M^2)$ and $M^1M^2 - \frac{1}{2}(M^1M^1+M^2M^2)$.

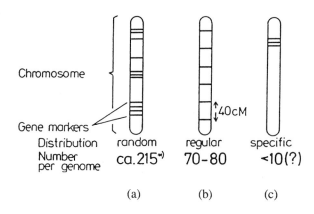

Chromosome

Gene markers

	random	regular	specific
Distribution	random	regular	specific
Number per genome	ca. 215*)	70–80	<10(?)

(a)   (b)   (c)

Fig. 20-13 Number of polymorphic marker genes applied for a genome with 2,500 cM.
*) According to Beckmann and Soller [70] as explained in the text.

Detection of Genomic Similarities and Differences in Gene Arrangements

Comparative gene mapping and the identification of conserved linkage groups [52, 54, 55] stimulated new research activities. Using sets of clones for RFLP mapping the degree to which chromosome content and gene order have been conserved was determined. To this end, gene clones from one species were mapped in different species. The high degree of linkage conservation found between species suggests that substitutions of chromosomes or chromosomal segments from one species to the other might be definable. For example, all members of the Bovidae family share the same basic number of chromosomal arms (29 autosomal arms per haploid set). If the gene contents and orders within arms were highly conserved among related species, single chromosomes or chromosomal segments might be substituted between species (e.g. by somatic cell hybridization [71]) to combine genes that are not available in the normal crossing range of a species.

On the other hand, different arrangements of genes have been described even within species [72]. Investigations on gene arrangements in different farm animal populations can achieve information on evolution or special effects of gene clusters and provide new parameters for breeding programs. For example, genomic differences may be associated with fertility and viability. This may become evident when cross-bred generations are formed between populations with large genetic distance, e.g. Zebu and European cattle.

20.3   Prerequisites for Genome Analysis Applied
   to Animal Breeding

In animal breeding research the use of molecular genetics is at its beginning. Only few approaches have reached the level of application and further basic studies are required. Thereby the complex approaches of animal breeding need specially adapted methods. To implement genome analysis, special laboratories, up to date training of personnel as well as coordination of comprehensive programs are necessary. Methods, data and equipment should be standardized. Moreover, special statistical procedures have to be utilized, e.g. for complex linkage analysis and for marker gene studies.

Since molecular genetic methods are paramount to biological research it should be possible to connect genome analysis in farm animals to other research branches which use similar methods. Thereby, the access to resources, such as reference families, DNA libraries, probes or vectors, can be organized efficiently. Moreover, the collection and distribution of proceedings and software may promote optimal utilization of methods and data.

It is thus urgently recommended to develop an infrastructure for interactions between the numerous projects and results. A central coordination implying initiatives to settle strategies adequate for large projects and international cooperation.

20.4   Possible Consequences of Genome Research
   in Animal Breeding

Considerable consequences of genome analysis are expected although they are often overemphasized. Questions arise relative to aim and purpose of results, and conceivable risks as well as effects impairing other interests are discussed. Therefore the possible consequences of the use of recombinant DNA technology may be divided into intended and unwanted ones to assess advantages and risks.

20.4.1 Prospective Advantages

Among the intended consequences, the useful innovations for animal breeding, an improved interference of men in the microevolution of domestic animal species is the crucial goal. New approaches of molecular genetics develop such interferences efficiently, reliably and conscientiously. The DNA technology leads to increasing independency from the conditions of special population or environment. Along with discriminative methods, single traits can be regarded more properly in farm animals. Thus, far-reaching programs are discussed to improve the quality of special animal products, to enlarge the spectrum of production and to increase fertility and viability. In future the use of DNA technology will help in future to breed new strains with superior and special abilities. Moreover, genome analysis will offer a control tool for the type and heterogeneity of genes in populations. Hence, the breeders can trace genetic developments, which are generated by a certain selection, faster and more precisely. This allows to undertake corrections in time. Furthermore, laws or decrees demand increasingly to test individuals and their products. Examples are the milk protein genotyping of A.I. sires required in case of semen export or the DNA fingerprint technique in order to identify individuals or carcass samples required at forensic requests.

The activities offer a large spectrum of innovations valuable for economic and social reasons. Therefore considerable activities are underway in all industrial countries with a high speed of transfers from methodical development to practical applications. It is well known that technical progress influences the economic situation as well as the structure in agriculture. In animal breeding, population genetics, artificial insemination, cross-breeding programs or, more recently, the embryo transfer are well-known examples. From the actual research in molecular genetics, adapted to farm animals several DNA sequences are offered which are usable for economic purposes, e.g. identification of gene variants in populations or gene transfer experiments. Commercial DNA probes, e.g. for screening superior or unfavorable gene variants, are already in use. Appropriate supplies and patents are existing and laboratories are starting to test for the interest of breeders' associations.

Thus, new technical developments are launched into practical animal breeding. The resulting innovations have to be proven and properly used. This rational progress should be ensured by government and administration.

20.4.2 Risks

In the media, active and increasing discussions concentrate on the risks of the new technology. There are misgivings that ecological husbandry is abandoned and that food quality declines. New technologies for animals are under the suspicion to be the stepping stone into human genetics. To assure public acceptance, active discussions have to be carried out on all levels concerned, not only for individual projects but also for regional or global animal breeding strategies. Rational argumentation on the activities of genome analysis should be emphasized by precise and comprehensive information.

New technologies - not only those of molecular genetics - actually force the breeding remarkably in respect to quantity as well as to quality. The present development may not be able to realize all expectations and may in fact produce unexpected phenomena and consequences.

For example, extreme breeding targets markedly change morphology and physiology of individuals. Being deformed the newly bred individuals often have a reduced capacity to regulate metabolism and develop a disposition for defects. Elimination of such individuals seems to be a difficult task in microorganisms or plants whilst domestic animals are fully controlled by the breeders. Undesired variants have always occurred and were prevented from propagation by castration or slaughtering. However, problems arise where extreme trait expressions are associated with pathological events, even though the individuals have superior economic or sentimental values, e.g. special growth or muscle dispositions (e.g. alleles for dwarfism, double muscling, creeper, achondroplasm). Presently this is already a problem in some cases. It might be aggravated by new techniques, such as improved screening and a selection on the level of gene variants. Thus, in breeding for special traits, the species specific conditions have to be considered. Breeding goals should not only be of advantage for human but also take into account the needs of the bred animals.

Moreover, the new methods are likely to accelerate the changes of allele frequencies in populations, promote small effective population size and further reduce the number of breeds. A rising loss of genetic variability follows. Finally the adaptability to changing environmental influences or to new commercial interests may be diminished by the increase in homogeneity of individuals within a breed. Thus, populations will loose their ability for production or will at least reduce their selection responses from generation to generation. In case of a decreased number of breeds, no substitutes or resource populations will be available in future.

Correspondingly, the responsibility of breeders is growing and a demand of legal regulations arises in order to avoid undesirable side effects. All considered, development promises reliable economical as well as ecological progress in animal breeding.

20.5 Summary

Biotechnology will have a significant impact on the future of animal improvement. Particular impulses and promises are expected by the methods of molecular genetics. Recombined DNA sequences, polymorphic DNA regions and gene maps deliver important informations for basic genome analysis. New methods and DNA probes arise which are usable for selection of breeding individuals, influencing gene expression, supporting gene transfer experiments and analyzing complex traits for genetic components. The technology is likely to be integrated into existing animal breeding programs allowing access, transfer and combination of genes at a rate and with a precision not possible previously.

Thus, in animal breeding, genome analysis does not only enhance basic knowledge but supports future practical breeding programs as well. Recent developments in molecular genetics suggest an approach by which breeders could reduce effects of environmental variation on selection processes and thus improve the genetic gain. This applies to various breeding purposes, e.g. the increase of efficiency, the improvement of animal product quality and the development of alternative products.

Activities of genome analysis in farm animals are widespread in several countries and are supported by far-reaching basic and applied experiments. Modern techniques open new venues to breeding that will eventually be beneficial to animals, environment and men. To approach these goals, combinations of conventional methods along with new technologies of genome analysis are required for optimal success.

20.6 References

[1] Hodges, J., *World Animal Rev. 1986, 59,* 2-10.

[2] Smidt, D., *Der praktische Tierarzt 1988, 9,* 12-19.

[3] Mercier, J.-C., Structure and function of milk protein genes. In: *Genome Analysis in Domestic Animals.* VCH, Weinheim, *1990.*

[4] Alexander, L.J., Stewart, A.F., Mackinlay, A.G., Kapelinskaya, T.V., Tkach, T.M., and Gorodetsky, S.T., *Eur. J. Biochem 1988, 178,* 395-401.

[5] Bonsing, J., and Mackinlay, A.G., *J. Dairy Res. 1987, 54,* 447-461.

[6] Andersson, L., Major histocompatibility genes in cattle and their significance for immune response and disease susceptibility. In: *Genome Analysis in Domestic Animals.* VCH, Weinheim, *1990.*

[7] Vaiman, M., Chardon, P., and Cohen, D., *Anim. Genet. 1986, 17,* 113-133.

[8] Ivell, R., The structure and function of peptide hormone genes. In: *Genome Analysis in Domestic Animals.* VCH, Weinheim, *1990.*

[9] Woychik, R.P., Camper, S.A., Lyons, R.H., Horowitz, S., Goodwin, E.C., and Rottman, F.M., *Nucleic Acid Res. 1982, 10,* 7197-7210.

[10] Atencio, E.J., et al., *Nucleotide sequences 1986/87.* Volume III, Other vertebrates and invertebrates. A compilation from the GenBank and EMBL data libraries. Academic Press, Orlando, *1987.*

[11] Shook, G.E., *J. Dairy Sci. 1989, 72,* 1349-1362.

[12] Gavora, J.S., and Spencer, J.L., *Animal Blood Grps. Biochem. Genet. 1983, 14,* 159-180.

[13] Patterson, D.F., Haskins, M.E., Jezyk, P.F., Giger, U., Meyers-Wallen, V.N., Aguirre, G., Fyfe, J.C., and Wolfe, J.H., *JAVMA 1988, 193,* 1131-1144.

[14] Östergaard, H., The influence of the MHC sytem in farm animals on disease resistance and fertility. *38th Ann. Meeting European Association Animal Production (Lisbon), 1987,* 10 pp.

[15] Brem, G., *Züchtungskunde 1988, 60,* 248-262.

[16] Hammer, R.E., Purcel, V.G., Rexroad, C.E., Wall, R.J., Bolt, D.J., Palmiter, R.D., and Brinster, R.L., *J. Anim. Sci. 1986, 63,* 269-278.

[17] Wagner, T.E., *Canad. J. Anim. Sci. 1985, 65,* 539-552.

[18] Ellendorff, F., Farries, E., Oslage, H.J., Rohr, K., and Smidt, D., (Eds.), *BST-Symposium.* Landbauforschung Völkenrode, Sonderheft 88, *1988.*

[19] Jeffreys, A.J., and Flavell, R., *Cell 1977, 12,* 429-439.

[20] Bufton, L., Mohandas, T.K., Magenis, R.E., Sheehy, R., Bestwick, R.K., and Litt, M., *Hum. Genet. 1986, 74,* 425-431.

[21] Jeffreys, A.J., Wilson, V., and Thein, S.L., *Nature 1985, 314,* 67-72.

[22] Jeffreys, A.J., Wilson, V., Thein, S.L., Weatherall, D.J., and Ponder, B.A., *Am. J. Hum. Genet. 1986, 39,* 11-24.

[23] Nakamura, Y., Leppert, M., O'Connell, P., Wolff, R., Holm, T., Culver, M., Martin, C., Fujimoto, E., Hoff, M., Kumlin, E., and White, R., *Science 1987, 235,* 1616-1622.

[24] Skolnick, M.H., and White, R., *Cytogenet. Cell Genet. 1982, 32,* 58-67.

[25] Prokop, C.-M., Informations on RFLPs and VNTRs in farm animals. In: *Genome Analysis in Domestic Animals*. VCH, Weinheim, *1990*.

[26] Geldermann, H., *Appl. Genet., 1975, 46,* 319-330.

[27] Geldermann, H., Pieper, U., and Roth, B., *Theor. Appl. Genet. 1985, 70,* 138-146.

[28] Smith, C., and Simpson, S.P., *J. Anim. Breed. Genet. 1986, 103,* 205-217.

[29] Stam, P., *Anim. Genet. 1987, 18, Suppl. 1,* 97-99.

[30] Soller, M., *Anim. Prod. 1978, 27,* 133-129.

[31] Soller, M., and Beckmann, J.S., *Theor. Appl. Genet. 1982, 67,* 25-33.

[32] Falconer, D.S., *Quantitative Genetics*. 2nd Edition. Longman, London & New York. *1981*.

[33] Tanksley, S.D., Young, N.D., Paterson, A.H., and Bonierbale, M.W., *Bio/Technology, 1989, 7,* 257-264.

[34] Leonard, M., Kirszenbaum, M., Cotinot, C., Chesné, P., Heyman, Y., Stinnakre, M.G., Bishop, C., Delouis, C., Vaiman, M., and Fellous, M., *Theriogenology 1987, 27,* 248.

[35] Bondioli, K.R., Ellis, S.B., Pryor, J.H., Williams, M.W., and Harpold, M.M., *Theriogenology 1989, 31,* 95-104.

[36] Ebensberger, C., Studer, R., and Epplen, J.T., *Hum. Genet. 1989,* in press.

[37] Saiki, R.K., Gelfand, D.H., Stoffel, S., Scharf, S.J., Higuchi, R., Horn, G.T., Mullis, K.B., and Erlich, H.A., *Science 1988, 239,* 487-491.

[38] Glodek, P., and Kräusslich, H., *Züchtungskunde 1988, 60,* 263-271.

[39] Robinson, J.L., Dombrowski, D.B., Harpestad, G.W., and Shanks, R.D., *J. Hereditiy, 1984, 75,* 277-280.

[40] Davies, W., Harbitz, I., Fries, R., Stranzinger, G., and Hauge, J.G., *Anim. Genet., 1988, 19,* 203-212.

[41] Gahne, B., and Juneja, R.K., *Anim. Blood Grps. Biochem. Genet., 1985, 16,* 265-284.

[42] Ellis, K.P., and Davies, K.E., *Biochem. J. 1985, 226,* 1-11.

[43] Hill, W.G., *Nature 1987, 327,* 98-99.

[44] Gill,, P., Lygo, J.E., Fowler, S.J., and Werrett, D.J., *Electrophoresis 1987, 8,* 38-44.

[45] Geldermann, H., Pieper, U., and Weber, E., *J. Anim. Sci. 1986, 63,* 1759-1768.

[46] Mueller, B., *Gerichtliche Medizin.* Bd. 2, 2. Auflage, Springer-Verlag, Berlin-Heidelberg-New York *1975*.

[47] Botstein, D., White, R.L., Skolnik, M., and Davis, R.W., *Am. J. Hum. Genet. 1980, 32,* 314-331.

[48] Morton, N.E., *Am. J. Hum. Genet. 1956, 8,* 80-96.

[49] Clerget-Darpoux, F., and Baur, M.P., Linkage studies and mapping. In: *Genome Analysis in Domestic Animals*. VCH, Weinheim, *1990*.

[50] Human Gene Mapping, Proc. 9th International Workshop Human Gene Mapping (Paris). *Cytogenet. Cell Genet. 1987, 46*.

[51] Human Gene Mapping, Update of the 9th International Workshop Human Gene Mapping (Paris). *Cytogenet. Cell Genet. 1988, 49,* 1-258.

[52] Lalley, P.A., Davisson, M.T., Graves, J.A.M., O'Brien, S.J., Roderick, T.H., Doolittle, D.P., and Hillyard, A.L., *Cytogenet. Cell Genet. 1988, 49,* 228-235.

[53] Fries, R., Beckmann, J.S., Georges, M., Soller, M., and Womack, J., *Anim. Genet. 1989, 20,* 3-29.

[54] O'Brien, B.J., Genetic analysis in mammals: past, present and future. *Genet. Engin. Anim.* (Eds. J. Warren and A. Hollaender) Plenum Press *1986*, 139-149.

[55] Stranzinger, G., *Anim. Genet. 1987, 18, Suppl. 1,* 111-116.

[56] Stranzinger, G., Stage of gene mapping in farm animal species. In: *Genome Analysis in Domestic Animals.* VCH, Weinheim, *1990.*

[57] Thompson, E.A., Kravitz, K., Hill, J., and Skolnick, M., Linkage and the power of a pedigree structure, in: *Genetic Epidemiology,* Eds. Morton, N.E., & C.S. Chung, New York, Academic Press, *1978,* pp 247-253.

[58] White, R., and Lalouel, J.M., *Sci. Amer. 1988, 258,* 20-28.

[59] Kluge, R., and Geldermann, H., *Theor. Appl. Genet. 1982, 62,* 1-4.

[60] Hetzel, D.J.S., Genomic DNA banks. *Workshop on RFLP Markers, Gene Mapping and Linkage Studies in Domestic Animals.* Mt. Victoria *1988,* 6 pp.

[61] Maijala, K., Gene mapping in cattle. *4th International Symposium "Progress in Cattle Breeding",* Czechoslovakia *1989,* Sept. 14-17.

[62] Soller, M., and Beckmann, S., Toward an understanding of the genetic basis for trypano-tolerance in the N'Dama cattle of West Africa. *Consultation Report, FAO, Rome, 1987.*

[63] Baker, B., Schell, J., Lörz, H., and Federaff, N., *Proc. Natl. Acad. Sci. USA 1986, 83,* 4844-4848.

[64] Orkin, S.H., *Cell 1986, 47,* 845-850.

[65] Bender, W., Spierer, P., and Hogness, D., *J. Supra Molec. Structure 1979, 10,* 32.

[66] Poustka, A., Rackwitz, H.-R., Frischauf, A.-M., Hohn, B., and Lehrach, H., *Proc. Natl. Acad. Sci. USA 1984, 81,* 4129-4133.

[67] Cooke, H., *Trends in Genetics 1987, 3,* 173-174.

[68] Poustka, A., Pohl, T.M., Barlow, D.P., Frischauf, A.-M., and Lehrach, H., *Nature 1987, 325,* 353-355.

[69] Poustka, A., and Lehrach, H., *Trends in Genetics 1986, 2,* 174-179.

[70] Beckmann, J.S., and Soller, M., *Theor. Appl. Genet. 1983, 67,* 35-43.

[71] Rossant, J., Croy, B.A., Chapman, V.M., Siracusa, L., and Clark, D.A., *J. Anim. Sci. 1982, 55,* 1241-1248.

[72] Mettler, L.E., and Gregg, T.G., *Population Genetics and Evolution.* Prentice-Hall, Englewood Cliffs, New Jersey, *1969.*

Index